The Institute of Mathematics
and its Applications
Conference Series

Previous volumes in this series were published by
Academic Press to whom all enquiries should be addressed.
Forthcoming volumes will be published by
Oxford University Press throughout the world.

Continued overleaf

Computer-aided Surface Geometry and Design

The Mathematics of Surfaces IV

Based on the proceedings of the fourth conference on The Mathematics of Surfaces, organized by the Institute of Mathematics and its Applications and held at the University of Bath in September 1990.

Edited by
ADRIAN BOWYER
School of Mechanical Engineering
University of Bath

CLARENDON PRESS · OXFORD
1994

Oxford University Press, Walton Street, Oxford OX2 6DP
Oxford New York Toronto
Delhi Bombay Calcutta Madras Karachi
Kuala Lumpur Singapore Hong Kong Tokyo
Nairobi Dar es Salaam Cape Town
Melbourne Auckland Madrid
and associated companies in
Berlin Ibadan

Oxford is a trade mark of Oxford University Press

Published in the United States by
Oxford University Press Inc., New York

© Institute of Mathematics and its Applications, 1994

A catalogue record for this book is available from the British Library

Library of Congress Cataloging in Publication Data
(data applied for)

ISBN 0 19 853648 8

Printed in Great Britain by
Bookcraft (Bath) Ltd
Midsomer Norton, Avon

Preface

The fourth IMA Mathematics of Surfaces Conference took place at Bath University from the tenth to the twelfth of September 1990. The programme committee were Professor Mike Pratt, Professor James Davenport, Dr Ralph Martin, and me.

The conference was attended by about eighty participants, and papers were presented on subjects ranging from finding the zeros of multivariate implicit polynomials to the shapes of yacht sails. Indeed, the conference looked at a very broad range of surface mathematics centred on — but not restricted to — the use of surfaces in computer-aided design.

I would like to thank the Conference Office at Bath University for their help with the organization of lecture rooms, meals and accommodation, and the IMA for dealing with many other aspects of the conference. In particular, thanks are due to Pam Irving and Donna Budd of the IMA who put in a great deal of work both before and during the conference.

These proceedings have taken me rather longer to edit than I expected. This was, in part, due to a need to do other things as well. But it was also due to the fact that there was much more to the editing than I had anticipated from previous experience of camera-ready-copy proceedings. These proceedings were prepared from author-produced LaTeX files, and I would like finally to thank Adèle Sharples and Debbie Brown of the IMA, who have done a great deal of work on the computer-preparation of them.

Adrian Bowyer
University of Bath

ACKNOWLEDGEMENTS

The Institute thanks the authors of the papers, the editor, Adrian Bowyer (University of Bath) and also Miss Adèle Sharples and Miss Debbie Brown for preparing and typing the papers.

CONTENTS

CONTRIBUTORS

S. ARNBORG; Department of Numerical Analysis and Computing Science, The Royal Institute of Technology, S-10044 Stockholm, Sweden.

R.P. BENNELL; Applied and Computational Mathematics Group, Royal Military College of Science (Cranfield), Shrivenham, Wiltshire, SN6 8LA.

M.I.G. BLOOR; Department of Applied Mathematical Studies, University of Leeds, Leeds, LS2 9JT.

M.S. BLOOR; Department of Mechanical Engineering, University of Leeds, Leeds, LS2 9JT.

J.M. BROWN; Department of Applied Mathematical Studies, University of Leeds, Leeds, LS2 9JT.

L.-D. CAI; AI Vision Research Unit, University of Sheffield, Western Bank, Sheffield, S10 2TN.

J.F. CANNY; Computer Science Division, University of California, Berkeley, CA 94720, USA.

C. COTTIN; Department of Mathematics, University of Duisburg, 4100 Duisburg 1, Germany.

U. CUGINI; Dip. di Ingegneria Industriale, Università degli Studi di Parma, Viale delle Scienze, 43100 Parma, Italy.

W. DEGEN; Mathematisches Institut B, Universität Stuttgart, Pfaffenwaldring 57, 70569 Stuttgart, Germany.

A. DOLENC; Institute of Industrial Automation, Helsinki University of Technology, Otakaari 1, Espoo 02150, Finland.

R. EKPETE; Computer Science Department, Arizona State University, Tempe, AZ 85287, USA.

G. FARIN; Department of Computer Science, College of Engineering and Science, Arizona State University, Tempe, AZ 85287-5406, USA.

R.T. FAROUKI; IBM Thomas J. Watson Research Center, P.O. Box 218, Yorktown Heights, New York 10598, USA.

T. FOLEY; Department of Computer Science, Arizona State University, Tempe, AZ 85287, USA.

R.J. GOULT; LMR Systems, 33 Filgrave, Newport Pagnell, Buckinghamshire, MK16 9ET.

H. HAGEN; FB Informatik, Universität Kaiserslautern, Postfach 3049, D-6750 Kaiserslautern, Germany.

F.D. HALES; Department of Transport Technology, Loughborough University of Technology, Loughborough, LE11 3TU.

T. HERMANN; Computer and Automation Institute, Hungarian Academy of Sciences, H-1518, Budapest, P.O. Box 63, Hungary.

C.M. HOFFMANN; Department of Computer Science, Purdue University, West Lafayette, IN 47907, USA.

J. HOLLMAN; Department of Mathematics, Stockholm University, Box 6701, S-11385 Stockholm, Sweden.

L. LANGEMYR; Wilhelm-Schickard-Institut für Informatik, Universität Tübingen, Auf dem Sand 13, Tübingen, W-7400, Germany.

M.H.E. LARCOMBE; Department of Computer Science, University of Warwick, Coventry, CV4 7AL.

M.C. LÓPEZ DE SILANES; Departamento de Matemática Aplicada, Centro Politécnico Superior, Universidad de Zaragoza, Maria de Luna 3, E-50015 Zaragoza, Spain.

G. LUKÁCS; Computer and Automation Institute, Hungarian Academy of Sciences, H-1518, Budapest, P.O. Box 63, Hungary.

D. MANOCHA; Department of Computer Science, CB 3175, Sitterson Hall, University of North Carolina, Chapel Hill, NC 27599-3175, USA.

R.R. MARTIN; Department of Computing Mathematics, Mathematics Institute, University of Wales College of Cardiff, Senghenydd Road, P.O. Box 916, Cardiff, CF2 4YN.

J.C. MASON; Applied and Computational Mathematics Group, Royal Military College of Science (Cranfield), Shrivenham, Wiltshire, SN6 8LA.

E. McLELLAN; LMR Systems, 33 Filgrave, Newport Pagnell, Buckinghamshire, MK16 9ET.

A.E. MIDDLEDITCH; Computer Aided Engineering Research Group, Physics Department, Brunel University, Uxbridge, Middlesex UB8 3PH.

P. MILNE; School of Mathematical Sciences, University of Bath, Claverton Down, Bath BA2 7AY.

M. NORDGREN; Institute of Industrial Automation, Helsinki University of Technology, Otakaari 1, Espoo 02150, Finland.

H. NOWACKI; Institut für Schiffs- und Meerestechnik, Technische Universität Berlin, Salzufer 17-19, D-1000 Berlin 10, Germany.

M.J. PRATT; Design and Manufacturing Institute, Rensselaer Polytechnic Institute, Troy, NY 12180-3590, USA.

S. RADI; KAEMaRT Group - CNR IMU, Via A.M. Ampère 56, 20131 Milano, Italy.

C. RIZZI; KAEMaRT Group - CNR IMU, Via A.M. Ampère 56, 20131 Milano, Italy.

P.A. ROACH; Department of Computing Mathematics, Mathematics Institute, University of Wales College of Cardiff, Senghenydd Road, P.O. Box 916, Cardiff, CF2 4YN.

M.A. SABIN; FEGS Limited, 5 Coles Lane, Oakington, Cambridge.

F.J. SERÓN; Departamento de Matemática Aplicada, Centro Politécnico Superior, Universidad de Zaragoza, Maria de Luna 3, E-50015 Zaragoza, Spain.

J.J. TORRENS; Departamento de Matemática Aplicada, Centro Politécnico Superior, Universidad de Zaragoza, Maria de Luna 3, E-50015 Zaragoza, Spain.

R. VAN DAMME; Department of Applied Mathematics, University of Twente, P.O. Box 217, 7500 AE Enschede, The Netherlands.

M.J. WILSON; Department of Applied Mathematical Studies, University of Leeds, Leeds, LS2 9JT.

C. WOODWARD; Institute of Industrial Automation, Helsinki University of Technology, Otakaari 1, Espoo 02150, Finland.

G.M. YOUNG; School of Electronic Systems Design, Cranfield Institute of Technology, Cranfield, Bedford, MK43 0AL.

Standards for Curves and Surface Data Exchange

R.J. Goult

LMR Systems, Newport Pagnell, Buckinghamshire

1 Introduction

During the past decade the growth of the use of computers in the design, analysis and manufacture of products has led to an increasing requirement to communicate product data between computer systems, particularly between CAD systems. Within a single company it is not unusual to find different systems used for initial design, detailed design, FE mesh generation, analysis and manufacturing. After taking account of different systems used by suppliers and sub-contractors the number of systems needing to communicate product design data can exceed 50. Two basic techniques have been used to solve this communication problem, the first involves the provision of dedicated translators between each pair of systems, the second uses an agreed standard neutral file interface and requires a pre- and a post-processor for each system. For a large number of systems there are obvious theoretical advantages with the neutral file approach, which requires $2n$ processors for n systems as opposed to the dedicated translator approach with its requirement for the development and maintenance of $n(n-1)$ translators for n systems. This paper will concentrate on the neutral file approach giving a brief description of existing neutral file standards and describing the considerations taken into account in developing the ISO Standard STEP (Standard for the Exchange of Product data). Particular emphasis will be given to those parts of the standards concerned with parametric curve and surface data where major differences exist between the forms of representation used by current CAD systems.

Ideally a neutral file interface for CAD data exchange should make it possible to communicate 100% of the data between any two systems without loss of accuracy or information. The effectiveness of this transfer depends not only upon the neutral file specification but also upon the quality of the pre- and post-processors provided. In the context of sculptured surface data, fundamental incompatibilities between the forms of surface representation used by different systems mean that no matter how good

1

the neutral file specification and the processors the objective of 100% interchange between all systems will never be attainable. The practical limit to the amount of data which can be fully exchanged is the intersection of the capabilities of the two systems, the neutral file specification and of the processors concerned. The design of the neutral file is important in that it defines the ultimate scope of the exchange and strongly influences the quality of pre- and post-processors.

2   Existing National Standards

At the present time there are 3 National Standards which make provision for the exchange of product design data; these are IGES, an ANSI standard [1], SET a French National Standard [2] and VDA-FS a DIN standard [3]. All of these are supported by major CAD system vendors and are used outside their country of origin.

The first and most widely used of these is IGES. When it first appeared in 1981 as version 1.0 it was described as an initial graphics exchange specification and was designed to enable the exchange of 2D drawings as ASCII character strings using fixed format 80 column card images. Later versions of IGES have increased the scope by adding new entities but the basic file format has remained unaltered in the interests of upward compatibility. IGES files can be very cumbersome and the file structure with its header, directory and data sections makes it virtually impossible to produce one pass translators. Geometrically the original IGES entities support lines, circular arcs, conic sections, composite curves and parametric curves and surfaces. In the earlier versions only explicit parametric polynomial curves and surfaces of degree less than or equal to 3 were supported, but since version 3.0 rational B-Spline curves and surface entities of arbitrary degree have been provided as well as bounded surface entities. Version 4.0, the last version released, provides entities to support the exchange of CSG solid models, the entities used here are logically very similar to those being proposed for STEP. Also included in the later versions are entities to support the exchange of electrical and finite element data. It was originally intended that version 5.0 of IGES, due to be released later this year would include provision for the exchange of B-Rep models but this may not be the case due to technical difficulties.

SET first appeared some years later than IGES and offers a similar scope. Important design objectives were to provide facilities for archiving as well as data exchange and to produce more compact data files. This has been achieved by abandoning the fixed 80 column format and in some cases providing different forms of representation for logically equivalent entities. Parametric curve and surface representation in SET was originally by explicit polynomial coefficients but unlike the simple IGES entities there is no limit to the degree and provision is made for rational forms. The latest version of SET has considerably expanded scope with provision for electrical and finite element data and the exchange of solid model data including BREP and CSG. Also included in this version are entities or 'blocks' to permit the exchange of rational B-spline curves and surfaces.

VDA-FS was originally designed to meet some very specific requirements within the German motor industry for the exchange of computer sculptured surface geometry. It was designed to be compatible with IGES in its file format and in Version 1.0 the only geometric entities were points and parametric polynomial curves and surfaces. The curves and surfaces used a simple explicit coefficient representation for each segment of the curve or patch of the surface. There is no limit to the degree but no provision is made for rational forms. Version 2.0 has added circular arc and bounded surface entities but unlike IGES and SET there is no intention to extend it to a full product data exchange specification and full German support is being given to the development of STEP.

3   The ISO Standard STEP

During 1984 the International Standards Organisation considered the need for a standard for the exchange of product data. The existing national standards were reviewed and it was decided that none was capable of being developed to meet all the requirements. A working group (ISO TC 184 SC4 WG1) was then established with the task of developing the new standard. The secretariat of this working group was given to the IGES organisation in the US where there was an existing research programme called PDES. The scope of the proposed standard is very wide and includes all aspects of product data. A large number of applications are to be provided for including electrical, finite element analysis, manufacturing, shipbuilding and architecture and civil engineering. In addition provision is made for inclusion from the start of conformance and test criteria.

The work has been a true example of international collaboration with major contributions from many countries including Britain, Germany, Japan and The Netherlands, also from the Esprit CAD*I project [4]. As an indication of the wholehearted US support the interpretation of the acronym PDES has changed during the project from Product Data Exchange Specification to Product Data Exchange using STEP.

The first release of STEP as a Draft Proposal came at the beginning of 1989 when the entire proposed standard, as it was then, was released for international review and criticism. None of those closely involved in the project expected the proposal to be accepted but it was felt that the final quality would be improved by a widespread critical review at this time. Since the middle of last year all sections of the document have been rewritten in response to the comments received and it has been decided that STEP is too large to be released as a single document, also it is unrealistic to expect to synchronise the development of all the sections. The decision has therefore been taken to subdivide the standard into Parts which can be released for review in related Groups. The first of these Parts (Part 11) has just been released to the national standards organisation and describes the formal specification language Express which is used in all the other technical parts. It is anticipated that Part 42 which contains the geometry and all aspects of shape representation will be released later this year. The logical content of the shape representation has changed very little since the 1989 Draft Proposal but the formal descriptions have changed in line with the evolution of Express. Important considerations affecting the development of Part 42 will be described in the remainder of this paper.

3.1   Design objectives

The major objectives in neutral file design can be loosely defined as breadth of scope, efficiency, stability of representation and the provision of unambiguous definitions. Of these objectives efficiency is both the hardest to define and the one which introduces conflicting requirements which ensure that any neutral file design is always a compromise.

A very narrow interpretation of efficiency is that the neutral file communicating any given set of product data should be as small as possible. The problem with this as a sole objective is that a very compressed physical file format is likely to complicate the development of pre- and post-processors and make human interpretation of the contents of the data file virtually impossible. The IGES specification includes provision for compact binary files but this feature is not supported by the majority of commercial pre- and post-processors. A broader interpretation of efficiency should take into account not only file size but also the ease with which pre- and post-processors can convert data between the CAD systems and the neutral file. One pass processing can be an objective and to enable this an important feature of the CAD*I neutral file specification is the avoidance of forward pointers. The breadth of scope of the entity set also influences the ease of processor development and the overall effectiveness of the exchange process. A large entity set with multiple representations of equivalent data makes it possible to match closely the internal representations of a variety of CAD systems and simplifies pre-processor development since direct mappings are provided. This approach, although ideal for archiving, will not however improve the overall efficiency of the exchange process since post-processor development is then more complex and there is a tendency for different systems to support only those entities close to their internal representations. The opposite approach is to insist that there is no redundancy in the physical file representation, but this in turn can produce complex representations of basically simple data occupying an unnecessarily large amount of file space. The advantage of a minimal entity set is that different systems are forced to use a common neutral file representation for equivalent data and there is less danger of producing representations not recognised by particular post-processors. In STEP the entity set is nearly minimal but there are a few exceptions which offer particularly compact representations. One example of this type of compromise is the polyline, simply represented by a sequence of points but essentially redundant since the same information could, with more complexity, be communicated as a B-Spline curve of degree 1, or as a composite curve.

Early investigations in the CAD*I project of IGES pre- and post-processors available with CAD systems showed that at least some of the communication problems were due to the different processor developers using a different interpretation of the entity definitions. This problem is always potentially present when a natural language like English is used for the standard specification. For STEP a computer sensible language EXPRESS has been developed and is included as part of the standard itself. This approach cannot of itself ensure that the definitions themselves are correct or appropriate but they are only capable of one interpretation and the syntax can be verified by suitable software tools. Examples of EXPRESS definitions for simple geometric entities are included in the next section.

3.2 Neutral file entities

In defining the individual entities to represent the shape and size of a product in a neutral file a primary consideration is the stability of the representation. Most geometric entities are represented in the neutral file essentially by a sequence of real numbers. During the communication which includes the pre-processor, the neutral file and the post-processor these real numbers may be rounded or truncated and hence the data communicated is rarely precise. A stable neutral file representation for a particular geometric entity should be robust in the presence of such small errors. Two particular aspects of this stability are:

- the post-processor should be able to unambiguously interpret the received data as being of the same geometric form as the original entity.

- for small transmission errors the geometry reconstructed by the receiving system should be close to that originally defined.

Not all entities in existing standards satisfy the above requirements for stability. As an example IGES [1] defines a circular arc (Entity Type 100) in the xy plane by its centre point together with start point and end point for the arc. It is not clear how this data is to be interpreted if the received data is such that the distance from the start point to the centre is not identical to the distance from the centre to the end point.

In selecting the geometric entity representations for STEP stability was a primary consideration. STEP contains entities to represent conic sections and simple surfaces such as sphere, cone, cylinder and torus, in addition to more general parametric curve and surface entities. The specification separates the position and orientation information from the essential geometric properties of the curve or surface being represented. For stability most of the data is in the form of points and directions.

In the STEP specification the ellipse entity definition is given by:

```
ENTITY ellipse
        SUBTYPE OF (conic);
        semi_axis_1  :  REAL;
        semi_axis_2  :  REAL;
        position     :  axis_2_placement;
    WHERE
        semi_axis_1 > 0.0;
        semi_axis_2 > 0.0;
        coordinate_space(ellipse)=coordinate_space(location);
    END_ENTITY.
```

Within this definition the axis_2 placement entity is the entity used to locate a point and a pair of mutually perpendicular axes in space. It is defined in terms of a point and two directions representing the axis (in this case normal to the plane of the ellipse) and a reference direction. The STEP definition for a 3D entity projects the reference direction onto the plane normal to the axis $\mathbf{a}$ to obtain a unit vector $\mathbf{x}$, then defines $\mathbf{y} = \mathbf{a} * \mathbf{x}$. For a 2D entity $\mathbf{x}$ is obtained by normalising the reference direction and $\mathbf{y}$ is perpendicular to $\mathbf{x}$. If $\mathbf{c}$ is the point defined in the axis_2 placement the parametric equation of the ellipse is defined in the STEP specification as:

$$\mathbf{r}(u) = \mathbf{c} + \text{semi\_axis\_1} \, cos(u) \, \mathbf{x} + \text{semi\_axis\_2} \, sin(u) \, \mathbf{y}$$

It should be noted that this entity provides a common format for the definition of an ellipse in either 2 dimensional space (coordinate_space = 2) or 3 dimensional space and the definition defines an unambiguous interpretation for all legal values of the data, even if rounding errors produce a reference direction not strictly in the plane of the curve.

The other simple curve and surface entities in STEP are defined in a similar way. One example of an entity where the definition has been modified from the usual one in order to improve the numerical stability is the conical surface. The geometric properties of such a surface would appear to be simply defined in terms of a vertex position, an axis and a semi-angle. In practice the entire infinite conical surface is rarely required and the region of interest may be some part of the surface at a considerable distance from the vertex of the cone. In such cases the effective radius and position of the surface section are very sensitive to the values of the semi-angle and the axis direction. The corresponding STEP entity is defined in terms of an axis_2_placement, which include the axis of the cone and a point on this axis, a radius and a semi-angle. The radius defines the radius of a frustrum of the cone from the axis point. Clearly this definition corresponds to the elementary one if the axis point is the vertex with corresponding radius of 0.0, but, for a surface section some distance from the vertex, a more stable representation is available, if the axis point is selected close to this section.

3.3 General parametric curve and surface representations

Amongst CAD systems a wide variety of formats is used to represent general parametric curves and surfaces. For parametric polynomial curves and surfaces these include power series coefficients, B-Spline, Bézier and Ferguson or Hermite. For a single segment curve of a given degree these representations are mathematically equivalent but the data content differs considerably. In the power series form the explicit polynomial coefficients are stored, in the B-Spline and Bézier representation the curve is defined in terms of a set of control points and the appropriate basis functions, whilst the Ferguson representation utilises end points and derivatives together with Hermite basis functions. For rational curves and surfaces the power series, B-spline and Bézier formats can be generalised by introducing homogeneous coordinates with a fourth 'weight' coordinate.

Which representation is 'best' for the set of parametric polynomial curves and surfaces is an issue which has yet to be resolved. Literature which supports one representation over another has been slow to appear.

Stability experiments

During the CAD*I project Lachance [5] conducted some experiments to compare the stability of different forms of parametric curve representation. We assume that both the sending and receiving systems employ explicit power basis coefficient representations of polynomials. We assume that there is some loss of precision due to the use of an ASCII format. For the purpose of these experiments, the errors were of three types: truncation to 6 significant figures, truncation to 12 significant figures, and truncation to 4 decimal place accuracy. It is further assumed that all local calculations are performed in double precision.

The polynomials $\mathbf{P}(u)$ to be transmitted could have been arbitrarily chosen. To give them some physical significance, they were chosen to be least squares approximates, of different degrees, to five separate curves in the xy-plane. Three of these were semi-circular arcs with different centers and radii, one was the upper half of the unit square, and the last was a mixed exponential and polynomial curve. Some of these curves are displayed in Fig. 1.

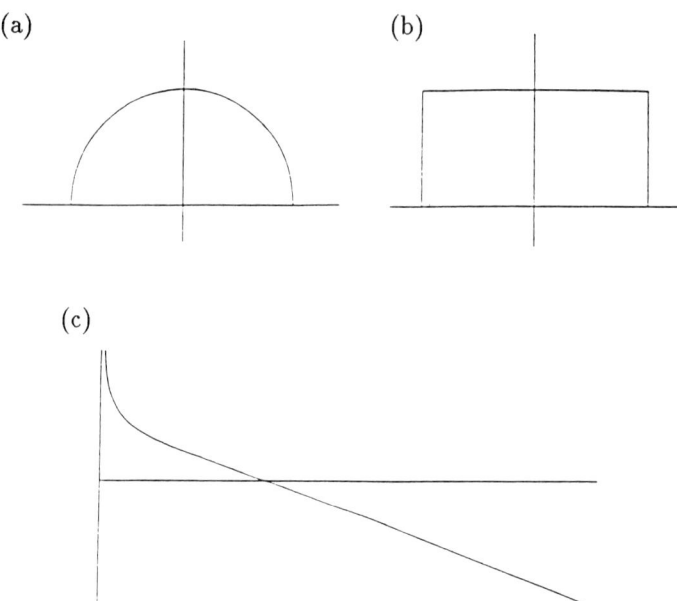

Figure 1.   Curves defining original polynomials (a) $\{(x,y) : x^2 + y^2 = 1, \quad y > 0\}$ (b) $\{(x,y) : \max(x,y) = 1, \quad y > 0\}$ (c) $\{(x,y) : x = e^{5u}, y = 1 - u(2 - u(3 - u(4 - u))), \quad 0 \le u \le 1\}$

Fig. 2 shows the various communication paths involved. The first path assumes that $\mathbf{P}(u)$ is communicated directly using explicit coefficients, in much the same way as the German automotive standard VDA-FS works today. However, the target system does not receive the polynomial $\mathbf{P}(u)$, but the truncated polynomial $\mathbf{TP}(u)$.

The second path assumes that the Bézier control points will be the medium of exchange. Thus the polynomial $\mathbf{P}(u)$ is transformed to its Bézier representation $\mathbf{B}(u)$, these control points are then truncated to yield $\mathbf{TB}(u)$, and then the receiving system reconstitutes these truncated Bézier control points into an explicit polynomial $\mathbf{RTB}(u)$.

The third path is similar to the second, except that uniform B-spline control points are employed. In this case the polynomial $\mathbf{P}(u)$ is transformed to its B-spline representation $\mathbf{S}(u)$, these control points are trun-

Sending System			Receiving System
a)			
Explicit polynomial representation	$\longrightarrow$		Truncated polynomial representation
b)			
Explicit$\longrightarrow$ polynomial representation	Bézier$\longrightarrow$ representation	Truncated$\longrightarrow$ Bézier	Reconstituted explicit polynomial
c)			
Explicit$\longrightarrow$ polynomial representation	B-spline$\longrightarrow$ representation	Truncated$\longrightarrow$ B-spline	Reconstituted explicit polynomial

Figure 2. a) Polynomial coefficients as exchange medium, b) Bézier control points as exchange medium, c) B-spline control points as exchange medium

cated to get $\mathbf{TS}(u)$, and then the truncated B-spline control points are reconstituted to form the explicit polynomial $\mathbf{RTS}(u)$. The splines used here are the normalised uniform B-splines $N_{i,n+1}(u)$ of order $n + 1$, with distinct integer knots $(i - n - 1, \ldots, i)$, where $i = 1, \ldots, n + 1$.

Analysis

It seems quite natural to exchange polynomial data by exchanging the actual polynomial coefficients. The only error that is incurred is a truncation error. That is, if $\mathbf{A} = (\mathbf{A}_0, \mathbf{A}_1, \ldots, \mathbf{A}_n)$ represents the coefficient vector of the original polynomial, then the target system will receive this coefficient vector with an associated error vector $\mathbf{a} = (\mathbf{a}_0, \mathbf{a}_1, \ldots, \mathbf{a}_n)$,

$$\mathbf{A} = \mathbf{A} + \mathbf{a}.$$

The approach which we are investigating here suggests that the coefficient vector $\mathbf{A}$ be first altered by a matrix transformation M. For example, the original polynomial might be alternatively represented by its Bézier or B-spline control points $\mathbf{V} = (\mathbf{V}_0, \mathbf{V}_1, \ldots, \mathbf{V}_n) = \mathbf{A}M$. The sequence of events in this case would be

$$\mathbf{A} \longrightarrow \mathbf{V} \longrightarrow \mathbf{V} + \mathbf{v} \longrightarrow \mathbf{A} + \mathbf{v}M^{-1},$$

where $\mathbf{v} = (\mathbf{v}_0, \mathbf{v}_1, \ldots, \mathbf{v}_n)$ represents the associated truncation error.

The errors which are introduced in these two transferring processes are basically of two types. The first and most pronounced is the error due to the truncation of the respective coefficients. The second is the error which results from matrix multiplication by the matrices M and M^{-1}, converting one representation into the other.

In this discussion the multiplication errors are assumed to be small relative to the truncation errors. It is felt that this assumption is reasonable because all local calculations are performed in double precision; this assumption is further supported by the fact that, in the Bézier case, the matrix and its inverse can be expressed explicitly, allowing the calculations to be performed optimally. Even in the B-spline case, where one numerical inversion must take place, the transformation matrices seem to be reasonably well-conditioned when the degree is small [5].

For these reasons we focus on the truncation vectors $\mathbf{a}$ and $\mathbf{v}M^{-1}$. For a polynomial $\mathbf{a}(u)$ parametrised over the interval $[0,1]$, with coefficient vector $\mathbf{a}$, it is easy to show that a uniform bound on $\mathbf{a}(u)$ is

$$\|\mathbf{a}(u)\| \leq (n+1) \max_{0 \leq i \leq n} \|\mathbf{a}_i\| \text{ for } 0 \leq u \leq 1,$$

where $\| \ \|$ denotes the Euclidian norm in three space.

Since the Bernstein polynomials and the B-spline basis functions form a partition of unity, it is also easy to see that for a polynomial $\mathbf{v}(u)$ with control polygon $\mathbf{v}$ we have a uniform bound

$$\|\mathbf{v}(u)\| \leq \max_{0 \leq i \leq n} \|\mathbf{v}_i\| \text{ for } 0 \leq u \leq 1.$$

At issue then, at least in the estimation of worst case behaviour, is the relative magnitudes of the vectors $\mathbf{a}$ and $\mathbf{v}$. Since these are truncation errors, their size depends upon the type of truncation and upon the relative sizes of the coefficient vectors. For example, if $\mathbf{A}$ and $\mathbf{V}$ were of the same size, one might expect from the above uniform bounds that the error terms associated with the Bézier or B-spline representations might be better behaved than that associated with the explicit polynomial coefficients, because of the factor of $n+1$ in the first inequality.

In these experiments it was discovered that the coefficients $\mathbf{A}$ and $\mathbf{V}$ are not generally of the same size. In fact, the relative sizes of the respective coefficients for each of the three representations can vary quite dramatically. To demonstrate this variation, we display in Fig. 3 the polynomial coefficients, and the Bézier and B-spline control points, for the fourth degree polynomial approximation to the unit semi-circle. For the sake of the illustration the consecutive coefficients of the explicit polynomial are joined to one another in a manner similar to the Bézier and B-spline control polygons. One rather striking feature of this figure is the relative sizes of the convex

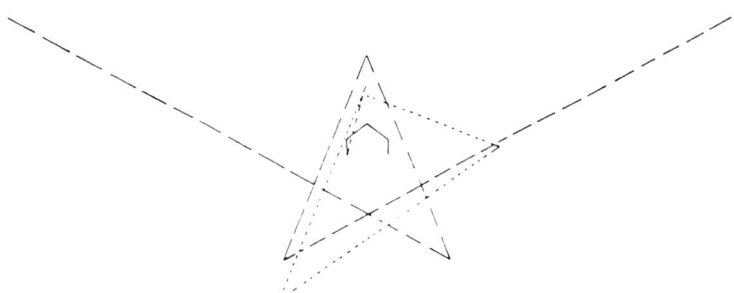

Figure 3. Quartic approximation to unit semi-circle. Solid line - Bézier control polygon, broken line - B-spline control polygon, dotted line - polynomial coefficients

hulls of these coefficients. Clearly, truncation of the larger coefficients will result in a greater loss of information than a corresponding truncation of smaller coefficients. Such a loss was observed in our experiments, and the behaviour was little changed as the polynomial was translated or scaled away from the origin.

It is well understood that the Bézier control polygon lies 'near' the curve which it describes, this being more so as the degree increases. The polynomial coefficients do not generally enjoy this property, exhibiting a more random and less intuitive orientation. As regards the B-spline control points, they share some of the properties of the Bézier control points, such as the convex hull property. It is also known that there is a certain 'laying-off' property associated with the uniform B-splines. That is the curve is not usually as good an approximation of the B-spline control polygon as for the Bézier case.

What is not very well publicized is the fact that for curves of relatively small degree, the uniform B-spline control polygon can be quite far from the curve which it represents. In Fig. 4 we illustrate this behaviour by displaying the uniform B-spline control polygon for the polynomial approximations to the unit semi-circle of degree 6.

Results

In our notation the truncated polynomial $\mathbf{TP}(u)$, the reconstituted Bézier polynomial $\mathbf{RTB}(u)$ and the reconstituted B-spline polynomial $\mathbf{RTQ}(u)$ are all images of the original polynomial $\mathbf{P}(u)$. To estimate the respective deviations of these polynomials from the original we use two measures of comparison: the uniform norm and the least squares norm denoted $\| \ \|_{L_\infty}$ and $\| \ \|_{L_2}$ respectively. The uniform norm estimate is based upon comparing a sample of 100 pairs of points corresponding to equally-spaced param-

Figure 4. Uniform B-spline control polygon for polynomial approximation of degree 6 to unit semi-circle

etric values in the interval [0,1], while the least squares estimate is based upon Romberg iterations applied to Simpson's Rule.

For the fourth degree polynomial example illustrated previously, we found that the uniform norms of the respective differences satisfied

$$\|\mathbf{P} - \mathbf{RTB}\|_{L_\infty} \approx 0.1426D - 05,$$
$$\|\mathbf{P} - \mathbf{TP}\|_{L_\infty} \approx 3.51\|\mathbf{P} - \mathbf{RTB}\|_{L_\infty},$$
$$\|\mathbf{P} - \mathbf{RTQ}\|_{L_\infty} \approx 1.34\|\mathbf{P} - \mathbf{RTB}\|_{L_\infty},$$

while the least squares norms satisfied

$$\|\mathbf{P} - \mathbf{RTB}\|_{L_2} \approx 0.1045D - 05,$$
$$\|\mathbf{P} - \mathbf{TP}\|_{L_2} \approx 2.14\|\mathbf{P} - \mathbf{RTB}\|_{L_2},$$
$$\|\mathbf{P} - \mathbf{RTQ}\|_{L_2} \approx 1.77\|\mathbf{P} - \mathbf{RTB}\|_{L_2}.$$

Note that in this instance the reconstituted truncated Bézier curve $\mathbf{RTB}(u)$ is 'closer' to the original curve $\mathbf{P}(u)$ than either $\mathbf{TP}(u)$ or $\mathbf{RTQ}(u)$, whether measured in the L_∞ or L_2 norms.

The comparisons illustrated above were made for the three types of truncation formats, for polynomials of degrees 2 through 15 approximating the five different planar curves. Table 1 displays the average ratios of the least squares norm comparing truncated explicit polynomial and truncated B-spline with truncated Bézier.

Conclusions

From these experiments it was concluded that the power series coefficients do not provide the most stable format for the communication of parametric polynomial curves and surfaces. A further disadvantage of this representation when applied to piecewise parametric curves and surfaces is that,

Table 1. Average ratio of least square norms

	Format F25.4=4d.p. D25.6=digits D25.12=12 digits	Least sq. $\dfrac{\|P\text{-}TP\|_2}{\|P\text{-}RTB\|_2}$	Least sq. $\dfrac{\|P\text{-}RTQ\|_2}{\|P\text{-}RTB\|_2}$
Semi-circle center: (0,0) radius:	F25.4 D25.6 1 D25.12	2.20 3.66 2.12	1.78 185.68 59.17
Semi-circle center: $(2000,500\pi)$ radius: 1	F25.4 D25.6 D25.12	2.10 1.65 1.90	1.10 8.35 3.20
Semi-circle center: (2000,0) radius: 2000	F25.4 D25.6 D25.12	3.27 9.34 8.27	1.42 196.06 45.97
Exponential - Polynomial Mix	F25.4 D25.6 D25.12	1.69 2.10 1.05	1.10 3350.19 8162.71
Unit Square	F25.4 D25.6 D25.12	3.03 705717.00 43731.13	10.92 95622484.00 836727671.41

unlike control point representations, it cannot guarantee that adjacent sections are indeed coincident at their common ends or edges. For the communication of high degree curves and surfaces the strictly uniform B-spline representation is also less stable than the Bézier representation or B-splines using coincident end knots. Both the CAD*I neutral file specification and STEP provide rational B-spline curve and surface entities which includes a special facility for the communication of Bézier control points. These entities are suitable for the communication of rational or polynomial curve or surface data of any degree.

3.4 The STEP B-spline curve entity

Special features of this definition when compared to the corresponding IGES entity are the inclusion of knots, knot multiplicities and weights as optional attributes, the inclusion of a uniform type indicator and the fact that a multiple knot can be included as a single real number and an integer multiplicity. This last feature enables multiple knots, which affect the level of continuity to be recognised easily rather than relying upon the questionable process of comparing real number values. The uniform type can take one of four values; non-uniform knots, uniform, quasi-uniform or piecewise_bezier_knots. Uniform knots implies that all knots are of multiplicity one and and equally spaced, quasi-uniform is similar except that the first and last knots have multiplicity degree + 1. A Bezier, or piecewise Bezier curve has a corresponding B-Spline representation with the same control points, first and last knots of multiplicity degree + 1 and any knots corresponding to segment break points of multiplicity degree. The Express NVL function is used in this definition to provide appropriate default values if the knot data is omitted. In particular STEP enables the direct communication of Bezier curve and surface data without the user being obliged to generate redundant knot information.

```
*)
ENTITY b_spline_curve
SUBTYPE OF (bounded_curve);
degree                : INTEGER;
upper_index_on_control_points :  INTEGER;
control_points        :  INTEGER;
control_points        :  ARRAY [0:upper_index_on_control_points]
OF cartesian_point;
uniform_data          : OPTIONAL uniform_type;
upper_ind_knots_data:  OPTIONAL INTEGER;
knot_mult_data        : OPTIONAL ARRAY [1:upper_index_on_knots] OF
INTEGER;
knots_data            : OPTIONAL ARRAY [1:upper_index_on_knots] OF
REAL;
weights_data          : OPTIONAL ARRAY [0:upper_index_on_control_points]
of REAL;
form_number           : OPTIONAL bspline_curve_form;
closed_curve          : OPTIONAL BOOLEAN;
self_intersect        : OPTIONAL BOOLEAN;
DERIVE
uniform               : uniform_type
:= NVL(uniform_data, nonuniform_knots);
upper_index_on_knots:  INTEGER
:= NVL(upper_ind_knots_data, default_bspl_knots_upper
(degree, upper_index_on_control_points, uniform));
knot_multiplicities : ARRAY [1:upper_index_on_knots] of INTEGER
:= NVL(knot_mult_data, default_bspl_knot_mult
(degree, upper_index_on_knots, uniform));
knots                 : ARRAY [1:upper_index_on_knots] of REAL
:= NVL(knots_data, default_bspl_knots
(upper_index_on_knots, uniform));
```

```
weights              :   ARRAY [0:upper_index_on_control_points] of
REAL
:=NVL(weights_data, default_bspl_curve_weights
(upper_index_on_control_points));
WHERE
WR1  :     constraints_param_bspl(degree, upper_index_on_knots,
upper_index_on_control_points,
knot_multiplicities, knots, uniform);
WR2  :     constraints_geom_bspl_curve(b_spline_curve);
END_ENTITY;
(*
```

Acknowledgement

Some of the work described in this paper was funded as part of Esprit
Project 322: CAD Interfaces.

The information given in this paper on STEP and existing standards was
correct at the date of the conference, there has been further evolution of
the standards since that date.

Bibliography

1. National Institute of Standards and Technology (1988). IGES, Initial
   Graphics Exchange Specification, Version 4.0.

2. AFNOR (1989). Z68-300-1 Specifications du Standard d'Echange et
   de Transfert (SET).

3. DIN (1984). Verbund der Automobilindustrie-Flächen Schnittstelle
   (VDA-FS), DIN 66301.

4. Schlechtendahl, E.G. (Ed.) (1988). *Specification of a CAD*I Neutral
   File for CAD Geometry, Version 3.3*, Springer-Verlag, Berlin.

5. Lachance, M.A. (1987). Some Experiments on Transferring Polynomial Data. Esprit Project 322: CAD*I, Status Report 3, Kern-
   forschungszentrum Karlsruhe.

The Geometry of the Helical Canal Surface

R. R. Martin

University of Wales College of Cardiff

1 Introduction

Sweeping has been used for some time as a method of generating surfaces in Computer Aided Design. Initially, straight line sweeps of laminae were used for generating prismatic objects, whilst other sweeps about circular paths were used to define surfaces of revolution which could be readily turned on a lathe [5, 13]. More recently, algorithms have been devised for sweeping three dimensional solid models along quite complex paths, with applications both in shape design and interference checking [7, 11].

The object of this paper is to consider a particular type of swept surface. The class of surfaces formed by sweeping a sphere was first investigated by Monge [9] in 1850, who named them *Canal Surfaces*. In the particular case when the path which the sphere is swept along is a helix as shown in Fig. 1, and the sphere has constant radius, the surface swept out may be called a *Helical Canal Surface*. This surface is of engineering interest, as it is the surface of an ideal helical spring. Although occasionally computer generated images of objects such as corkscrews are seen in the literature which appear to include portions of a helical canal surface, such objects are usually a cheat—the apparent helical canal surface is not what it seems, but instead is several half or quarter tori joined together with a slight angle between their successive axes [8], or even many pieces of circular cylinder [2]. It would thus seem that there is a need to provide some details of the helical canal surface.

Canal surfaces may be generated either by sweeping a sphere along a path, or by sweeping a particular circular cross-section of the sphere along the same path. This fact was first shown by Monge [9] in the case where the path is a planar curve; Salmon [12] extended the proof to cover space curves.

More recently, Pegna [11] has pointed out that when it comes to machining canal surfaces, the swept sphere representation is a more natural model when a ball-ended cutter is being used, whilst the swept disk description is more appropriate when a 5-axis milling machine is being used with a disk cutter.

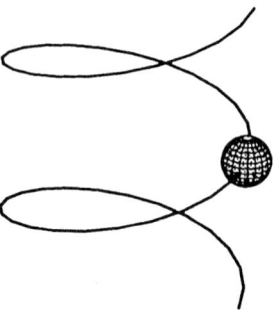

Figure 1. Sphere moving along a helical path

These two methods of sweeping a sphere and sweeping a disk will be discussed for the helical canal surface, showing how they lead to implicit and parametric formulae for this surface.

As can be seen, a helical canal surface is of infinite extent, and will intersect certain straight lines parallel to its axis an infinite number of times, which implies that the helical canal surface is not an algebraic surface. Most surfaces discussed for the purposes of Computer Aided Design have been assumed to be either polynomial or rational polynomial parametric surfaces, or more recently, implicit algebraic surfaces. This paper shows that more general surface types may be needed to describe other surfaces of real-world interest. It also shows how some of the methods devised for coping with the more restricted classes of surfaces may also be of use for these more general surfaces.

2   Sweeping and Envelopes

This section briefly reviews the mathematics of sweeping; for fuller details, the reader should consult [7]. Before proceeding to the details, however, we need to distinguish between two methods of defining sweeping. In the first case, we take a solid object, and move it through space. The set of all points which the solid occupies at some time during its motion gives the swept volume, whose boundary is the desired swept surface. In the second case, a *cross-section curve* (usually, but not necessarily, planar) is swept along a space curve, or *spine*. If the cross-section curve is planar, it is usually, but not necessarily, kept perpendicular to the spine. Here, the swept surface is the union of all instances of the cross-section curve taken at all instants of time.

In general, it is possible for the swept surface to intersect itself. We will ignore this possibility here.

2.1 Sweeping a solid volume

The theory of envelopes from classical differential geometry [14] provides us with the means of finding the swept surface in the first case. For simplicity here we will consider the object to be bounded by a single closed surface. The swept surface is then found by sweeping this bounding surface. The original surface may be expressed in either implicit or parametric form.

Sweeping an implicit surface

A surface in space may be written in implicit form as

$$f(x, y, z) = 0. \tag{2.1}$$

If we now allow this surface to move, we have a family of surfaces

$$F(x, y, z, t) = 0 \tag{2.2}$$

where a particular member of the family is obtained by chosing a specific value for t. (Often t is chosen to represent time, although it may also represent any other convenient function of time.)

The envelope to this family is the swept volume, which touches each member of the family. It can also be thought of as the surface which is the locus of intersections of two adjacent members of the family at times t and $t + dt$ in the limit dt tends to zero. Thus, the envelope is the surface which simultaneously satisfies

$$F(x, y, z, t) = 0, \qquad \frac{\partial F(x, y, z, t)}{\partial t} = 0, \tag{2.3}$$

and may be found by eliminating t from these two equations.

To describe the path of the object, we assume that the moving object is defined in terms of a set of coordinate axes which moves with the object, and which rotate as the object rotates. We then specify where the origin of these coordinates is at time t, and the relative orientation of these axes to the global coordinate system. Thus, given a description of the initial static object, and the path, we may find the implicit equation for the family of surfaces as follows. A point $\mathbf{p}(x, y, z)$ in the object will be at a point $\mathbf{p}(t)(x(t), y(t), z(t))$ at a time t, where $\mathbf{p}(t)$ is given by

$$\mathbf{p}(t) = \mathbf{p}R(t) + T(t), \tag{2.4}$$

and $R(t)$ is a 3×3 orthogonal matrix describing the rotation and $T(t)$ is a row vector giving the translation. The equation for the family of surfaces, $F(x, y, z, t) = 0$, can simply be found by inverting this equation to give

$$\mathbf{p} = (\mathbf{p}(t) - T(t)) R^{-1}(t), \tag{2.5}$$

and substituting into the equation $f(x, y, z) = 0$ for the original surface.

Sweeping a parametric surface

Alternatively, the original surface may have been given in parametric form

$$\mathbf{r} = (x(u,v),\ y(u,v),\ z(u,v))\,. \tag{2.6}$$

In this case, the family of surfaces can be described as a parametric hyper-surface in the form

$$\mathbf{R} = (x(u,v,t),\ y(u,v,t),\ z(u,v,t),\ t)\,, \tag{2.7}$$

found by substituting $\mathbf{r}$ for $\mathbf{p}$ in (2.4) to obtain the x, y, z components of $\mathbf{R}$; the fourth is simply t.

It can be shown [7] that in the parametric case, the condition for finding the common points between successive members of the family is that the determinant below is zero:

$$\begin{vmatrix} \dfrac{\partial x}{\partial t} & \dfrac{\partial y}{\partial t} & \dfrac{\partial z}{\partial t} \\[2mm] \dfrac{\partial x}{\partial u} & \dfrac{\partial y}{\partial u} & \dfrac{\partial z}{\partial u} \\[2mm] \dfrac{\partial x}{\partial v} & \dfrac{\partial y}{\partial v} & \dfrac{\partial z}{\partial v} \end{vmatrix} = 0. \tag{2.8}$$

This gives us a relationship between u, v, t, which in principle can be used to eliminate either u or v from $x - x(u, v, t) = 0$, and the corresponding relations for y and z. Ideally, this will lead to a parametric form for the envelope as either

$$\mathbf{R}\,(x(v,t),\ y(v,t),\ z(v,t)) \tag{2.9}$$

or

$$\mathbf{R}\,(x(u,t),\ y(u,t),\ z(u,t))\,. \tag{2.10}$$

Unfortunately, in practice, things may not be quite so simple. Instead of having $x = x(u, t)$, etc., a more complicated relationship may exist between x, u, and t, where to find points on the surface with given parametric values of u and t, we must solve a non-linear equation to obtain the corresponding values of x. The surface form we end up with

$$g(x, u, t) = 0$$
$$h(y, u, t) = 0 \tag{2.11}$$
$$k(z, u, t) = 0$$

is neither really a parametric form nor an implicit form. We will refer to it as an *implicit parametric form* henceforth.

This form may be converted to implicit form if desired by eliminating u and t. However, if we started with a parametric form for our original surface, it is likely that we would prefer a parametric form for the envelope.

2.2   Sweeping a cross-section curve

Let us now consider the second possibility of generating a surface by sweeping a cross-section curve, using a second curve as an axis. As might be expected, this is somewhat easier. It is simplest here to suppose that the cross-section curve is described in parametric form, rather than as the intersection of a plane and another surface. In this case it will have the form

$$\mathbf{r} = (x(u), y(u), z(u)),\qquad\qquad(2.12)$$

where u varies as we go round the curve; u may, for example, be arc-length.

We may describe the motion of this cross-section along the spine again using (2.4). A reference point on the spine at each moment in time gives the $T(t)$ component of the motion. One way of defining $R(t)$ if the cross-section is a planar curve is to insist that it remains perpendicular to the spine; a further constraint is needed to complete the description, which gives the rotation of the plane about the spine. One possibility is to chose a reference direction in the plane to be always aligned with the binormal of the spine curve [6, 10].

The parametric form of the swept surface may now be easily found by substituting $\mathbf{r}$ for $\mathbf{p}$ in (2.4) to obtain its x, y, z components directly as $(x(u,t), y(u,t), z(u,t))$.

An interesting observation which can be made by comparing this result with the one given for sweeping a parametric surface is as follows:

Theorem 1. Any *sweep of a parametric surface which can also be expressed as the sweep of a parametric cross-section curve (planar or nonplanar) along a parametric spine curve may always be expressed as a new parametric surface of the form* $\mathbf{R}(x(u,t), y(u,t), z(u,t))$ *rather than in implicit parametric form.*

As the mixed form is computationally more tricky to work with, this suggests that wherever possible swept surfaces should be generated by sweeping cross-sections in preference to sweeping surfaces. This is also desirable from the point of view that less algebra is involved when starting from a cross-section, as will be demonstrated later.

3   The Helical Canal Surface

We will now consider how each of the three approaches described above can be used to find descriptions of a helical canal surface.

When considering such a surface, there are three free parameters which we are at liberty to choose, which can most readily be expressed in terms of the generating sphere and helix. The first is the radius of the sphere, r. The second is the radius of the helix, a (*i.e.* the perpendicular distance

from a point on the axis of the helix to a point on the helix). The third is
the pitch of the helix, $2\pi b$ (*i.e.* the distance between successive turns of the
helix, measured parallel to the axis). Using these definitions, the equation
of the helix which the sphere moves on is

$$T(t) = (a\cos(t),\ a\sin(t),\ bt)\,, \qquad (3.1)$$

where the axis of the helix is the z-axis.

   In what follows, we will consider the three approaches described above
in a slightly different order, as this will make it easier to discuss the relationships between the results obtained from these methods.

3.1   Generation by sweeping an implicit sphere

If we allow each of r, a and b to be a variable in what follows, the algebra
becomes tedious in the extreme, requiring an unreasonable time to work
through even for a computer algebra system. However, we can simplify
matters by assuming that we are using units of length such that the radius
of the sphere is one unit, and by measuring a and b in these units. At the
end of our calculations, we can obtain the general solution by substituting
x/r for x, y/r for y, and z/r for z in the final result, where r is the radius
of the sphere. At the same time we must, of course, also substitute a/r for
a and b/r for b.

   To decide where points on the sphere are after a time t using (2.4) we
need $R(t)$ as well as $T(t)$. Obviously, as rotating a sphere makes no change
to it, we can just choose $R(t)$ to be a unit matrix.

   We can now use the method of Section 2.1 to obtain $F(x, y, z, t) = 0$
using the equation for a unit sphere centered at the origin, and $T(t)$ and
$R(t)$ as above:

$$(x - a\cos(t))^2 + (y - a\sin(t))^2 + (z - bt)^2 - 1 = 0. \qquad (3.2)$$

Differentiating and cancelling, we obtain for $\partial F/\partial t = 0$

$$ax\sin(t) - ay\cos(t) - bz + b^2t = 0. \qquad (3.3)$$

   The presence of trigonometric functions here makes direct elimination
of t from these two equations rather tricky. The best way of proceeding is
to replace the trigonometric functions by new variables

$$s = \sin(t), \qquad c = \cos(t), \qquad (3.4)$$

and to introduce a third equation linking them,

$$s^2 + c^2 - 1 = 0. \qquad (3.5)$$

We now have three algebraic equations, from which we wish to eliminate
the three algebraic variables s, c and t. Clearly, we can only eliminate

any two variables from three such equations. The outline of the procedure we adopt is as follows. Firstly, we eliminate s and c to give a polynomial equation only containing t, and secondly we eliminate t and c to give another polynomial equation for s. We can then make use of the relationship $s = \sin(t)$ to equate corresponding solutions of the two polynomial equations to finally arrive at the desired result, an implicit surface form containing neither s nor t.

Elimination of the variables from algebraic equations can be *systematically* performed in several ways, for example by using Multivariate Resultants [1], or Gröbner Bases [3, 4]. These particular methods have the advantages that no spurious solutions are introduced as an artifact of the method of performing the elimination. Gröbner Basis methods were used to perform the calculations described in this paper, using the *MACSYMA* computer algebra system.

Eliminating s and c from (3.2), (3.3) and (3.5) gives the following quartic polynomial in t, after simplification:

$$
\begin{aligned}
b^4 t^4 &- 4b^3 z t^3 + 2b^2(3z^2 + y^2 + x^2 + 2b^2 + a^2 - 1)t^2 \\
&- 4bz(z^2 + y^2 + x^2 + 2b^2 + a^2 - 1)t \\
&+ z^4 + 2(y^2 + x^2 + 2b^2 + a^2 - 1)z^2 \\
&+ (y^2 + x^2 - a^2 - 2a - 1)(y^2 + x^2 - a^2 + 2a - 1) = 0.
\end{aligned}
\tag{3.6}
$$

Similarly, eliminating c and t from (3.2), (3.3) and (3.5) gives the following quartic polynomial in s:

$$
\begin{aligned}
a^4(y^2 + x^2)^2 s^4 &+ 4a^3 b^2 y(y^2 + x^2)s^3 \\
- 2a^2(b^2 y^4 &+ a^2 y^4 + a^2 x^2 y^2 - 2b^4 y^2 + a^2 b^2 y^2 \\
- b^2 y^2 &- b^2 x^4 - 2b^4 x^2 - a^2 b^2 x^2 + b^2 x^2)s^2 \\
- 4ab^2 y(b^2 y^2 &+ a^2 y^2 + b^2 x^2 + 2a^2 x^2 + a^2 b^2 - b^2)s \\
+ (b^2 y^2 &+ a^2 y^2 + b^2 x^2 - 2ab^2 x + a^2 b^2 - b^2) \times \\
(b^2 y^2 &+ a^2 y^2 + b^2 x^2 + 2ab^2 x + a^2 b^2 - b^2) = 0.
\end{aligned}
\tag{3.7}
$$

We must now solve these quartic equations for t and s, and equate the corresponding solutions.

The four solutions for t are

$$
t_1 = \frac{z - \sqrt{-y^2 - x^2 - 2w^2 - 2b^2 - a^2 + 1}}{b},
\tag{3.8}
$$

$$
t_2 = \frac{z + \sqrt{-y^2 - x^2 - 2w^2 - 2b^2 - a^2 + 1}}{b},
\tag{3.9}
$$

$$t_3 = \frac{z - \sqrt{-y^2 - x^2 + 2w^2 - 2b^2 - a^2 + 1}}{b}, \tag{3.10}$$

$$t_4 = \frac{z + \sqrt{-y^2 - x^2 + 2w^2 - 2b^2 - a^2 + 1}}{b}, \tag{3.11}$$

where w^2 is defined as the positive root of

$$w^4 = (b^2 + a^2)(y^2 + x^2) + (a^2 + b^2 - 1)b^2. \tag{3.12}$$

For a non-self-intersecting helical canal surface, a must be greater than one. If we apply this restriction, it can be seen that the first two roots must be complex, and thus can be discarded.

This may be explained from geometrical considerations. If we take suitable values of x and y for a point lying on the envelope surface, and the corresponding value for t, on substituting these into (3.10) and (3.11), we see that this gives two values for z. These correspond to points on the upper and lower hemispheres of the generating sphere each with the given values of x and y.

The four solutions for s are

$$s_1 = -\frac{(w^2 + b^2)y + bx\sqrt{-y^2 - x^2 - 2w^2 - 2b^2 - a^2 + 1}}{ay^2 + ax^2} \tag{3.13}$$

$$s_2 = -\frac{(w^2 + b^2)y - bx\sqrt{-y^2 - x^2 - 2w^2 - 2b^2 - a^2 + 1}}{ay^2 + ax^2} \tag{3.14}$$

$$s_3 = \frac{(w^2 - b^2)y - bx\sqrt{-y^2 - x^2 + 2w^2 - 2b^2 - a^2 + 1}}{ay^2 + ax^2} \tag{3.15}$$

$$s_4 = \frac{(w^2 - b^2)y + bx\sqrt{-y^2 - x^2 + 2w^2 - 2b^2 - a^2 + 1}}{ay^2 + ax^2} \tag{3.16}$$

where w is defined before. Again, in this case, it can be seen that the first two roots are complex, and can be discarded.

We now need to determine how the possibilities for s are to be paired with the possibilities for t.

On eliminating c between (3.2), (3.3) and (3.5), using Gröbner Bases or otherwise, we may obtain the following simple relationship which is linear in s, and quadratic in t:

$$2as(y^2 + x^2) - b^2 t^2 y + 2bt(yz + bx)$$
$$- y(z^2 + y^2 + x^2 + a^2 - 1) - 2bxz = 0. \tag{3.17}$$

If we now substitute in turn the four different combinations of (s_3, t_3), (s_3, t_4), (s_4, t_3), and (s_4, t_4) into this relationship, we find that the pairs

(s_3, t_4) and (s_4, t_3) give identically zero results while the other two pairs do not.

Thus, we can now state that the relations which give the two branches of the surface are

$$s_3 = \sin(t_4), \qquad s_4 = \sin(t_3). \tag{3.18}$$

Hence the overall helical canal surface is

$$(s_3 - \sin(t_4))\,(s_4 - \sin(t_3)) = 0, \tag{3.19}$$

or, expanding this out and ignoring the denominators which are always positive,

$$
\begin{aligned}
\left(a\,\left(y^2 + x^2\right)\sin\left((z - v)/b\right) + \left(b^2 - w^2\right)y - b\,v\,x\right) &\quad \times \tag{3.20}\\
\left(a\,\left(y^2 + x^2\right)\sin\left((z + v)/b\right) + \left(b^2 - w^2\right)y + b\,v\,x\right) &\quad = 0
\end{aligned}
$$

where v is given by the positive root of

$$v = \sqrt{-y^2 - x^2 + 2\,w^2 - 2\,b^2 - a^2 + 1}. \tag{3.21}$$

Finally, if we allow for a sphere of radius r rather than unit radius using the observations made earlier, we obtain the following implicit form for the general helical canal surface:

$$
\begin{aligned}
\left(a\,\left(y^2 + x^2\right)\sin\left((z - v')/b\right) + \left(b^2 - w'^2\right)y - b\,v'\,x\right) &\quad \times \tag{3.22}\\
\left(a\,\left(y^2 + x^2\right)\sin\left((z + v')/b\right) + \left(b^2 - w'^2\right)y + b\,v'\,x\right) &\quad = 0
\end{aligned}
$$

where w'^2 is the positive root of

$$w'^4 = \left(b^2 + a^2\right)\left(y^2 + x^2\right) + \left(a^2 + b^2 - r^2\right)b^2 \tag{3.23}$$

and v' is given by

$$v' = \sqrt{-y^2 - x^2 + 2\,w^2 - 2\,b^2 - a^2 + r^2}. \tag{3.24}$$

3.2 Generation by sweeping a disk

In this approach, the helical canal surface will be found by sweeping a disk, which we shall keep perpendicular to the generating helix at all times. If we project a single turn of the helix onto the x-y plane, we can see that it goes round a distance $2\pi a$ in a single turn while it goes up a distance $2\pi b$. Thus the helix makes a constant angle α to the x-y plane, where $\tan\alpha = b/a$.

If we take a disk in the x-z plane, of radius r, its parametric form in a local coordinate system centred on $T(0) = (a, 0, 0)$ and aligned with the global coordinate system is

$$(r \cos \theta, \, 0, \, r \sin \theta), \qquad (3.25)$$

where θ measures the angle round the disk.

If we now tip this disk over by an angle α about the x-axis, so that it is perpendicular to the helix at this point, its parametric form becomes

$$\mathbf{p} = (r \cos \theta, \, -r \sin \theta \sin \alpha, \, r \sin \theta \cos \alpha). \qquad (3.26)$$

As the disk sweeps out the helical canal surface, it rotates about the global z-axis at unit speed. Thus $R(t)$ is given by

$$R(t) = \begin{pmatrix} \cos t & \sin t & 0 \\ -\sin t & \cos t & 0 \\ 0 & 0 & 1 \end{pmatrix}. \qquad (3.27)$$

We may now directly put these values for $\mathbf{p}$, $R(t)$ and $T(t)$ into (2.4) to obtain the following parametric form for the helical canal surface:

$$
\begin{aligned}
x(\theta, t) &= r \sin \alpha \sin \theta \sin t + r \cos \theta \cos t + a \cos t & (3.28) \\
y(\theta, t) &= -r \sin \alpha \sin \theta \cos t + r \cos \theta \sin t + a \sin t & (3.29) \\
z(\theta, t) &= r \cos \alpha \sin \theta + bt. & (3.30)
\end{aligned}
$$

The relative ease with which a parametric form for the helical canal surface has been obtained by sweeping a planar cross-section rather than the solid sphere should be carefully noted.

Fig. 2 shows a facetted representation bounded by constant parameter lines of a helical canal surface generated by sweeping a disk. The *Mathematica* computer algebra system was used to generate the image.

Finally, let us take the three components of the parametric form of the surface as given by (3.28)–(3.30), replace $\sin \alpha$ and $\cos \alpha$ by $b/\sqrt{a^2 + b^2}$ and $a/\sqrt{a^2 + b^2}$, and choose the special case of $r = 1$. Let us further replace $\sin t$ and $\cos t$ by s and c, and replace $\sin \theta$ and $\cos \theta$ by s_θ and c_θ. Let us also take the complementary equations $s^2 + c^2 = 1$ and $s_\theta^2 + c_\theta^2 = 1$. Then, if we algebraically eliminate s, c, s_θ and c_θ from these five equations, we arrive once more at (3.7), linking x, y, z and t, while if we eliminate t, c, s_θ and c_θ the result is (3.8) between x, y, z and s, showing agreement between this parametric form and the implicit form obtained earlier.

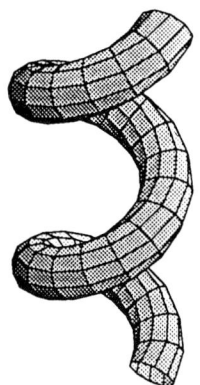

Figure 2. Helical canal surface generated by sweeping a disk

3.3 Generation by sweeping a parametric sphere

Using polar coordinates aligned with the z-axis

If we start with the usual polar form for a parametrically defined sphere

$$\mathbf{r} = (r \cos \phi \sin \psi,\ r \sin \phi \sin \psi,\ r \cos \psi) \tag{3.31}$$

then on using the method of Section 2.1, and $R(t)$ and $T(t)$ as in the previous Section, we obtain the following form for $\mathbf{R}(\psi, \phi, t)$:

$$
\begin{aligned}
x(\psi, \phi, t) &= -r \sin \phi \sin t \sin \psi + r \cos \phi \cos t \sin \psi + a \cos t & (3.32) \\
y(\psi, \phi, t) &= r \cos \phi \sin t \sin \psi + r \sin \phi \cos t \sin \psi + a \sin t & (3.33) \\
z(\psi, \phi, t) &= r \cos \psi + bt. & (3.34)
\end{aligned}
$$

On evaluating the determinant in (2.8), we obtain

$$r^2 \sin \psi\, (a \sin \phi \sin \psi + b \cos \psi) = 0. \tag{3.35}$$

We may ignore the solution $\sin \psi = 0$, which we can immediately see is not on the desired envelope, leaving us with

$$a \sin \phi \sin \psi + b \cos \psi = 0. \tag{3.36}$$

By introducing new algebraic variables for $\sin \psi$ and $\cos \psi$ as was done previously, we may use the equation above to eliminate ψ from (3.32)–(3.34). This gives us the following quadratics for $x(\phi, t)$, $y(\phi, t)$ and $z(\phi, t)$:

$$(a^2 \sin^2 \phi + b^2) \, x^2 + (-2 \, a^3 \sin^2 \phi - 2 \, b^2 \, a) \cos t \, x$$
$$- \, b^2 \sin^2 \phi \, r^2 \sin^2 t + 2 \, b^2 \cos \phi \sin \phi \, r^2 \cos t \sin t$$
$$+ (-b^2 \cos^2 \phi \, r^2 + a^4 \sin^2 \phi + b^2 \, a^2) \cos^2 t \quad = \quad 0, \quad (3.37)$$

$$(a^2 \sin^2 \phi + b^2) \, y^2 + (-2 \, a^3 \sin^2 \phi - 2 \, b^2 \, a) \sin t \, y$$
$$+ (-b^2 \cos^2 \phi \, r^2 + a^4 \sin^2 \phi + b^2 \, a^2) \sin^2 t$$
$$- \, 2 \, b^2 \cos \phi \sin \phi \, r^2 \cos t \sin t - b^2 \sin^2 \phi \, r^2 \cos^2 t \quad = \quad 0, \quad (3.38)$$

$$(b^2 + a^2 \sin^2 \phi) \, z^2 + (-2 \, b^3 - 2 \, a^2 \sin^2 \phi \, b) \, t \, z$$
$$+ (b^4 + a^2 \sin^2 \phi \, b^2) \, t^2 - a^2 \sin^2 \phi \, r^2 \quad = \quad 0. \quad (3.39)$$

The roots of these are found to be, after trigonometric simplification,

$$x_1 \quad = \quad a \cos t + \frac{b \, r \cos (t + \phi)}{\sqrt{w}} \qquad (3.40)$$

$$x_2 \quad = \quad a \cos t - \frac{b \, r \cos (t + \phi)}{\sqrt{w}} \qquad (3.41)$$

$$y_1 \quad = \quad a \sin t - \frac{b \, r \sin (t + \phi)}{\sqrt{w}} \qquad (3.42)$$

$$y_2 \quad = \quad a \sin t + \frac{b \, r \sin (t + \phi)}{\sqrt{w}} \qquad (3.43)$$

$$z_1 \quad = \quad b \, t - \frac{a \, r \sin \phi}{\sqrt{w}} \qquad (3.44)$$

$$z_2 \quad = \quad b \, t + \frac{a \, r \sin \phi}{\sqrt{w}}, \qquad (3.45)$$

where w is given by

$$w = a^2 \sin^2 \phi + b^2. \qquad (3.46)$$

From the definition of polar coordinates, ψ is in the range $0 \le \psi \le \pi$, and hence $\sin \psi$ is always positive. If we also take $\sqrt{w}$ to be the positive root, we can immediately see that by comparing the above solutions with (3.32) and (3.33) that we must choose $x = x_1$ and $y = y_2$. Finally, if we use (3.36) to replace $\cos \psi$ by $\sin \psi$ in z_1 and z_2, we can also similarly determine that $z = z_1$. We thus obtain the following parametric form for the helical canal surface:

$$x(\phi, t) \quad = \quad a \cos t + \frac{b \, r \cos (t + \phi)}{\sqrt{w}} \qquad (3.47)$$

Figure 3. Helical canal surface generated by sweeping a sphere

$$y(\phi,t) \;=\; a \sin t + \frac{b\, r\, \sin(t+\phi)}{\sqrt{w}} \tag{3.48}$$

$$z(\phi,t) \;=\; b\,t - \frac{a\, r\, \sin \phi}{\sqrt{w}}. \tag{3.49}$$

A facetted representation of a helical canal surface generated using this parametric form is shown in Fig. 3. Note the different appearance of the parameter lines compared to Fig. 2. This is a different parametrisation of the helical canal surface to the one obtained by sweeping the disk.

Nevertheless, we can find a connection between these two parametrisations. If we expand out the trigonometric functions in the above equations, and compare them with the parametrisation obtained by sweeping a disk, we may equate the coefficients in $\cos t$, $\sin t$, t and 1. On doing this, five non-trivial equations result. It can be shown that they can all be satisfied provided that we choose

$$\cos \theta \;=\; \frac{b \cos \phi}{\sqrt{w}}, \tag{3.50}$$

$$\sin \theta \;=\; \frac{-\sqrt{a^2 + b^2}\, \sin \phi}{\sqrt{w}}, \tag{3.51}$$

which also satisfy $\cos^2 \theta + \sin^2 \theta = 1$ as required.

Using polar coordinates based on sweeping a disk

If we return to Section 3.2, and look at the parametric form for the disk being swept, repeated here for convenience,

$$\mathbf{p} = (r \cos \theta,\; -r \sin \theta \sin \alpha,\; r \sin \theta \cos \alpha), \tag{3.52}$$

this represents a circle in space as θ is varied. However, if instead of regarding α as a fixed constant, we regard α as variable, then the same equation is the parametric equation for a sphere using polar coordinates initially aligned with the x-axis.

Thus, we may also consider the generation of a helical canal surface by sweeping a parametric sphere defined by the above Equation using the method of Section 2.1, using $R(t)$ and $T(t)$ as defined in Section 3.2. On evaluating the determinant in (2.8), we obtain after simplification

$$\tan \alpha = \frac{b}{a} \tag{3.53}$$

which is how α was defined originally in Section 3.2. Using the notation of Section 2.1, the relationship between the variables u, v, and t, or in this case α, θ, and t is a particularly simple one — α is a constant, immediately allowing us to eliminate it and arrive at (3.28)–(3.30) again as a parametric form of the helical canal surface:

$$\begin{aligned}
\mathbf{R}(\theta, t) \quad = \quad & (r \sin \alpha \sin \theta \sin t + r \cos \theta \cos t + a \cos t, \\
& -r \sin \alpha \sin \theta \cos t + r \cos \theta \sin t + a \sin t, \qquad (3.54) \\
& r \cos \alpha \sin \theta + bt).
\end{aligned}$$

This particular result illustrates the following general observation, which goes somewhat further than Salmon's proof [12] that any canal surface can be generated either by sweeping a sphere or by sweeping a disk:

Theorem 2. *Generating a canal surface by sweeping a disk is equivalent to generating a canal surface by sweeping a sphere under the following circumstances. Let the sphere be parameterized in such a way that, and $R(t)$ and $T(t)$ be chosen such that, the disk is a great circle which stays as the same fixed constant parameter line on the sphere as the sphere moves. Then if this is so, the two methods of generation are equivalent in the sense that the same parametric form is obtained for the resulting canal surface.*

4   Conclusions

This paper has discussed swept envelopes in general. It has also shown several different ways of generating helical canal surfaces in particular, giving both parametric and implicit forms and showing the relationships between them. Finding the parametric form by sweeping a disk was much the simplest method.

Unlike most surfaces currently considered in computer aided geometric modelling, the helical canal surface does not have a rational polynomial form, nor is it an algebraic surface. However, techniques applicable to such surfaces have proved to be practical in elucidating the details of canal surfaces, using the approach of introducing extra variables which are removed at the end of the calculations. In particular, computer algebra systems utilising algorithms for computing Gröbner Bases have been used extensively throughout this paper for systematic elimination of variables. Perhaps this paper will encourage the consideration of other non-algebraic surfaces which may also be of interest in computer aided geometric design.

References

1. Bajaj, C., Garrity, T. and Warren, J. (1988). *On the Applications of Multi-Equational Resultants.* Technical Report CSD-TR-826, Computer Sciences Department, Purdue University.

2. Blinn, J.F. (1989). Optimal tubes. *IEEE Computer Graphics and its Applications, 9*, (5).

3. Buchberger, B. (1983). Gröbner Bases: an algorithmic method in polynomial ideal theory. in *Recent Trends in Multidimensional Systems*, Ed. N. K. Bose, D. Reidel, 16–28.

4. Buchberger, B. (1987). Applications of Gröbner Bases in non-linear computational geometry. *Proc. Workshop on Scientific Software, Minneapolis, 1987.* IMA Volumes in Mathematics and its Applications 14, Springer Verlag.

5. Cary, C.A.G. (1979). *BUILD User's Guide* C. A. D. Group Document No. 102, Cambridge University Computer Laboratory.

6. Klok, F. (1986). Two moving coordinate frames for sweeping along a 3D trajectory. *Computer Aided Geometric Design, 3*, 217–229.

7. Martin, R.R. and Stephenson, P.C. (1989). Swept volumes in solid modellers. *The Mathematics of Surfaces III*, Ed. D. C. Handscomb, Oxford University Press.

8. Middleditch, A.E. (1989). Personal Communication.

9. Monge, G. (1850). *Application de l'Analyse à la Géométrie* Bachelier, Paris.

10. Nutbourne, A.W. and Martin, R.R. (1988). *Differential Geometry Applied to Curve and Surface Design Volume 1: Foundations.* Ellis Horwood, Chichester.

11. Pegna, J. (1988). *Variable sweep geometric modeling.* Ph.D. Thesis, Stanford University.

12. Salmon, G. (1882). *A Treatise on the Analytic Geometry of Three Dimensions (4th Edition)*. Hodges and Figgs, Dublin.

13. Shiroma, Y., Okino, N. and Kakazu, Y. (1982). Research on 3-D geometric modelling by sweep primitives. *Proc. CAD '82 Conference*, Butterworths.

14. Weatherburn, C.E. (1927). *Differential Geometry of Three Dimensions*. Cambridge University Press.

Efficient Multivariate Approximation Methods for Lines and Curves of Data

J.C. Mason and R.P. Bennell

Applied and Computational Mathematics Group, Royal Military College of Science (Cranfield)

1 Introduction

There are important practical applications in which measurements of a dependent variable z are available along a number of curved paths in the plane of the independent variables x and y. For example, soundings of ocean depths may be taken from a vessel which follows a number of curved tracks across a relevant area, or heights and related readings may be recorded by a reconnaissance aircraft passing along curved paths over a region. In Fig. 1 we show a set of such data, namely ocean depth measurements (the decimal point denoting each measurement position). Clearly there is a structure to such data and it would be inefficient in principle to treat the data as scattered.

Handscomb and MacCarthy [1] consider ocean-depth data on curves and use local techniques to refer it first to a rectangular mesh before using this to form an approximation to the sea-bed surface $z(x,y)$. A mesh of data, as in Fig. 2, may be fitted extremely rapidly, and indeed, lines of data, where ordinates occur in a variety of positions on each line as in Fig. 3 and Fig. 4 may also be fitted efficiently. Hayes [2] gives an excellent discussion of both these cases and we summarise the situation in §2 and §3 below, but no comparably efficient algorithm has, to our knowledge, been specifically tailored to a family of curves of data as in Fig. 5 and Fig. 7. One of the aims of the present paper is to provide a simple approach to curves of data, based on the generation of lines of data from the given curves. Indeed the algorithm which we present for curves has an operations count which is essentially twice that of the Clenshaw-Hayes algorithm [3] for lines of data (but in the context of a more general measure of the degree of approximation), and this would therefore appear to be a very efficient procedure.

Other types of data are also considered here; we consider firstly data on two families of perpendicular lines (as in Fig. 8), and secondly data on

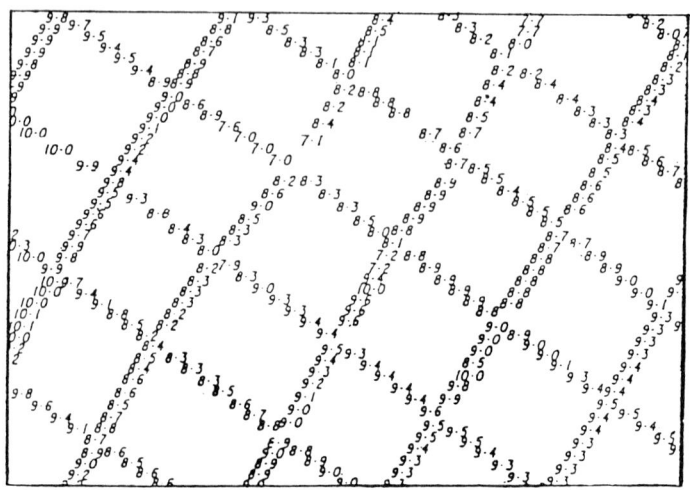

Figure 1. Data on two families of curves - ocean depth measurements

two intersecting families of curves (as in Fig. 1). We shall also consider techniques for dealing with a variety of boundaries to the data, and in particular consider the possibility of transforming boundaries to yield a rectangular region or alternatively of fitting on a fixed interior rectangle.

Most of the discussion and numerical tests are based here on the use of Chebyshev polynomial techniques. However, B-splines are also considered, and indeed our general approach is, in principle, applicable to any linear approximating form.

2 Meshes of Data - the Forsythe-Clenshaw Algorithm

If data are available, as in Fig. 2, on a mesh of points (x_i, y_i), namely the tensor product of abscissae

$$x = x_1, x_2, \ldots, x_{m_1} \text{ and } y = y_1, y_2, \ldots, y_{m_2}$$

then a least-squares bivariate polynomial approximation may be obtained by exploiting a one-dimensional procedure.

Following Forsythe [4], orthogonal polynomials $\{P_i(x)\}$, (where P_i is of degree i in x), on $x_1, x_2, \ldots, x_{m_1}$, may be used to approximate $z(x, y)$ on each of the lines $y = y_1, y_2, \ldots, y_{m_2}$ in the form

$$z \approx \sum_{i=1}^{n_1} a_i(y) P_{i-1}(x) \tag{2.1}$$

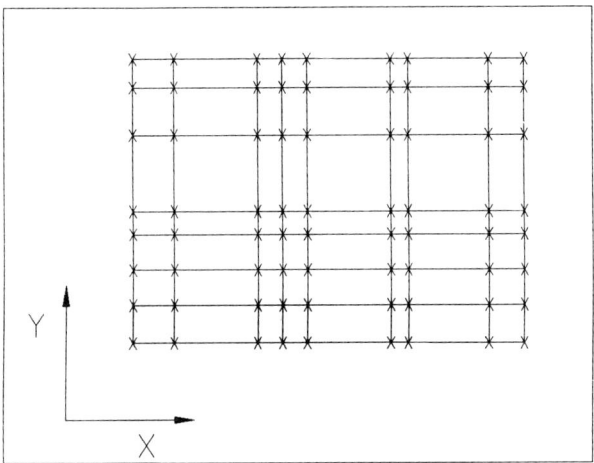

Figure 2. Rectangular mesh of data

where it is assumed that the range of the data is adjusted to $[-1, 1]$ in both x and y. The first approximation (2.1) takes $O(m_1 n_1^2)$ operations, while each subsequent approximation takes $O(m_1 n_1)$, so that the total operations count is,

$$O(m_1 m_2 n_1) + O(m_1 n_1^2). \tag{2.2}$$

Clenshaw [5] noted that it was more convenient in practice to generate (2.1) directly in the Chebyshev form,

$$z \approx \sum_{i=1}^{n_1} b_i(y) T_{i-1}(x) \tag{2.3}$$

where $T_{i-1}(x)$ is the Chebyshev polynomial of the first kind of degree $i-1$, and gave an algorithm for which the operations count was also of the form (2.2).

The next step is to use the same least-squares algorithm as above, but now in the y variable, to fit each coefficient $b_i(y)$ by a polynomial in y of the form,

$$b_i(y) \approx \sum_{j=1}^{n_2} c_{i,j} T_{j-1}(y) \qquad i = 1, 2, \ldots, n_1 \tag{2.4}$$

based on orthogonal polynomials $Q_{j-1}(y)$ which are defined on the point set $y_1, y_2, \ldots, y_{m_2}$, to give a final approximation

$$z \approx \sum_{i=1}^{n_1} \sum_{j=1}^{n_2} c_{i,j} T_{i-1}(x) T_{j-1}(y). \tag{2.5}$$

The additional operations involved in generating (2.3) amount to $O(m_2 n_2^2)$ for the first $b_i(y)$ and $O(m_2 n_2)$ for each subsequent $b_i(y)$, giving an additional count of,

$$O(m_2 n_1 n_2) + O(m_2 n_2^2). \tag{2.6}$$

The sum of the operations in (2.2) and (2.6) is thus

$$O(m^2 n) + O(mn^2) \tag{2.7}$$

if we assume that $m_1 = m_2$ and $n_1 = n_2$, and, moreover, if $m \propto n$, this becomes $O(n^3)$ operations.

Clearly the algorithm is dominated by the $O(m_1 m_2 n_1)$ operations in the first least-squares fit, and so for greater efficiency the axes x and y should be chosen (or interchanged) so that $n_1 \leq n_2$.

The algorithm of Clenshaw [5] is implemented in the NAG Library routine E02ADF, but no routine is provided to carry out the bivariate algorithm of this section. The coding of this procedure is not a trivial application of E02ADF, since, to achieve $O(m^2 n)$ operations it is necessary to have access to, or code, the inner product procedures from the one-dimensional algorithm.

3   Lines of Data - the Clenshaw-Hayes Algorithm

3.1   Data bounded by lines ($x = \pm 1$)

Suppose that lines of data x are given, as in Fig. 3, on $y = y_1, y_2, \ldots, y_{m_2}$, and that the x data change from one y-line to another (so that the data do not form a rectangular mesh), then it is no longer appropriate to use the orthogonal representation (2.1) on each y-line, since $\{P_{i-1}(x)\}$ will in fact vary with y (as the data x vary). However, following Clenshaw and Hayes [3], we may use the Chebyshev representations (2.3) and (2.4) to fit z in the bivariate form (2.5). Here (2.3) is the least-squares approximation in x to z on $y = y_1, y_2, \ldots, y_{m_2}$, based on a different orthogonal polynomial system for each y, and then (2.4) is the least-squares approximation in y to the coefficient $b_i(y)$ over the data points $y_1, y_2, \ldots, y_{m_2}$.

This algorithm does not yield the true least-squares approximation over the complete data set, but in practice it produces an approximation which is close to this.

Clearly this algorithm is more expensive than that of §2, as a consequence of the need to recompute orthogonal polynomials. Thus (2.3) involves m_2 Forsythe-Clenshaw procedures each of $O(m_1 n_1^2)$, and (2.4) then involves n_1 Forsythe-Clenshaw procedures each of $O(m_2 n_2^2)$, giving a total of

$$O(m_1 m_2 n_1^2) + O(m_2 n_1 n_2^2). \tag{3.1}$$

If we assume that $m_1 = m_2$ and $n_1 = n_2$, this gives

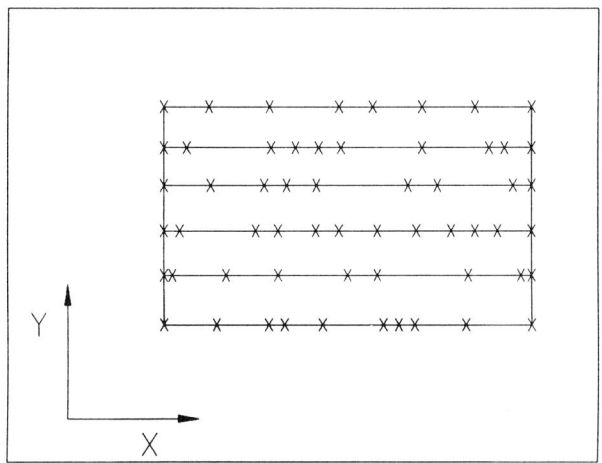

Figure 3. Lines of data - bounded by lines

$$O(m^2n^2) + O(mn^3) \tag{3.2}$$

operations, namely $m + n$ Forsythe-Clenshaw least-squares procedures. If in addition $m \propto n$, then this becomes $O(n^4)$ operations.

We note that an extra factor of n (or n_1 in (3.1)) is introduced into this algorithm compared with the mesh-based algorithm of §2. The above algorithm has been implemented in a slightly more general form in the NAG library routine E02CAF.

3.2 Data bounded by curves

If, as in Fig. 4, the data lines are bounded at the extremes of x by two curves $x = A(y)$ and $x = B(y)$, rather than by $x = \pm 1$, then following Hayes [2] we may transform these to parallel lines $\bar{x} = \pm 1$ by writing

$$\bar{x} = \frac{2x - A(y) - B(y)}{B(y) - A(y)}. \tag{3.3}$$

The algorithm then proceeds as before with the new variables $\bar{x}$ and y and leads to a bivariate polynomial approximation in $\bar{x}$ and y of the form

$$z \approx \sum_{i=1}^{n_1} \sum_{j=1}^{n_2} c_{i,j} T_{i-1}(\bar{x}) T_{j-1}(y). \tag{3.4}$$

The evaluation of z for a given (x, y) is then reasonably straightforward from (3.4) and (3.3). However, in some applications (e.g. in using some

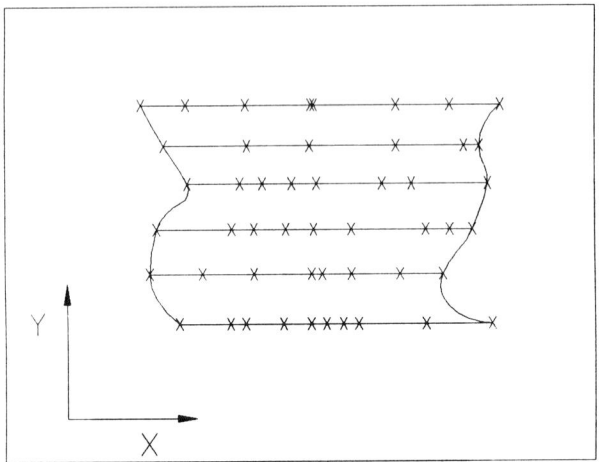

Figure 4. Lines of data - bounded by curves

surface generation computer packages), it is not too convenient that a rectangular mesh of (x, y) points does not correspond to a rectangular mesh in $(\bar{x}, y)$.

An alternative approach is to use Chebyshev polynomials on a fixed range of x interior to all pairs of curves $x = A(y), B(y)$; this actually leads to (at least) two possible procedures.

In the first procedure, on any given line y of data, all data between $A(y)$ and $B(y)$ are adopted, an approximation

$$z \approx \sum_{i=1}^{n_1} \bar{b}_i(y) T_{i-1}(\bar{x}) \tag{3.5}$$

(where $\bar{x}$ is given by (3.3)), is formed by least-squares, and (3.5) is then re-expressed in terms of a fixed set of Chebyshev polynomials $T_{i-1}(X)$, where

$$X = \frac{2x - x_A - x_B}{x_B - x_A}$$

in the form,

$$z \approx \sum_{i=1}^{n_1} b_i(y) T_{i-1}(X). \tag{3.6}$$

Here x_A and x_B might be chosen, for example, so that

$$x_A = \max_y A(y) \quad \text{and} \quad x_B = \min_y B(y).$$

For tidiness, the range $[x_A, x_B]$ is then translated to $[-1, 1]$ so that $T_{i-1}(X)$ becomes the standard Chebyshev polynomial $T_{i-1}(x)$. This yields, on fitting $b_i(y)$ by least-squares over y, to a bivariate approximation (2.5) in (x, y) with standard Chebyshev polynomials. This approximation may then be used over the whole (non-rectangular) region of the data, and it may also be readily used with computer packages to give surface fits over any rectangular domain, such as $[-1, 1] \times [-1, 1]$, of (x, y). Clearly this representation is going to be most effective if the domain of the data is not dramatically different from a rectangular one.

In the second procedure, we simply use only that part of the original data which lies in a chosen (fixed) rectangular region and then proceed with the algorithm of §3.1 on this new region.

In summary there are (at least) four possible algorithms for lines of data, which have been described in §3.1 and §3.2 above, corresponding to the following cases:

  (i)   lines of data with rectangular boundary (§3.1)
  (ii) lines of data with curved boundary - transformed to rectangle
  (iii) lines of data with curved boundary - using Chebyshev
        polynomials referred to an interior rectangle
  (iv) lines of data with curved boundary - using interior rectangle
        only

where each of these algorithms is a variant of the Clenshaw-Hayes algorithm of §3.1.

The algorithm of the next section builds further on these four algorithms. Cases (i) and (ii) above have already been implemented in the NAG library routine E02CAF.

4   Curves of Data - A General Approach

It is surprising that, although 25 years have passed since the Clenshaw-Hayes algorithm was published, there does not appear to have been an algorithm developed for curves of data of comparable efficiency to their early algorithm for lines of data. This is possibly because previous attempts have tended to focus on techniques for processing the curves of data themselves. However, in our new approach, which is a very simple but effective one, we process the curves of data by interpolating them onto a family of parallel secant lines, these lines being so directed as to ensure a good distribution of data. The problem then reduces to that of fitting lines of data, and the algorithms of §3 may be used directly. The latter algorithms may be used particularly efficiently, since the choice of secant lines is open to the user and may be optimised in order to reduce the computation.

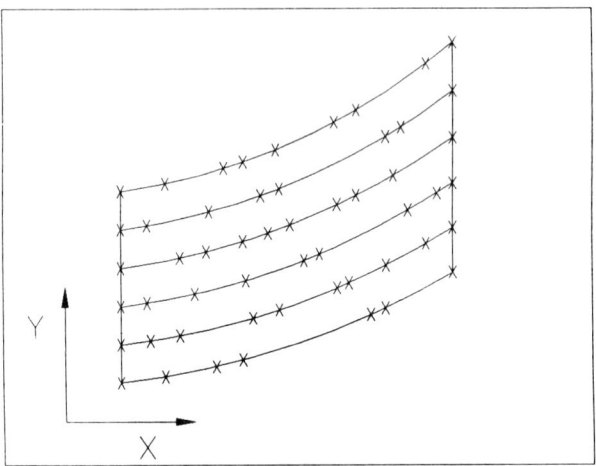

Figure 5. Curves of data - bounded by lines

4.1 Curves of data - straight line boundaries in x

Assume, as in Fig. 5 that the data are given on m_2 curves,

$$y = y_1(x), y_2(x), \ldots, y_{m_2}(x) \tag{4.1}$$

with at most m_1 data occurring on any given curve, and suppose that the x-abscissae of the data on $y_i(x)$ are $x_j^{(k)}$ for $k = 1, 2, \ldots, m_1^{(j)}$ where these all lie between the straight lines $x = \pm 1$. Assume also that the x- and y-axes are so orientated that each line parallel to the y-axis has no more than one intersection with any particular data curve and intersects as many curves as possible. This last assumption is not strictly necessary; the practical requirement is that the resulting lines of data should be well covered with data and should realistically represent the original data.

Linear forms of approximation $y_j^*(x)$ and $z_j^*(x)$ are now obtained to $y_j(x)$ and $z_j(x) \equiv z(x, y_j(x))$ respectively on $\{x_j^{(k)}\}$ by least-squares, and these are interpolated on chosen secant lines

$$x = x_1, x_2, \ldots, x_l \tag{4.2}$$

which leads to *new* sets of data (y, z) say, on each of the latter lines. This process is illustrated in Fig. 6, where the original curves of data are denoted by *crosses* and the newly formed line of data are *ringed*. Values of y and z, and hence data of z as a function of y, are thus now available on the lines (4.2). Note that these data are bounded by the pair of curves $y = C(x)$

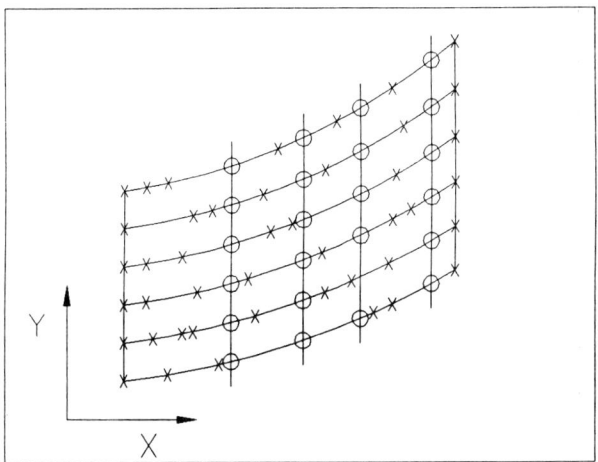

Figure 6. Curves of data - referred to secant lines

and $y = D(x)$ where $C(x) \equiv y_1(x)$ and $D(x) \equiv y_{m_2}(x)$, namely the first and last curves on which data were originally given.

The problem thus reduces to that of §3.2 of fitting data on lines bounded by curves, and we may, for example, adopt the approach of transforming y to $\bar{y}$,

$$\bar{y} = \frac{2y - C(x) - D(x)}{D(x) - C(x)} \tag{4.3}$$

so that the extreme abscissae y_1 and y_{m_2} become $\bar{y} = \pm 1$. Generalising the approach of §3.1, we then form approximations to the data of z as a function of $\bar{y}$ on each line $x = x_i$ by least-squares in the form

$$z^* \approx \sum_{j=1}^{n_2} a_j^{(i)} \psi_j(\bar{y}) \tag{4.4}$$

where $\{\psi_i\}$ is a chosen basis in y, e.g. Chebyshev polynomials, B-splines, etc. Finally, the coefficients $a_j^{(i)}$ are approximated over the points (4.2) in the form

$$a_j^{(i)} \approx \sum_{i=1}^{n_1} c_{i,j} \phi_i(x) \tag{4.5}$$

where $\{\phi_j\}$ is a chosen basis in x, to yield finally

$$z^* \approx \sum_{i=1}^{n_1} \sum_{j=1}^{n_2} c_{i,j} \phi_i(x) \psi_j(\bar{y}). \tag{4.6}$$

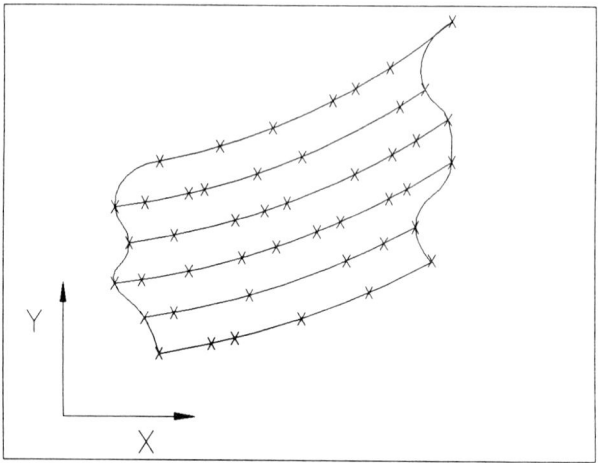

Figure 7. Curves of data - bounded by curves

The latter is based on an interpolation procedure, if there are precisely $l = n_1$ data lines, or a least-squares procedure if $l > n_1$.

In practice (compare §3.2) it is often easier to avoid the change of variables (4.3), and adopt a basis $\{\psi_j\}$ appropriate to a fixed interior interval $[y_A, y_B]$.

4.2 Curves of data - curved boundaries (in x)

If the boundaries in x, say $x = A(y)$ and $x = B(y)$, are curves rather than straight lines, as in Fig. 7, then these may be initially transformed to $\bar{x} = \pm 1$ by the transformation (3.3). The data are then of the type described in §4.1 and so the algorithm proceeds as before.

Alternatively, for example, the original geometry could be retained and secant curves used only strictly within the bounding curves $x = A(y)$ and $x = B(y)$, so that the final fit was restricted to an interior region bounded by the first and last secant lines.

To clarify matters further for specific forms, we now explain the use of Chebyshev polynomials in some detail, before discussing the case of splines.

4.3 Curves of data - using Chebyshev polynomials

It is natural to start with Chebyshev polynomials, since these were used by Clenshaw and Hayes [3], resulting techniques are extremely simple and such approximations are far more versatile than is generally credited.

Assume that the case of §4.1 is first considered, namely that of data $z(x, y)$ on a family of curves (4.1) with abscissae $x_j^{(k)}$ all between $x = \pm 1$. The basis functions are then chosen to be

$$\phi_i(x) = T_{i-1}(x) \text{ and } \psi_j(\bar{y}) = T_{j-1}(\bar{y}) \tag{4.7}$$

in the final approximation (4.6), based on one-dimensional approximations (4.4) and (4.5) after the transformation (4.3).

A rapid and most convenient technique is to choose as secant lines the zeros $x = x_1, x_2, ..., x_{n_1}$ of $T_{n_1}(x)$, and then to fit (4.5) by (Chebyshev) interpolation at these abscissae.

It remains to convert the initial curves of data into lines of data, and this is amplified in the following summary of the steps in the algorithm together with relevant NAG Library subroutines:

Step (i) : Form polynomial approximations y_j^* and z_j^* in x of order n_0 to $y_j(x)$ and $z_j(x) \equiv z(x, y_j)$ respectively, by least-squares over the data abscissae $x_j^{(k)}$ using the Forsythe-Clenshaw least-squares algorithm. (This provides a continuum representation of the curves of data, and may be effected, for example, by using E02ADF)

Step (ii) : Evaluate y_j^* and z_j^* at $x = x_1, x_2, ..., x_{n_1}$. (This involves evaluation of a Chebyshev sum at Chebyshev zeros and may be effected, for example, using E02AEF)

Step (iii) : Fit a polynomial approximation in $\bar{y}$, (4.4) with (4.7), on $x = x_i$ for the data obtained in *Step* (ii), using the Forsythe-Clenshaw least-squares algorithm. (This provides a Chebyshev approximation on each secant line, and may be effected using E02ADF).

Step (iv) : Fit a polynomial approximation, (4.5) with (4.7), to each coefficient in *Step* (iii) by interpolation at the zeros of $T_{n_1}(x)$. (This involves n_2 Chebyshev interpolations - which may be carried out by a Fast (discrete) Fourier Transform technique - see Canuto et al [6]).

The multiplications count for the above algorithm is as follows, where for simplicity we assume that $m_1^{(j)} = m_1$, so that there are m_1 data on every curve:

Step (i) : $2m_2$ Forsythe-Clenshaw procedures of degree n_0 on m_1 points :
$O(m_1 m_2 n_0^2)$

Step (ii) : $2m_2$ polynomial evaluations at n_1 points : $O(m_2 n_1^2)$

Step (iii) : n_1 Forsythe-Clenshaw procedures of degree n_2 on m_1 points:
$O(m_1 n_1 n_2^2)$

Step (iv) : n_2 Chebyshev interpolations of order n_1 : $O(n_1 n_2 \log n_1)$

If we assume that $m_1 = m_2$ and $n_0 = n_1 = n_2$, and neglect the lower order calculations of *Step* (ii) and *Step* (iv), this gives

$$O(m^2 n^2) + O(mn^3) \text{ operations} \qquad (4.8)$$

namely $2m+n$ Forsythe-Clenshaw least-squares procedures. If, in addition, $m \propto n$, then this becomes $O(n^4)$ operations.

On comparison with the Clenshaw-Hayes algorithm count (3.2), we observe that the same order of operations is involved in terms of m and n, but that the essential difference is that the number of Forsythe-Clenshaw least-squares procedures is increased from $m + n$ to $2m + n$.

The assumption here that n_0 may be taken equal (or nearly so) to n_1 and n_2 was found to be a reasonable one in the numerical examples we considered in §5 below, but there may be occasions when a somewhat higher degree of n_0 is needed.

Thus, the new algorithm is about twice as expensive as the Clenshaw-Hayes algorithm. Since curves of data essentially introduce a new variable into the problem, the comparison appears to be rather satisfactory.

4.4   Curves of data - B-splines

The procedure of §4.1 may be readily based on B-splines. For example, considering for simplicity the case of cubic splines, assume that a rectangular subregion of the data is used for the definition of a family of bivariate B-splines of form (4.6) with

$$\phi_i(x) \equiv B_i^{(1)}(x) \text{ and } \psi_j(\bar{y}) \equiv B_j^{(2)}(y) \qquad (4.9)$$

where $\{B_i^{(1)}(x)\}$ and $\{B_j^{(2)}(y)\}$ are families of cubic B-splines appropriate respectively to the fixed knots

$$x = \xi_1, \xi_2, ..., \xi_{k_1} \qquad (4.10)$$

and,

$$y = \eta_1, \eta_2, ..., \eta_{k_2} \qquad (4.11)$$

where $n_1 = k_1 + 4$ and $n_2 = k_2 + 4$. Here, 8 additional (exterior or endpoint) knots are introduced in both x and y in order to provide a full family of

B-splines (one per knot in x and one per knot in y), but excluding 4 outside knots in each case.

The proposed spline procedure, in the context of the curves of data defined in §4.1, is then as follows (with relevant NAG Library subroutines shown in brackets):

Step (i) : Form spline approximations to $y_j(x)$ and $z_j(x)$ on the curves by least-squares over the data. (E02BAF)

Step (ii) : Evaluate y_j^* and z_j^* at n_1 specified x values x_i *i.e.* secant lines. (E02BBF)

Step (iii) : Fit a least-squares spline approximation, (4.4) with (4.9), on $x = x_i$ for the data obtained in *Step* (ii). (E02BAF)

Step (iv) : Interpolate a spline approximation, (4.5) with (4.9), to each coefficient in *Step* (iii) at $x = x_i$. (E02BAF)

It remains to specify the values x_i of the secant lines, and these might well be chosen, for example, as the knots together with 4 other x values, such as 4 of the additional knots if these are suitable. Often in practice the additional knots are chosen to be four coincident knots at each interval end-point, and in that case the four additional points might, for example, be chosen as coincident pairs at each end-point, provided that in *Step* (ii) of the algorithm y^* and z^* and their first derivatives in x are evaluated at these points.

Clearly a number of alternative procedures may be proposed by modifying the choices of knots, interpolation procedures, and so on; however, the basic formulation is clear.

Another possibility, in connection with splines and curves of data, might be to attempt to place knots at positions on the curves rather than on a rectangular mesh, using techniques in the spirit of Hayes [7] and Walton [8], who adopt curvilinear knot meshes.

It is important to note that in the above discussions there is an implicit assumption that there are enough data on each curve or secant line in relation to the number and placement of knots, so that all spline approximations are well defined. In particular in Step (iv), for interpolation to be valid the standard Schoenberg-Witney conditions must be satisfied. Ideally, therefore, a more sophisticated algorithm should be developed which automatically guarantees that such requirements are met. Such an algorithm has recently been developed for data interpolation in the special case of lines of data, based on the introduction of a "common knot set" for all lines, by Anderson, Cox and Mason [9]. Moreover data near to a set of lines are also treated in [9] by an iterative procedure.

5   Two Families of Lines or Curves of Data

It is not uncommon for data to be defined on a pair of families of curves. For example, a survey ship may carry out soundings in two pairs of *directions*, roughly *North-South* and *East-West*, as in Fig. 1, but with the axes suitably rotated. Using the techniques of §4, these might well then be converted respectively into two families of lines of data in the directions of the two major axes x and y. The key requirement therefore reduces to that of fitting data on two families of perpendicular lines. How might the Clenshaw-Hayes algorithm be modified or extended to cover this?

Suppose that data are given on two families of lines:

$$m_A \text{ data on } x = x_j \text{ for } j = 1, 2, ..., m_1 \tag{5.1}$$

and,

$$m_B \text{ data on } y = y_k \text{ for } k = 1, 2, ..., m_2 \tag{5.2}$$

then, the same device as that used for curves of data in §4 may be used to convert the first set of data, on $x = x_j$, to a new set of data on $y = y_k$. Specifically, it suffices to approximate z on each $x = x_j$ by a polynomial, for example, and to interpolate the values of this at $y = y_k$ for each k.

Thus we obtain an augmented set of $m_1 + m_B$ data on $y = y_k$, comprising the original data on (5.2) together with the additional set of m_1 data obtained by converting the data on (5.1). This idea is illustrated in Fig. 8, where *crosses* denote the original data and *rings* denote the *new* data generated on $y = y_k$ from the data on $x = x_j$. The data fitting problem now reduces to that of fitting a (larger) set of data on one of the families of lines (5.2).

This technique, although not perhaps a pretty one at first sight, since it does not treat the data in a symmetrical way, is nevertheless tidy in that it reduces an awkward two-family problem into a tractable one-family problem. Moreover, we do not believe that there is normally a significant loss of information involved in converting the data on (5.1) into further data on (5.2).

6   Numerical Results

6.1   One family of curves

Since, in the case of Chebyshev polynomials (§4.3), the secant lines are the (typically irrational) zeros of a Chebyshev polynomial, there is no fundamental simplification in taking, in our model problem, data abscissae which are equally spaced in x on $[-1, 1]$. Consider the family of exponential curves,

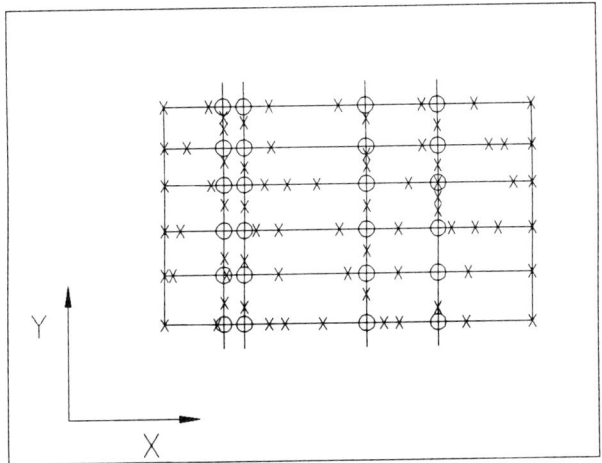

Figure 8. Two families of lines of data - referred to one family

$$y = 0.5(p+1)\exp[0.1(x+1)]$$
$$+0.05(x+1)\{\exp[0.5(p+1)] - 1\} + 0.5(p-1) \qquad (6.1)$$

where $p = -1, -1+h, -1+2h, \ldots, 1$ and $h = \frac{2}{m_2-1}$, thus providing m_2 curves $y = y_j$ for $j = 1, 2, \ldots, m_2$.

Note that, conveniently for a first model's purposes,

$$y = -1 \text{ on } p = -1$$

and,

$$y = \exp[0.1(x+1)] \text{ on } p = +1$$

so that y is a straight line on the lowest curve and a gently growing exponential on the highest curve. Moreover, the curves (6.1) have intercepts of $y = p$ on $x = -1$, so that the parameter p may be thought of simply as the initial y value on the curve.

The data then consist of values of z on the curves (6.1) for equally spaced values of x. The procedures of §4 work beautifully for such a smooth family of data abscissae.

For the purposes of testing, data were chosen as exact values of the function,

$$z(x,y) = \exp[x+y] \qquad (6.2)$$

and numerical tests were carried out successfully for a variety of values of m_1, m_2, n_1, n_2. We give results here for $m_1 = m_2 = 11$, and $n_0 = n_1 =$

Table 1. One family of curves - Chebyshev polynomial fits

	Errors on Secant Lines		Final Errors on Mesh	
Problem	rms in y	rms in z	rms in z	max in z
(A) Curves (6.1)	0.79×10^{-9}	0.53×10^{-3}	0.16×10^{-2}	0.51×10^{-2}
(B) Curves (6.1) exact data on x_i	0.0	0.0	0.16×10^{-2}	0.56×10^{-2}
(C) Straight Lines $y = p$	0.74×10^{-16}	0.19×10^{-3}	0.84×10^{-3}	0.32×10^{-2}
(D) Straight Lines exact data on x_i	0.0	0.0	0.86×10^{-3}	0.30×10^{-2}

$n_2 = 5$ so that the x-abscissae are -1 in steps of 0.2 to $+1$. A square mesh of points at steps of 0.2 in both x and y was used for comparing the final bivariate approximation to (6.2); results are shown in Table 1. Four types of data were considered: (A) and (B) use data (6.1), (C) and (D) use straight lines of data $y = p$, while (A) and (C) use original data but (B) and (D) use exact data at the interpolation points on the secant lines $x = x_i$, for test purposes only. The errors shown are, firstly interpolation errors occurring from *Step* (i) and *Step* (ii) in referring the data onto secant lines, and secondly, the final error from *Step* (iii) and *Step* (iv), both r.m.s and maximum absolute errors being given.

Note from these results that all four cases give fairly comparable accuracy. Thus the interpolation process onto secant lines has in these cases added no perceptible error to the final result, while the use of curves of data is almost as effective as the use of lines of data (in spite of the fact that the least-squares fit for the curves of data is carried out over a larger region with a curved boundary).

In all cases above, the data were fitted on the whole region of concern but Chebyshev polynomials were referred to the fixed interval $[-1, 1]$ in both x and y. Results for other values of m_1, m_2, n_1, n_2 were very similar.

6.2 Two families of curves

As a simple model for two families, we first considered the data on (6.1) together with data on the corresponding family in y, for a parameter q:

$$\begin{aligned} x \;=\; & 0.5(q+1)\exp[0.1(y+1)] \\ & +0.05(y+1)\{\exp[0.5(q+1)]-1\}+0.5(q-1) \end{aligned} \quad (6.3)$$

for the function (6.2), and adopted the Chebyshev polynomial procedures of §5 above, using Chebyshev polynomials referred to fixed intervals $[-1, 1]$ in both x and y, and taking as secant lines the Chebyshev zeros. In (6.3), $q = -1, -1+h, -1+2h, \ldots, 1$ and $h = \frac{2}{m_1 - 1}$, thus providing m_1 curves $x = x_k$ for $k = 1, 2, \ldots, m_1$.

Table 2. Two families of curves - Chebyshev polynomial fits

	Errors on Secant Lines		Final Errors on Mesh	
$z(y)$	rms in y	rms in z	rms in z	max in z
	0.79×10^{-9}	0.53×10^{-3}	0.16×10^{-2}	0.48×10^{-2}
$z(x)$	rms in x	rms in z		
	0.79×10^{-9}	0.53×10^{-3}		

Taking 11 curves of 11 data for each of (6.1) and (6.3), with equally spaced y and x abscissae respectively, and orders $n_1 = n_2 = 5$ and $n_0 = 5$ (in each variable), the problem was converted to that of fitting 16 data on each of one family of 5 secant lines. These 16 data points on $x = x_i$ for $i = 1, 2, ..., n_1$ comprise 11 points obtained by interpolating one family of 11 curves of data together with 5 extra points at the Chebyshev zeros in y, obtained by interpolating the corresponding data on the secant lines in y. We obtained the values for the errors which are given in Table 2, and note that these results are barely distinguishable from the corresponding results (A) of Table 1. Thus, in this example, the second family of data curves has served only to confirm the results obtained for one family alone. Nevertheless it is clear that the algorithm has been applied successfully.

Acknowledgements

This work was carried out with the academic support of the Applied and Computational Mathematics Group at RMCS (Cranfield), Shrivenham, the Mathematics Departments at the Australian Defence Force Academy (University of New South Wales) and University of Queensland, and the Centre for Mathematical Analysis at the Australian National University.

We are indebted to the staff at all those Universities for their encouragement and for the use of a variety of computing facilities which severely tested the portability of the software! We also acknowledge the encouragement and interest of Mr J. G. Hayes of the National Physical Laboratory, Teddington.

References

1. Handscomb, D.C. and MacCarthy, B. (1989). The approximation of hydrographic survey data using tensor product B-spline surfaces. In *Mathematics of Surfaces III*, D. C. Handscomb (Ed.), Oxford University Press, 373–389.

2. Hayes, J.G. (Ed.), (1970). Fitting data in more than one variable. Chapter 7 in *Numerical Approximation to Functions and Data*. The Athlone Press.

3. Clenshaw, C.W. and Hayes, J.G. (1965). Curve and surface fitting. *J. Inst. Maths Applics.*, *1*, 164–183.

4. Forsythe, G.E. (1957). Generation and use of orthogonal polynomials for data-fitting with a digital computer. *J. SIAM*, *5*, 74–88.

5. Clenshaw, C.W. (1960). Curve fitting with a digital computer. *Comput. J.*, *2*, 170–173.

6. Canuto, C., Hussaini, M.Y., Quarteroni, A. and Zang, T.A. (1988). *Spectral Methods in Fluid Dynamics*. Springer-Verlag, London.

7. Hayes, J.G. (1974). New shapes from bi-cubic splines. In *Proc. CAD74*. IPC Business Press, Guildford. Also in NPL Report NAC58.

8. Walton, D.J. (1987). Terrain modelling with B-spline type surfaces defined on curved knot lines. *Image and Vision Computing*, *5*, 1, 37–42.

9. Anderson, I.J., Cox, M.G. and Mason, J.C. (1993). Tensor-product spline interpolation to data on or near a family of lines. In *Algorithms for Approximation 3*, M.G. Cox and J.C. Mason (Eds), Special Issue of *Numerical Algorithms*, in press.

Shape Improvements for Curves and Surfaces

Gerald Farin

Arizona State University, USA

1 Introduction

We will study two instances of the concept of curve and surface fairing. We will call a curve or a surface fair if a derived construct exhibits a high amount of fairness; these constructs are the curvature plot for a curve and the reflection line pattern for surfaces. While such a definition of fairness sounds circular, it is actually very practical; both curvature plots and reflection lines are widely used in industry.

2 Curve Fairing

This first part of this paper addresses the problem of generating aesthetically pleasing, or fair, curves in the B-spline form.

For a detailed description of cubic B-splines, as well as their conversion to the piecewise Bézier form, see [10].

It is known from differential geometry that a planar curve is uniquely defined by its curvature [1]. We may visualize curvature by plotting curvature *vs* arc length or *vs* parameter, generating the so-called *curvature plot*. The curvature plot is an extremely accurate tool in providing information about the shape of a curve. In fact, two curves may not be visually distinguishable and yet may exhibit quite different curvature plots (see the examples below). These differences, subtle as they may be, may be crucial in an environment where aesthetics or aerodynamics are important, i.e., in the automotive, aircraft, or naval industries.

To a practitioner in the field, it is obvious that the analysis tool "curvature plot" is essential, but this insight is only slowly gaining acceptance in the academic community. In fact, almost all definitions of the (aesthetic) fairness of a curve rely on the notion of a curvature plot. We use the following:

A curve is fair if its curvature plot consists of relatively few monotone pieces.

I would like to add a personal note: There are definitions of curve fairness that do not address the shape of the curvature plot. The most common one is: *a curve is fair if it minimizes* $\int[\kappa(s)]^2 ds$, where $\kappa(s)$ denotes curvature as a function of arc length. This minimum property is satisfied by the elastic beam. I fail to see what an elastic beam has to do with aesthetic appearance. The reason for the minimum property definition of fairness is that it lends itself to mathematical treatment, in particular when used in the austere version that replaces curvature by second derivative (see Lee [2] for more details).

2.1   Knot Removal Fairing

A typical problem in the design process is that of *digitizing errors*: data points have been obtained from some digitizing device (a tablet being the simplest), and a fair curve is sought through them. In many cases, the digitized data are inaccurate, and this presence of digitizing error manifests itself in a "rough" curvature plot of an interpolating spline curve. Splines that are obtained from interactive adjustment of control polygons usually exhibit rough curvature plots as well.

In [3], [4] and [5], we have presented algorithms that fair B-spline curves in the sense that they improve their curvature plots. In this paper, those algorithms are referred to as *knot removal fairing algorithms*. Specifically, those algorithms reduce the curvature discontinuities between adjacent cubic pieces in the following way: the control point $\mathbf{d}_i$ of a C^2 cubic B-spline curve is associated with the parameter value u_i. If the B-spline curve were three times differentiable at u_i instead of just twice, it would not have a curvature discontinuity. Thus we try to move $\mathbf{d}_i$ to a new position $\hat{\mathbf{d}}_i$ such that the new curve is now C^3 at u_i.

After some calculation (equating the left and the right third derivative of the new spline curve), one verifies that the new vertex $\hat{\mathbf{d}}_i$ is given by

$$\hat{\mathbf{d}}_i = \frac{(u_{i+2} - u_i)\mathbf{l}_i + (u_i - u_{i-2})\mathbf{r}_i}{u_{i+2} - u_{i-2}}, \tag{2.1}$$

where the auxiliary points $\mathbf{l}_i, \mathbf{r}_i$ are given by

$$\mathbf{l}_i = \frac{(u_{i+1} - u_{i-3})\mathbf{d}_{i-1} - (u_{i+1} - u_i)\mathbf{d}_{i-2}}{u_i - u_{i-3}}$$

and

$$\mathbf{r}_i = \frac{(u_{i+3} - u_{i-1})\mathbf{d}_{i+1} - (u_i - u_{i-1})\mathbf{d}_{i+2}}{u_{i+3} - u_i}.$$

In practice, the improved vertex $\hat{\mathbf{d}}_i$ may be further away from the original vertex $\mathbf{d}_i$ than a prescribed tolerance allows. In that case, one restricts a realistic $\hat{\mathbf{d}}_i$ to be in the direction towards the optimal $\hat{\mathbf{d}}_i$, but within tolerance to the old $\mathbf{d}_i$.

Knot removal fairing can be made more sophisticated than described here; for instance, N. Sapidis devised an algorithm to enforce convexity in the fairing process, see [6] and [5].

Other methods for curve fairing exist. We mention Kjellander's method [7] that moves a data point to a more favorable location and then interpolates the changed data set with a C^2 cubic spline. This method is global. Methods that aim at the smoothing of single Bézier curves are discussed by Hoschek [8], [9].

2.2 Degree reduction fairing

The idea behind the algorithm (2.1) is to equate third derivatives from left and right at u_i. If the Bézier polygons of the two involved cubics are denoted by $\mathbf{b}_{3i-3}, \mathbf{b}_{3i-2}, \mathbf{b}_{3i-1}, \mathbf{b}_{3i}$ and $\mathbf{b}_{3i}, \mathbf{b}_{3i+1}, \mathbf{b}_{3i+2}, \mathbf{b}_{3i+3}$, the condition for equal third derivatives at u_i becomes (see [10]):

$$\frac{\Delta^3 \mathbf{b}_{3i-3}}{(\Delta_{i-1})^3} = \frac{\Delta^3 \mathbf{b}_{3i}}{(\Delta_i)^3}, \tag{2.2}$$

where the Δ^3 denote third forward differences and again $\Delta_i = u_{i+1} - u_i$.

The idea behind *degree reduction fairing* is not to equate third differences, as in (2.2), but rather to make each individual third difference $\Delta^3 \mathbf{b}_{3i}$ small. Recall that

$$\Delta^3 \mathbf{b}_{3i} = \mathbf{b}_{3i} - 3\mathbf{b}_{3i+1} + 3\mathbf{b}_{3i+2} - \mathbf{b}_{3i+3} = 0 \tag{2.3}$$

means that the cubic defined by $\mathbf{b}_{3i}, \mathbf{b}_{3i+1}, \mathbf{b}_{3i+2}, \mathbf{b}_{3i+3}$ is actually a quadratic curve, i.e., a parabola. It seems reasonable to hope for a better curve shape if all cubic pieces are close to being parabolic. The rationale is that parabolas are simpler in shape than cubics, and when we try to combat digitizing errors, we also combat an overabundance of information in the curve shape.

We could rewrite (2.3) in terms of the B-spline coefficients $\mathbf{d}_i$. This would lead to a global linear system for new control points $\hat{\mathbf{d}}_i$. There is a simpler geometric method that achieves the same goal. In addition, it will have the property of being local.

A cubic may be approximated by a quadratic by the process of *degree reduction*, see [11], [10], [12]. The approximating quadratic, with Bézier points $\mathbf{c}_{2i}, \mathbf{c}_{2i+1}, \mathbf{c}_{2i+2}$ may be obtained as

$$\mathbf{c}_{2i} = \mathbf{b}_{3i}, \quad \mathbf{c}_{2i+1} = \mathbf{m}_i, \quad \mathbf{c}_{2i+2} = \mathbf{b}_{3i+3},$$

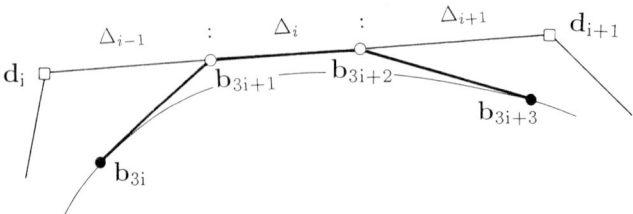

Figure 1. Cubic B-spline curves: the relationship between B-spline control polygon and Bézier control points

where

$$\mathbf{m}_i = \frac{1}{2}(\frac{3}{2}\mathbf{b}_{3i+1} - \frac{1}{2}\mathbf{b}_{3i}) + \frac{1}{2}(\frac{3}{2}\mathbf{b}_{3i+2} - \frac{1}{2}\mathbf{b}_{3i+3}). \qquad (2.4)$$

This quadratic may be brought back to cubic form by the process of *degree elevation*.

The involved Bézier points may be expressed in terms of the B-spline vertices $\mathbf{d}_{i-1}, \mathbf{d}_i, \mathbf{d}_{i+1}, \mathbf{d}_{i+2}$ according to the formulas given above. Let us agree to keep $\mathbf{d}_{i-1}$ and $\mathbf{d}_{i+2}$ fixed and to only change the other two control vertices. This simplification leads to the following algorithm to fair the ith curve segment:

step 1 Given the initial B-spline curve, compute the Bézier points $\mathbf{b}_{3i}, \mathbf{b}_{3i+1}, \mathbf{b}_{3i+2}, \mathbf{b}_{3i+3}$.

step 2 Compute the quadratic approximation to the cubic defined by these Bézier points.

step 3 Degree elevate that quadratic, resulting in a cubic control polygon $\mathbf{b}_{3i}, \hat{\mathbf{b}}_{3i+1}, \hat{\mathbf{b}}_{3i+2}, \mathbf{b}_{3i+3}$.

step 4 The new $\hat{\mathbf{d}}_i$ and $\hat{\mathbf{d}}_{i+1}$ are on the straight line through $\hat{\mathbf{b}}_{3i+1}, \hat{\mathbf{b}}_{3i+2}$. The involved ratios are known; they are illustrated in Fig. 1.

It should be kept in mind that the above algorithm does not produce a curve which consists of quadratic (i.e., degree reduced) segments only:
a) only $\mathbf{d}_i$ and $\mathbf{d}_{i+1}$ are changed, whereas also $\mathbf{d}_{i-1}$ and $\mathbf{d}_{i+2}$ would have to be changed in order to build a degree reduced cubic, and
b) the result of the $(i-1)^{\text{st}}$ step is distorted in the ith step.

3 Comparison

Before we describe differences between knot removal fairing and degree reduction fairing, it is interesting to point to a common theme that both adhere to. Both methods aim at a simplification in curve shape, or at a removal of superfluous information. Removing a knot from a B-spline curve changes its information content, and so does the reduction of the polynomial degree of a segment. So both schemes have the same high level description: Remove information from the curve by suitably eliminating some of its degrees of freedom, and then formally rewrite it to be compatible with the initial curve representation.

Both knot removal fairing (2.1) and the new degree reduction methods were implemented and tested on data that are a reasonable simulation of real-world design data.

Although the methods are local, we applied them to every segment of a given spline curve, assuming that all B-spline coefficients were in error as the result of digitizing. An automatic localization for knot removal smoothing has been implemented by N. Sapidis [5]. It looks for the knot where two cubics join with the largest discontinuity in the slope of the curvature plot and then fairs only there.

An automatic localization for degree reduction fairing might look for the cubic segment with the largest third derivative, and then fair only there.

In the following examples, (see Figures 2-7) we compare degree reduction fairing to knot removal fairing. In all examples, we did not impose a limit on the maximal deviation between old and new curve, in order to see what each algorithm deems the most appropriate change. In practice, such limits will be necessary. In most examples, it was not possible to tell the curves apart as displayed on the screen; their plots are omitted.

The following observations were made:

Magnitude of change: Knot removal fairing changes a curve more than does degree reduction fairing. The change was measured by simply (and somewhat naively) finding the largest distance $\|\mathbf{d}_i - \hat{\mathbf{d}}_i\|$.

Fairing speed: Typically, one has to fair a curve several times until the curvature plot looks satisfactory. Knot removal fairing needs about two to three iterations, while degree reduction fairing needs about five iterations. Even so, degree reduction fairing changes the curve less.

Dimensionality: Both methods may be applied to space curves, but this has not been tested.

The rational case: Both methods may be applied to rational B-spline curves (NURBS). Knot removal fairing would be carried out in homogeneous space, and the result will hopefully be a smoother curve. Degree reduction fairing will assume a different flavor: instead of aiming to come close to quadratics, i.e., parabolas, one might try to come close to conics. Conics of "optimal shape" are those of minimum eccentricity. We

are currently experimenting with these ideas; the results will be reported
elsewhere.

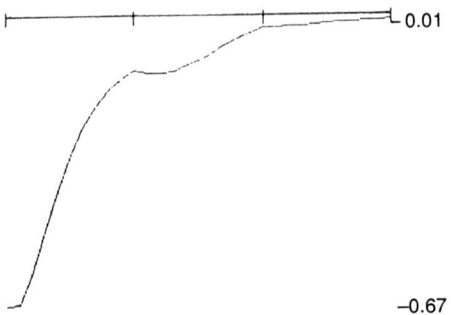

Figure 2. Original curvature plot

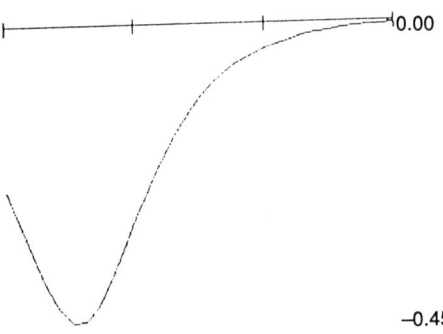

Figure 3. Curvature plot after two applications of knot removal fairing

4   The Surface Case

The above curve fairing methods may be applied to parametric tensor product surfaces in a straightforward way, see [5]. We will not discuss this matter here, but rather point out a method that has recently been developed
for piecewise triangular surfaces, in joint work with P. Kashyap. In particular, we address the problem of fairing a given surface that was obtained
as a scattered data interpolant using the Clough-Tocher element; we only
address the problem for functional surfaces of the form $z = f(x, y)$. Clough-
Tocher interpolants were invented as a tool for the finite element method

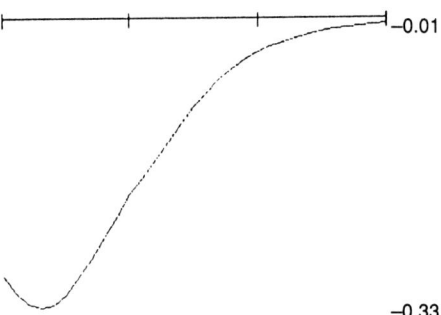

Figure 4. After five applications of degree reduction fairing

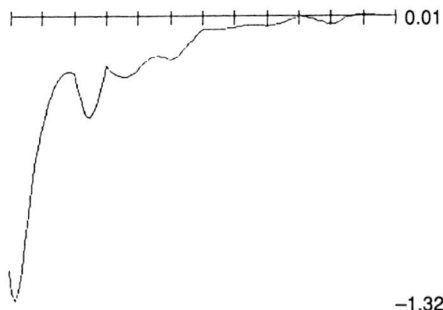

Figure 5. Original curvature plot

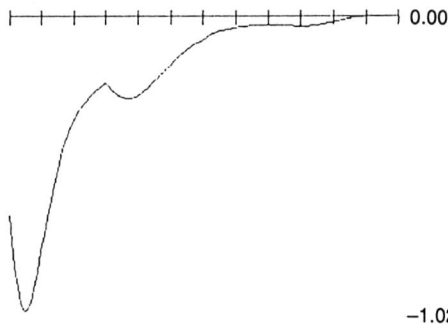

Figure 6. Curvature plot after two applications of knot removal fairing

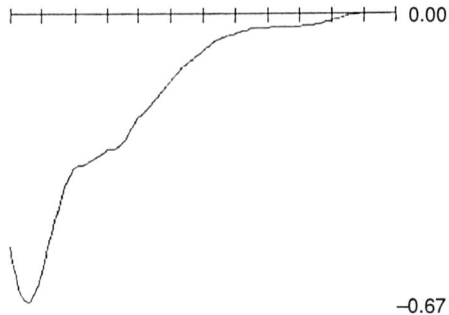

Figure 7. After five applications of degree reduction fairing

[13], but for some years they have been used in the field of CAGD (Computer Aided Geometric Design) in the area of scattered data interpolation, see [14], [15], [16].

A Clough-Tocher interpolant produces a C^1 piecewise polynomial surface, defined over a triangulation of scattered data sites. Previously the original interpolant was modified, [15], to increase the fairness of the overall interpolant by using the available degrees of freedom. In this paper we use an iterative scheme to further increase the smoothness of the overall interpolant.

The iterative scheme is then compared to the other methods by using an interrogation technique which simulates reflection lines [15].

4.1 Continuity conditions for triangular Bézier patches

For the theory of Bernstein–Bézier polynomials defined over triangles the reader is referred to [17], [10]. An nth degree Bernstein–Bézier triangular patch is of the form

$$b^n(\mathbf{u}) = \sum_{|\mathbf{i}|=n} b_{\mathbf{i}} B_{\mathbf{i}}^n(\mathbf{u}) \tag{4.1}$$

where the Bernstein polynomials $B_{\mathbf{i}}^n(\mathbf{u})$ are defined by

$$B_{\mathbf{i}}^n(\mathbf{u}) = \binom{n}{\mathbf{i}} u^i v^j w^k; \quad |\mathbf{i}| = n$$

the $b_{\mathbf{i}}$ are Bézier ordinates which form the control net of the triangular patch; $\mathbf{u} = (u, v, w)$ are the barycentric coordinates of the domain triangle, and $\mathbf{i} = (i, j, k)$ is a multiindex with $|\mathbf{i}| = i + j + k$.

Now consider two adjacent domain triangles and Bézier nets defined on each of them; we want to find the conditions which the nets must satisfy in order for the surface patches defined by the two nets to be C^1 or C^2.

For the C^1 case Fig. 8 (a) illustrates the result: the condition for C^1 continuity of the interpolant is that the shaded pair of triangles should be coplanar.

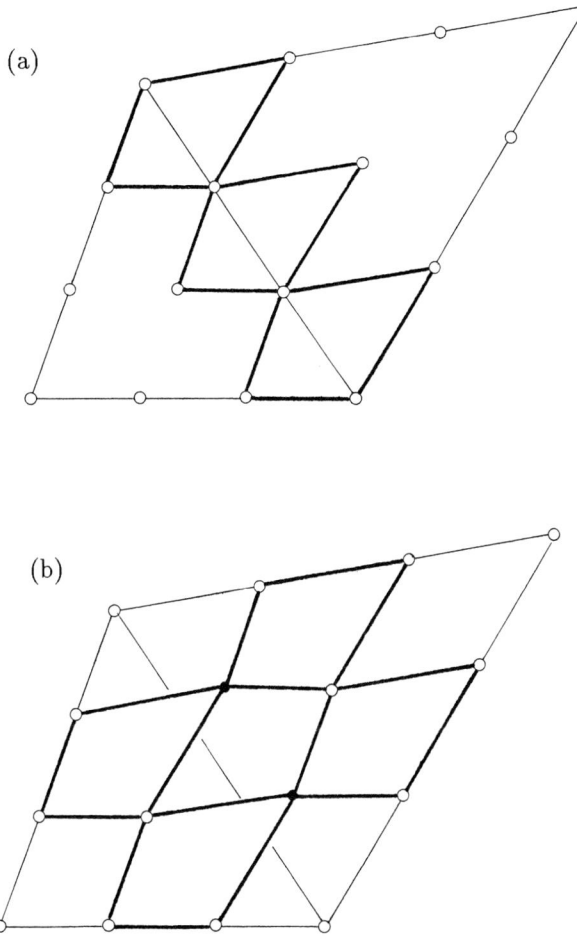

(a)

(b)

Figure 8. Continuity conditions of Bézier triangular patches: (a) C^1 continuity, (b) additional C^2 continuity

For the C^2 case: the shaded pairs of triangles in Fig. 8 (b) are constructed to be coplanar. Each of the extension points (marked points) as found from the two triangles would have a different z–value. The condition for C^2 continuity is that these two values be identical. These extension points are analogous to the extension points used to define C^2 conditions for spline curves in Bézier form, see [18] or [10].

4.2   The Clough-Tocher interpolant

Given the z–values and gradients over a set of triangulated data points, we want a "good" (in terms of continuity) piecewise cubic interpolant over the data set. As a first approximation, the 9-parameter piecewise cubic Bernstein-Bézier interpolant [17] can be used. Its nine boundary ordinates (all $b_{i,j,k}$ except $b_{1,1,1}$) are determined from the data by univariate cubic Hermite interpolation, the remaining ordinate is given by

$$b_{1,1,1} = \frac{1}{4}(b_{2,0,1}+b_{1,0,2}+b_{0,2,1}+b_{0,1,2}+b_{2,1,0}+b_{1,2,0}) - \frac{1}{6}(b_{3,0,0}+b_{0,3,0}+b_{0,0,3})$$

This choice of $b_{1,1,1}$ ensures quadratic precision.

This interpolant is only C^0 in general, and needs to be modified if one desires overall C^1 smoothness. This can be done by splitting each triangle in the given triangulation into three minitriangles (for an algorithm see [17]). This subdivided domain now has enough degrees of freedom (twelve per subdivided triangle instead of ten before subdivision) to allow for C^1 continuity of the overall interpolant.

In Fig. 9, let C, P_2, P_3 and C', P_3, P_2 be the vertices of two adjacent minitriangles (coming from two different macro triangles). Expressing the center points C and C' in terms of barycentric coordinates of the opposite triangles C', P_3, P_2 and C, P_2, P_3, respectively :

$$C = u'C' + v'P_3 + w'P_2$$

and

$$C' = uC + vP_2 + wP_3.$$

The C^1 conditions are fulfilled by the subtriangle pair formed by c_6, c_9, c_{10} and c'_4, c'_7, c'_8 and pair c_4, c_7, c_8 and c'_6, c'_9, c'_{10} (same tangent plane). So the only condition for a C^1 patch is that the middle pair of subtriangles should be coplanar, i.e.,

$$c'_5 = uc_5 + vc_8 + wc_9. \tag{4.2}$$

This can be achieved by following the given scheme: Chose a direction **l** (the components l_1, l_2, l_3 are its barycentric representation in terms of triangle C, P_2, P_3), not parallel to the triangle edge P_2P_3. Then the directional derivative of mini-cubic $\mathcal{P}_1$ defined over the minitriangle C, P_2, P_3 is a univariate quadratic Bézier polynomial with Bézier ordinates

$$3(l_1c_6 + l_2c_9 + l_3c_{10}), \quad 3(l_1c_5 + l_2c_8 + l_3c_9), \quad 3(l_1c_4 + l_2c_7 + l_3c_8).$$

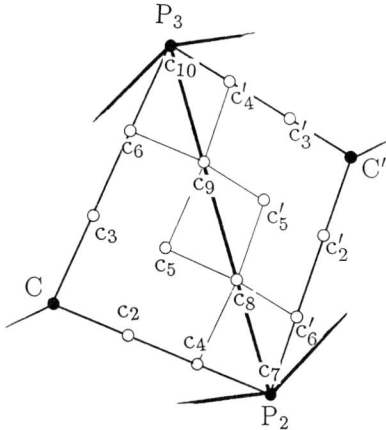

Figure 9. Cross-boundary derivatives: the Bézier ordinates that are needed to formulate C^1 and C^2 conditions between two adjacent triangular patches

We can fix the unknown c_5 by choosing a linear variation for the directional derivative. This choice can be expressed as

$$(l_1 c_6 + l_2 c_9 + l_3 c_{10}) - 2(l_1 c_5 + l_2 c_8 + l_3 c_9) + (l_1 c_4 + l_2 c_7 + l_3 c_8) = 0, \quad (4.3)$$

see [19]. The unknown c_5 may be found from (4.2), or by an analogous procedure for the other mini-cubic $\mathcal{P}_2$ over the minitriangle C', P_3, P_2. It is imperative that the l denotes the same direction both in $\mathcal{P}_1$ and $\mathcal{P}_2$. One way of doing this is by choosing l to be perpendicular to edge P_2, P_3, although it makes the interpolant affinely variant (perpendicular lines, in general, do not map onto perpendicular lines in an affine transformation).

After finding all three center points of the macro triangle, we can compute the rest of the interior points by applying the C^1 conditions four times.

4.3 Fairing the interpolant

The C^1 condition (4.2) has two unknowns, one of which can be fixed by chosing the linear cross boundary derivative condition (4.3). This appears to be quite arbitrary and so in this section we describe conditions which would improve the smoothness of the surface [15]. In section 3, we observed that C^1 cubics over split triangles enjoy an extra degree of freedom (this is manifested in (4.2) which has two unknowns in one equation). We can improve upon condition (4.3) by trying to achieve

$$uc_3 + vc_5 + wc_6 = u'c'_2 + v'c'_4 + w'c'_5 \tag{4.4}$$

and

$$uc_2 + vc_4 + wc_5 = u'c'_3 + v'c'_5 + w'c'_6, \tag{4.5}$$

thereby hopefully minimizing the jump in the second derivative across the boundary.

So we have a constrained minimization problem: Minimize the sum of errors in (4.4), (4.5) constrained by (4.2). We use the standard Lagrange multiplier method to obtain

$$c'_5 = (-us_1 - ua_{12} - u^2s_2 - r_3a_{11})/D$$

where

$$s_1 = 2(vr_1 + wr_2), \qquad s_2 = -2(w'r_1 + v'r_2)$$

and

$$
\begin{aligned}
r_1 &= u'c'_2 + v'c'_4 - uc_3 - wc_6 \\
r_2 &= u'c'_3 + w'c'_6 - uc_2 - vc_4 \\
r_3 &= vc_8 + wc_9 \\
a_{11} &= 2(v^2 + w^2) \\
a_{12} &= -2(vw' + wv') \\
a_{22} &= 2(w'^2 + v'^2)
\end{aligned}
$$

and the denominator

$$D = -2ua_{12} - u^2a_{22} - a_{11}.$$

The unknown c_5 then can be found from (4.2). After fixing the center points of the macro triangle, the inner points are recomputed using the C^1 conditions.

4.4   Iterative improvement

In the previous section, the C^2 error across the macro triangle was reduced by adjusting the center points c_5 and c'_5; based on the same idea, we now present an iterative scheme which further improves the smoothness of the overall interpolant by changing the inner points. These points can be found either by:

a) Form the control net over the macro triangle by using the 9-parameter interpolant, then subdivide it at the centroid into three subnets (use de-Casteljau algorithm). The inner points are Bézier ordinates neighboring the centroid. This interpolant is C^∞ over the macro triangle but only C^0 across them.

b) Find the center points using the linearized cross boundary derivative condition (4.2); the inner points are then found by using the C^1 conditions in the macro triangle. The resulting global interpolant is C^2 at the centroid of the macro triangle and C^1 all over.

The iterative idea is based on the fact that after application of the C^2 error minimization scheme, the C^2 error is reduced, hence the new inner points are "better" than the old ones. Now using these "better" inner points and once again applying the minimization scheme we can reduce the C^2 errors further. This method is due to P. Kashyap [20].

The iterative improvement weakens the locality of the standard and the modified Clough-Tocher schemes: suppose just one data value was nonzero. Using the standard Clough-Tocher scheme, all triangular patches sharing the corresponding data site would be nonzero. Using the modified method, all neighboring patches would be nonzero as well. With the iterative method, one further layer of triangles is added per iteration. This gradual loss of locality may explain the apparent shape improvement of the resulting surfaces; the less local a scheme is, the more potential exists for "ironing out" shape imperfections.

5 Reflection Lines

In order to judge the performance of different interpolation schemes, one may print out errors relative to known test functions, one may compare perspective views, or one may inspect contour plots. We have found that a third method is far more powerful: this is the use of reflection lines. The idea comes from the automotive industry. Here designers judge the aesthetic appearance of a car body by placing it under parallel fluorescent light bulbs. These reflect in the car surface, and instead of judging the car body directly, one judges how the light sources reflect. Tiny imperfections are detectable with this method. We use this method of quality inspection to evaluate our new interpolants. For literature on reflection lines etc., see [17], [21], [3], [4].

As shown in [17], the problem of finding reflection lines of surfaces of the form $z = f(x, y)$ amounts to contouring a directional derivative of that surface. Since we are using triangular cubic patches, its directional derivatives are quadratic patches, the contours of which are conic sections [22].

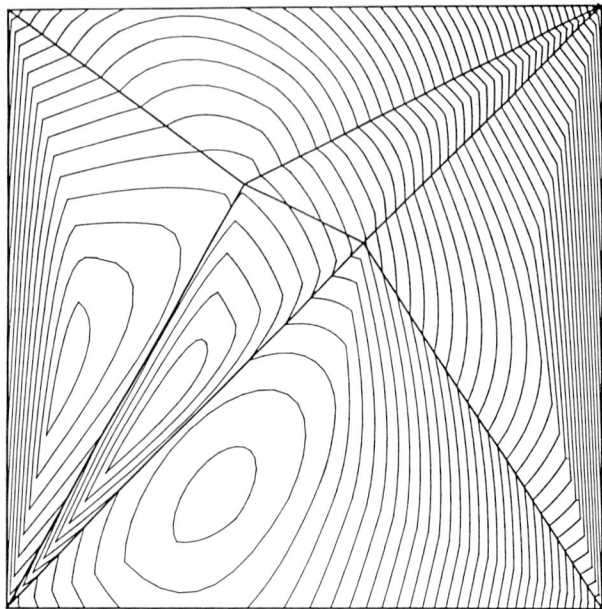

Figure 10. Reflection lines: the original Clough-Tocher interpolant with linearized cross-boundary derivatives

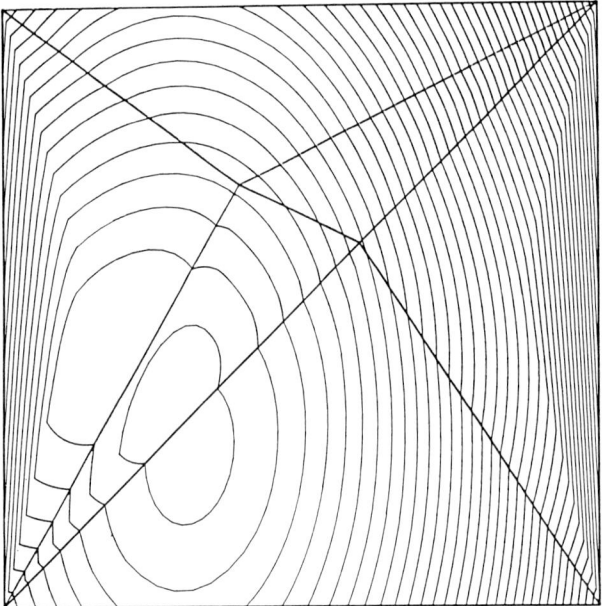

Figure 11. Reflection lines: the modified Clough-Tocher interpolant with no iterations

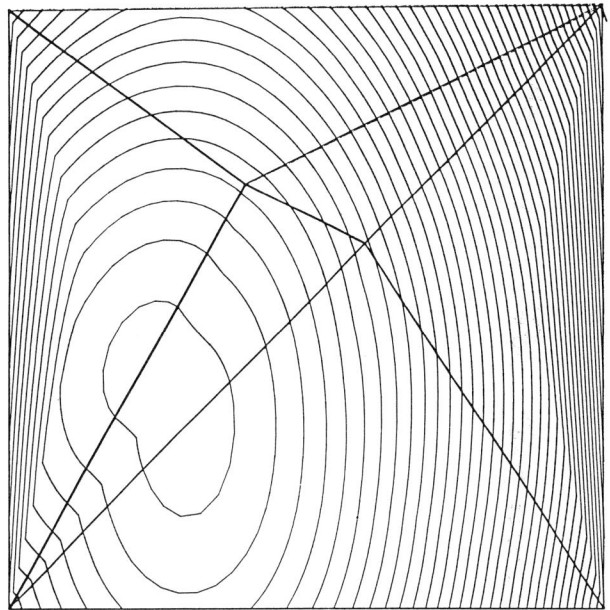

Figure 12. Reflection lines: the iterated Clough-Tocher interpolant after 5 iterations

The test function that we use is :

$$f(x,y) = (x - 0.3)^3 + x(y - 0.3)^2 - 0.1x$$

with exact gradients. The data sites consist of the corners of the unit square with two additional points $(0.4, 0.7)$ and $(0.6, 0.6)$.

In Figs. 10, 11 and 12, we can see that the boundaries of the mini-triangles are much more visible in the reflection line pattern of the older method, and in general the new scheme produces visually more pleasing (smoother) surfaces.

6   Conclusions

We have demonstrated how to use the Bézier technique to derive visually more pleasing curves and surfaces. Both methods are based on the concept of removing superfluous information from the original curve or surface. In the curve case, this could be done explicitly; in the surface case we had to observe additional constraints and attempt a least squares approach.

In both cases, the effect of the fairing operation was measured not by inspection of the generated curve or surface, but by the use of more sophisticated interrogation tools: curvature plots for the curve case, and reflection lines for the surface case.

Acknowledgements

This research was supported by the National Science Foundation grant DCR-8502858 and the Department of Energy contract DE-FG02-87ER25041, both awarded to Arizona State University.

References

1. do Carmo, M. (1976). *Differential Geometry of Curves and Surfaces.* Prentice Hall, Englewood Cliffs.

2. Lee, E. (1990). Energy, fairness, and a counterexample. *Computer Aided Design, 22,* 37–40.

3. Farin, G., Rein, G., Sapidis, N. and Worsey, A. (1987). Fairing cubic B-spline curves. *Computer Aided Geometric Design, 4,* 91–104.

4. Farin, G. and Sapidis, N. (1989). Curvature and the fairness of curves and surfaces. *IEEE Computer Graphics and Applications, 9,* 52–57.

5. Sapidis, N. and Farin, G. (1990). Automatic fairing algorithm for B-spline curves. *Computer Aided Design, 22,* 121–129.

6. Sapidis, N. (1987). Algorithms for locally fairing B-spline curves. Msc thesis, Univ. of Utah.

7. Kjellander, J. (1983). Smoothing of cubic parametric splines. *Computer Aided Design, 15,* 175–179.

8. Hoschek, J. (1984). Detecting regions with undesirable curvature. *Computer Aided Geometric Design, 18,* 183–192.

9. Hoschek, J. (1985). Smoothing of curves and surfaces. *Computer Aided Geometric Design, 2,* 97–105.

10. Farin, G. (1988). *Curves and Surfaces for Computer Aided Geometric Design.* Academic Press. Third edn. 1992.

11. Forrest, A. (1972). Interactive interpolation and approximation by Bézier polynomials. *Computer J., 15,* 71–79.

12. Watkins, M. and Worsey, A. (1988). Degree reduction for Bézier curves. *Computer Aided Design, 20,* 398–405.

13. Clough, R. and Tocher, J. (1965). Finite element stiffness matrices for analysis of plates in blending. In *Proceedings of Conference on Matrix Methods in Structural Analysis*.

14. Barnhill, R. (1977). Representations and approximation of surfaces. In *Mathematical Software III*, J. R. Rice, Editor, Academic Press.

15. Farin, G. (1985). A modified Clough-Tocher interpolant. *Computer Aided Geometric Design, 2*, 19–27.

16. Worsey, A. and Farin, G. (1987). An n-dimensional Clough-Tocher element. *Constructive Approximation, 3*, 99–110.

17. Farin, G. (1986). Triangular Bernstein-Bézier patches. *Computer Aided Geometric Design, 3*, 83–128.

18. Boehm, W., Farin, G. and Kahmann, J. (1984). A survey of curve and surface methods in CAGD. *Computer Aided Geometric Design, 1*, 1–60.

19. Barnhill, R. and Farin, G. (1981). C^1 quintic interpolation over triangles: two explicit representations. *Int. J. Num Methods in Engineering, 17*, 1763–1778.

20. Farin, G. and Kashyap, P. (1992). An iterative Clough-Tocher interpolant. *Mathematical Modelling and Numerical Analysis, 26*, 201–209.

21. Barnhill, R., Farin, G., Fayard, L. and Hagen, H. (1988). Twists, curvatures, and surface interrogation. *Computer Aided Design, 20*, 341–346.

22. Worsey, A. and Farin, G. (1990). Contouring a bivariate quadratic polynomial over a triangle. *Computer Aided Geometric Design, 7*, 337–352.

Watch Your (Parametric) Speed!

Rida T. Farouki

IBM Thomas J. Watson Research Center

1   Preamble

Computer–aided design systems rely almost exclusively on parametric curve and surface representations, based on polynomial or rational functions. While parameterizations frequently play a key rôle in this context, facilitating the computation and display of various curve and surface properties, we must concede that from a broader perspective their status is little more than that of an expedient intermediary. Thus, geometry–based engineering applications such as numerical control machining or tolerance analysis usually care little about parameterizations, and indeed often prefer to remain entirely oblivious of them.

Certainly it would be very undesirable that the accuracy, efficiency, or reliability of a surface intersection algorithm (for example) be sensitively dependent on the parameterizations of the surface patches that it receives. As long as flagrant misbehaviour in this respect is rather infrequent, we are inclined to adopt a somewhat indulgent attitude concerning the vagaries of parametric representations — any misgivings are relegated to a deep recess of the mind, to be recalled only when our curve or surface algorithms prove utterly incapable of disgesting some parametric outrage.

In this informal survey we shall attempt a frontal assault on some of the recalcitrant problems that parameterization presents. Specifically, we shall focus on plane curves, for which the central dilemma of parameterization can be concisely phrased: the absurd difficulty of constructing curves of constant — or at least nearly constant — "parametric speed" from a class of simple (*i.e.*, computationally tractable) functions. While our endeavour is perhaps predestined to rather limited success, we shall nevertheless encounter a variety of interesting problems and results along the way. Let us therefore proceed, undaunted:

Given a differentiable parametric curve $\mathbf{r}(t) = \{x(t), y(t)\}$, the element $\mathrm{d}s$ of its arc length corresponding to an infinitesimal increment $\mathrm{d}t$ in the parameter is given by

$$\mathrm{d}s^2 = \mathrm{d}x^2 + \mathrm{d}y^2, \tag{1.1}$$

where $\mathrm{d}x = x'(t)\mathrm{d}t$, $\mathrm{d}y = y'(t)\mathrm{d}t$, and primes denote derivatives. Hence, the rate of change ds/dt of the arc length s with respect to the parameter t is the function

$$\sigma(t) = \sqrt{x'^2(t) + y'^2(t)}. \tag{1.2}$$

We will call this function the *parametric speed* of $\mathbf{r}(t)$. It is, by definition, non–negative (s being measured always in the sense of increasing t).

By way of example, if $\mathbf{r}(t)$ is a polynomial Bézier curve of degree n,

$$\mathbf{r}(t) = \sum_{k=0}^{n} \mathbf{p}_k \binom{n}{k} (1-t)^{n-k} t^k \quad \text{for } t \in [0,1], \tag{1.3}$$

with control points $\mathbf{p}_k = (x_k, y_k)$ for $k = 0, \ldots, n$, then the argument of the radical in (1.2) is the polynomial

$$\sum_{k=0}^{2n-2} s_k \binom{2n-2}{k} (1-t)^{2n-2-k} t^k \tag{1.4}$$

of degree $2n - 2$, whose Bernstein coefficients $\{s_k\}$ are given by

$$s_k = \sum_{j=\max(0,k-n+1)}^{\min(k,n-1)} \frac{\binom{n-1}{j}\binom{n-1}{k-j}}{\binom{2n-2}{k}} n^2 (\Delta x_j \Delta x_{k-j} + \Delta y_j \Delta y_{k-j}) \tag{1.5}$$

for $k = 0, \ldots, 2n - 2$ (where $\Delta x_j = x_j - x_{j-1}$ and $\Delta y_j = y_j - y_{j-1}$).

Ideally, we would like the polynomial (1.4) to degenerate to a constant, but it is not difficult to convince oneself that this is impossible unless $n = 1$, *i.e.*, the curve in question is really a straight line. In fact, on differentiating (1.2) we have

$$\frac{d\sigma}{dt} = \frac{x'(t)x''(t) + y'(t)y''(t)}{\sigma(t)}, \tag{1.6}$$

and we remark that since the numerator of (1.6) is of *odd* degree (namely, $2n - 3$) it must have at least one real root, and $\sigma(t)$ must therefore have at least one real extremum, whenever $n \geq 2$. (We assume that $x'(t)$ and $y'(t)$ do not vanish simultaneously at any real value of t, so that $\sigma(t) \neq 0$ for all real t — see, however, §3 below.)

Another "simple" parametric form in common use is the rational curve

$$x(t) = \frac{X(t)}{W(t)}, \quad y(t) = \frac{Y(t)}{W(t)} \tag{1.7}$$

where $X(t)$, $Y(t)$, and $W(t)$ are polynomials of degree n such that

$$\text{GCD}(X, Y, W) = 1 \quad \text{and} \quad W(t) \neq \text{constant}. \tag{1.8}$$

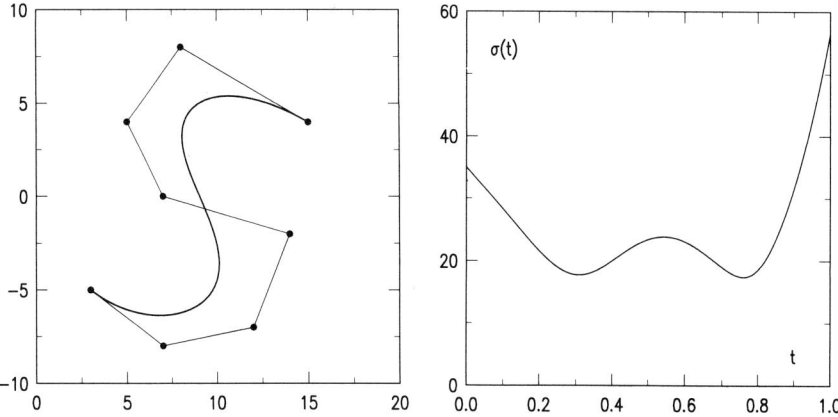

Figure 1. Parametric speed variation along a Bézier curve

In this case, the parametric speed $\sigma(t)$ is obtained by substituting the *rational functions*

$$x' = \frac{WX' - W'X}{W^2} \quad \text{and} \quad y' = \frac{WY' - W'Y}{W^2} \qquad (1.9)$$

into (1.2), and it is now a far more subtle matter to establish that $\sigma(t)$ can be a constant only when equations (1.7) define a straight line (see §5 below). Note that in the rational case the parametric speed increases without bound as we approach the parameter values defined by the roots of $W(t)$, *i.e.*, the "points at infinity" of the curve (1.7).

Since most algorithms used in the processing and display of parametric curves rely on the parameterization (rather than the intrinsic geometry) of a curve, it is important to note that even for apparently innocuous curves $\sigma(t)$ may exhibit substantial variation — see Fig. 1. If we wish to render the Bézier curve (1.3), for example, we are usually content to select a sequence of parameter values $t_0, t_1, \ldots, t_N$ corresponding to a uniform increment $\Delta t = t_k - t_{k-1}$. But these points will be evenly distributed along the locus of (1.3) only to the extent that $\sigma(t)$ does not vary significantly.

Another instance in which uneven parametric flow adversely affects the outcome of geometric calculations is the use of the *de Casteljau subdivision algorithm* (in computing intersections of Bézier curves, for example). This proceeds as follows: choosing a parameter value $\tau \in [0, 1]$ we iterate the sequence of linear interpolations:

$$\mathbf{p}_k^{(r)} = (1 - \tau)\,\mathbf{p}_{k-1}^{(r-1)} + \tau\,\mathbf{p}_k^{(r)} \quad k = r, r+1, \ldots, n, \qquad (1.10)$$

on the control points of the curve (1.3) for $r = 1, \ldots, n$. This gives

$$\mathbf{p}_0^{(0)}\,\mathbf{p}_1^{(1)}\,\cdots\,\mathbf{p}_n^{(n)} \quad \text{and} \quad \mathbf{p}_n^{(n)}\,\mathbf{p}_n^{(n-1)}\,\cdots\,\mathbf{p}_n^{(0)}, \qquad (1.11)$$

for the control points of the subsegments $t \in [0,\tau]$ and $t \in [\tau,1]$ of the original curve (see [4]). If we were to choose $\tau = \frac{1}{2}$, for example, we would usually wish the subsegments defined by (1.11) to be of commensurate length — but this is not so when $\sigma(t)$ exhibits considerable variation over $[0,1]$.

Finally, note that when a rational curve is written in Bézier form

$$\mathbf{r}(t) = \frac{\displaystyle\sum_{k=0}^{n} w_k \mathbf{p}_k \binom{n}{k} (1-t)^{n-k} t^k}{\displaystyle\sum_{k=0}^{n} w_k \binom{n}{k} (1-t)^{n-k} t^k}, \qquad (1.12)$$

we can no longer regard the shape of the control polygon $\mathbf{p}_0, \ldots, \mathbf{p}_n$ as a fair indicator of the shape of the curve that it defines. In fact, a given polygon may be associated with many different curves by simply varying the projective coordinates or "weights" $w_0, \ldots, w_n$. Conversely, a given curve can have many different (but equivalent) representations, corresponding to different values for $w_0, \ldots, w_n$. Each choice of the weights gives a different parametric flow across the curve; we will exploit this property in §8.

2   Keep an Eye on your Hodograph

The *hodograph* of a parametric curve $\mathbf{r}(t) = \{x(t), y(t)\}$ is the curve defined by its parametric derivative or "velocity vector," $\mathbf{r}'(t) = \{x'(t), y'(t)\}$. The notion of a hodograph has only occasionally been discussed, and applied to practical problems, in the computer–aided design literature (see [1], [14], [25], and [26]).

The parametric speed is evidently just the scalar function that gives the Euclidean length of the velocity vector:

$$\sigma(t) = \|\mathbf{r}'(t)\|. \qquad (2.1)$$

Thus, the hodograph describes not only how fast we move along a curve, but also our instantaneous *direction of motion* at each point.

Any polynomial curve clearly has a polynomial hodograph. Conversely, any polynomial hodograph will define — modulo translations — a unique polynomial curve. This one–to–one correspondence does not carry over to rational curves and hodographs, however. Whereas every rational curve has a rational hodograph, there are rational forms for $x'(t)$ and $y'(t)$ that are *not* the hodographs of rational curves (the integral of a rational function is not necessarily rational; see [13]).

Ideally, the hodograph of a curve would be simply (a segment of) the unit circle — the parametric speed would then be unity, *i.e.*, the curve would be parameterized by its arc length. The rate of rotation of the velocity vector about this circular segment tells us how rapidly the orientation of the tangent vector changes as we traverse the curve; this is none other than the curvature — see §5 below. We shall encounter in §6 a few curious curves that exhibit circular hodographs. Unfortunately, they do not seem to be of much use for practical free–form design problems.

To the extent that the hodograph of a curve does not stray too far from the unit circle, we would probably consider its parameterization to be quite "reasonable," and might expect our curve algorithms to accommodate its modest deviations from uniform parametric speed. But when the hodograph ventures dangerously close to the origin, or races off to infinity at finite values of the parameter, it is usually a sign that great trouble lies ahead.

We shall find that the notion of a hodograph proves useful in a variety of circumstances below: in analyzing the "irregular" points of parametric curves (see §3); in identifying a class of curves whose parametric speed, while not constant, is at least of a much simpler functional form than (1.2) (see §4); and in demonstrating the essential incompatibility of two equally desirable aspects of parameterizations, namely, constant parametric speed and simple (*i.e.*, rational) functional forms (see §5).

3 On Encountering a Red Light

We should take heed when the hodograph of a polynomial curve $\mathbf{r}(t)$ passes through the origin, since if t_0 is a value of the parameter such that $x'(t_0) = y'(t_0) = 0$, we come to a complete stop at the point t_0 on the curve!

But the fact that the parametric speed drops to zero at t_0 is perhaps the least of our worries in negotiating that point. For, as we shall soon see, such points generally imply a rather awkward *geometry* in their vicinity, and it is not possible to remedy this merely by re–parameterization. We shall call the points where $\mathbf{r}'(t) = \mathbf{0}$ the *irregular points* of $\mathbf{r}(t)$.

For simplicity, let us consider the polynomial curve

$$x(t) = \sum_{k=0}^{n} a_k t^k \ , \ y(t) = \sum_{k=0}^{n} b_k t^k \ . \tag{3.1}$$

By the Euclidean algorithm [29] we may compute the polynomial

$$\phi(t) = \mathrm{GCD}(x'(t), y'(t)) \tag{3.2}$$

of greatest degree that simultaneously divides the hodograph components $x'(t)$ and $y'(t)$. Evidently, the (real) irregular points of $\mathbf{r}(t)$ are identified by the (real) roots of $\phi(t)$. We are interested in deducing the geometric nature of the irregular point t_0, given its *multiplicity* as a root of $\phi(t)$.

In order to accomplish this, we first recall that the unit tangent vector $\mathbf{T}(t)$ and signed curvature $\kappa(t)$ of the curve $\mathbf{r}(t)$ are defined by[1]

$$\mathbf{T}(t) \; = \; \frac{\mathbf{r}'(t)}{\|\mathbf{r}'(t)\|} \quad \text{and} \quad \kappa(t) \; = \; \frac{[\mathbf{r}'(t) \times \mathbf{r}''(t)] \cdot \mathbf{z}}{\|\mathbf{r}'(t)\|^3}, \tag{3.3}$$

where $\mathbf{z}$ is a unit vector orthogonal to the plane of $\mathbf{r}(t)$ [18].

Noting that $\mathbf{r}'(t_0) = \mathbf{0}$, we will expand $\mathbf{r}'(t)$ in Taylor series about t_0 for positive and negative increments in t,

$$\mathbf{r}'(t_0 + \Delta t) \; = \; \mathbf{r}''(t_0)\Delta t + \frac{1}{2}\mathbf{r}^{(3)}(t_0)(\Delta t)^2 + \cdots,$$

$$\mathbf{r}'(t_0 - \Delta t) \; = \; -\mathbf{r}''(t_0)\Delta t + \frac{1}{2}\mathbf{r}^{(3)}(t_0)(\Delta t)^2 - \cdots. \tag{3.4}$$

Consider first a *simple* root t_0 of (3.2), so that $\phi(t_0) = 0 \neq \phi'(t_0)$ (*i.e.*, $x'(t_0) = y'(t_0) = 0$, but $x''(t_0)$ and $y''(t_0)$ are not *both* zero). Then at least one of the polynomials $x'(t)$, $y'(t)$ must change sign as we pass through $t = t_0$. In order to determine the limiting behavior of the tangent vector $\mathbf{T}(t)$ as we approch the parameter value t_0 from above and below, we need only keep in (3.4) the non–vanishing terms of lowest order in Δt. We then find that

$$\lim_{\Delta t \to 0} \mathbf{T}(t_0 - \Delta t) \; = \; -\lim_{\Delta t \to 0} \mathbf{T}(t_0 + \Delta t), \tag{3.5}$$

i.e., there is an abrupt reversal of the tangent vector on passing through t_0. Such a point is called a *cusp*.

Suppose next that t_0 is a *double* root of (3.2), *i.e.*, $\phi(t_0) = \phi'(t_0) = 0 \neq \phi''(t_0)$. Then we have $x'(t_0) = y'(t_0) = 0$ and $x''(t_0) = y''(t_0) = 0$, but the third derivatives $x^{(3)}(t_0)$ and $y^{(3)}(t_0)$ are not both zero. Since in this case the non–vanishing terms of lowest order in Δt in (3.4) are quadratic, the tangent $\mathbf{T}(t)$ is *continuous* as we pass through t_0,

$$\lim_{\Delta t \to 0} \mathbf{T}(t_0 - \Delta t) \; = \; \lim_{\Delta t \to 0} \mathbf{T}(t_0 + \Delta t). \tag{3.6}$$

All is not well, however. Although the curve is indeed tangent–continuous at $t = t_0$, an examination of the behavior of the curvature $\kappa(t)$ as we approach t_0 indicates that

$$\lim_{\Delta t \to 0} |\kappa(t_0 \pm \Delta t)| \; = \; \frac{2}{3} \lim_{\Delta t \to 0} \frac{\|\mathbf{r}^{(3)}(t_0) \times \mathbf{r}^{(4)}(t_0)\|}{(\Delta t)^2 \|\mathbf{r}^{(3)}(t_0)\|^3} = \infty \tag{3.7}$$

---

[1]Forrest [14] describes an interesting graphical method for identifying inflections from the hodograph: if at any point the tangent to the hodograph passes through the origin, the corresponding curve point must have zero curvature, since the first and second parametric derivatives are evidently parallel there, and the cross product in (3.3) vanishes.

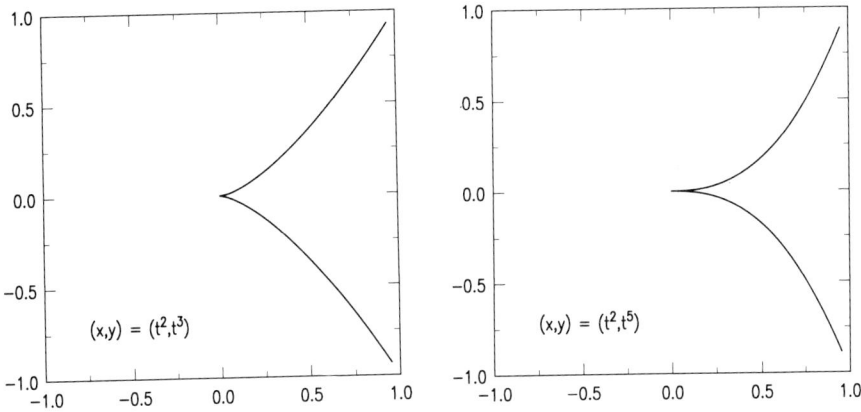

Figure 2. Examples of cusps

whenever $\mathbf{r}^{(3)}(t_0) \times \mathbf{r}^{(4)}(t_0) \neq \mathbf{0}$. Thus, the magnitude of $\kappa(t)$ increases without bound as we approach t_0!

In general, we have a cusp exhibiting the sudden tangent reversal (3.5) whenever t_0 is a root of $\phi(t)$ of *odd* multiplicity. If t_0 is a root of *even* multiplicity, the curve point $\mathbf{r}(t_0)$ is tangent continuous, but not always of infinite curvature — the behavior of $\kappa(t)$ in the vicinity of t_0 will depend on the lowest–order terms in the Taylor expansions (3.4) that make a non–zero contribution to the cross product in (3.3).

We can get a "feel" for the nature of irregular points by examining the behavior of polynomial curves of the form

$$x(t) = t^m , \quad y(t) = t^n \quad \text{where } n > m \geq 2 \tag{3.8}$$

at the origin (we should also ensure that $\mathrm{GCD}(m,n) = 1$, *i.e.*, that m and n are *relatively prime*). For such curves, $t = 0$ is clearly a root of $\phi(t) = \mathrm{GCD}(x'(t), y'(t))$ of multiplicity $\min(m,n) - 1$. Thus, with $m = 2$, we obtain cusps such as those shown in Fig. 2 for $n = 3$ and $n = 5$. With $m = 3$, we observe tangent–continuous points of infinite curvature; Fig. 3 shows the cases $n = 4$ and $n = 5$.

The reader may care to examine also the case $(m,n) = (3,7)$, for which the point $t = 0$ appears well–behaved at first sight, since it is both tangent–continuous and of finite (zero) curvature. However, it is not difficult to verify that the *rate of change*

$$\frac{\mathrm{d}\kappa}{\mathrm{d}s} = \frac{\|\mathbf{r}(t)\|^2 \left[\mathbf{r}'(t) \times \mathbf{r}^{(3)}(t)\right] \cdot \mathbf{z} - 3\,\mathbf{r}'(t) \cdot \mathbf{r}''(t) \left[\mathbf{r}'(t) \times \mathbf{r}''(t)\right] \cdot \mathbf{z}}{\|\mathbf{r}'(t)\|^6} \tag{3.9}$$

of the curvature with respect to arc length is infinite there!

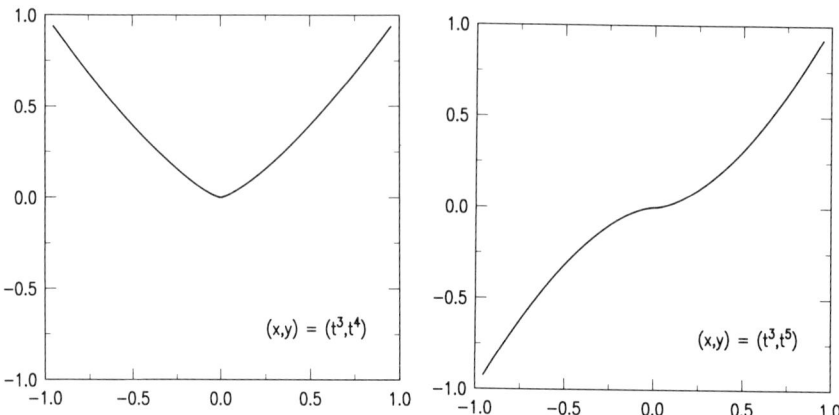

Figure 3. Tangent–continuous points of infinite curvature

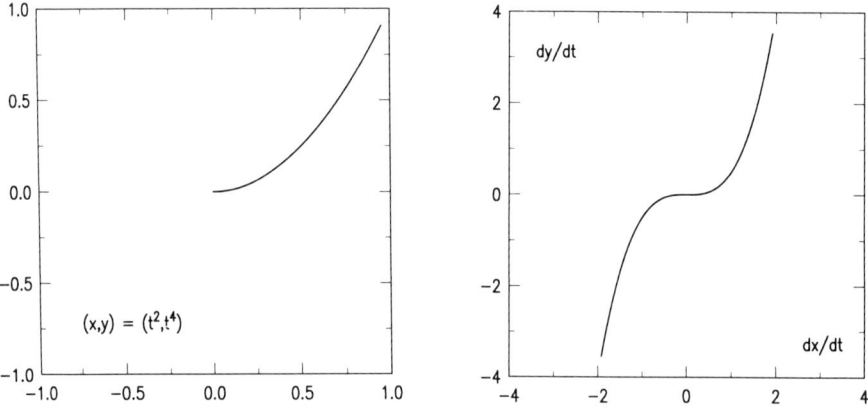

Figure 4. An "improperly" parameterized curve and its hodograph

(We avoid cases where m and n have a common factor because they
define *multiply–traced* curves. Fig. 4 shows $(x,y) = (t^2,t^4)$ — as t increases
from $-\infty$ to 0 we approach the origin along the segment of a parabola in
the positive quadrant, and as t passes through 0 and increases to $+\infty$ we
"reverse" and backtrack along our incoming path! This behavior is reflected
in the hodograph, which is symmetric with respect to inversion through the
origin. In fact, the same point locus could be defined by setting $u = t^2$ to
give $(x,y) = (u, u^2)$ where $0 \le u \le +\infty$. With the original parameterizat-

ion, each point of the curve has *two* parameter values $t = \pm u$! We shall return to such "improper" parameterizations in §7 below.)

In the case of a rational curve, it is possible to come to a complete halt and then start up again in a perfectly "smooth" manner (*i.e.*, the tangent and the curvature and all its derivatives are continuous). A familiar example is the point $t = \pm \infty$ on the unit circle,

$$x(t) = \frac{1 - t^2}{1 + t^2}, \quad y(t) = \frac{2t}{1 + t^2}. \tag{3.10}$$

In general, if $p = \max(\deg(X), \deg(Y))$ and $q = \deg(W)$ in (1.7) we can see from (1.9) and (1.2) that for a rational curve, $\sigma(\pm \infty) = \infty$ or $\sigma(\pm \infty) = 0$, according to whether $p > q + 2$ or $p < q + 2$. To determine whether or not the tangent and curvature are well–behaved at $t = \pm \infty$ in the latter case, we can make the substitution $t = 1/u$ in $X(t)$, $Y(t)$, $W(t)$ and perform Taylor expansions for increments $\pm \Delta u$ about $u = 0$.

Irregular points also arise generically on certain "procedurally defined" curves such as offsets [7], [8] (which do *not*, in general, have polynomial or rational parameterizations). Further discussion of these matters may be found in [2].

4 Pythagoras on the Road

Pythagoras was, presumably, not acquainted with polynomials — nor with curves parameterized in terms of them. But it is perhaps not unreasonable to assume that, if he had been in a position to ponder the matter of parametric speed, he would have been particularly appreciative of a class of polynomial curves whose parametric speed is just a *polynomial* function of the parameter, rather than the radical form (1.2).

If $\mathbf{r}(t) = \{x(t), y(t)\}$ is a polynomial curve of degree n, we say it belongs to the family of *Pythagorean–hodograph curves* [11] having the desired property if there exists a polynomial

$$\sigma(t) = \sum_{k=0}^{n-1} \sigma_k \binom{n-1}{k} (1 - t)^{n-1-k} t^k \tag{4.1}$$

of degree $n - 1$ such that

$$x'^2(t) + y'^2(t) \equiv \sigma^2(t). \tag{4.2}$$

It can be shown [19] that the Pythagorean identity (4.2) is satisfied if and only if the components of the hodograph of $\mathbf{r}(t)$ are of the form

$$x'(t) = w(t)\left[u^2(t) - v^2(t)\right] \quad \text{and} \quad y'(t) = 2\,w(t)u(t)v(t) \tag{4.3}$$

for real, non-zero polynomials $u(t)$, $v(t)$, $w(t)$ such that $u(t)$ and $v(t)$ have no factors in common and are not both constants. The parametric speed of $\mathbf{r}(t)$ is then the *polynomial*

$$\sigma(t) = w(t)\,[\,u^2(t) + v^2(t)\,]\,, \qquad (4.4)$$

and we say that $x'(t), y'(t), \sigma(t)$ form a "Pythagorean triple."

To avoid introducing irregular points, we shall usually want to choose the polynomial $w(t)$ to be simply a constant. This yields odd–degree curves with parametric speed $\sigma(t) = u^2(t) + v^2(t) > 0$ for all real values of t. Thus, the simplest Pythagorean–hodograph curves are cubics, obtained by choosing linear polynomials of the form

$$u(t) = u_0(1 - t) + u_1 t\,, \quad v(t) = v_0(1 - t) + v_1 t\,, \qquad (4.5)$$

substituting into expressions (4.3), and integrating to give $\mathbf{r}(t) = \{x(t), y(t)\}$. It is readily verified that when $\mathbf{r}(t)$ is expressed in the Bézier representation (1.3), its control points must be of the form

$$
\begin{aligned}
\mathbf{p}_1 &= \mathbf{p}_0 + \frac{1}{3}(u_0^2 - v_0^2, 2u_0v_0)\,, \\[2mm]
\mathbf{p}_2 &= \mathbf{p}_1 + \frac{1}{3}(u_0u_1 - v_0v_1, u_0v_1 + u_1v_0)\,, \\[2mm]
\mathbf{p}_3 &= \mathbf{p}_2 + \frac{1}{3}(u_1^2 - v_1^2, 2u_1v_1)\,, \qquad (4.6)
\end{aligned}
$$

for real values of u_0, u_1 and v_0, v_1 ($\mathbf{p}_0$ being chosen at will). In this case, the coefficients of $\sigma(t)$ in (4.4) are

$$
\begin{aligned}
\sigma_0 &= u_0^2 + v_0^2\,, \\
\sigma_1 &= u_0u_1 + v_0v_1\,, \\
\sigma_2 &= u_1^2 + v_1^2\,. \qquad (4.7)
\end{aligned}
$$

In fact, it turns out that when we discount freedoms of rigid motion and uniform scaling there is just *one* Pythagorean–hodograph cubic — it is a classical curve identified (in a rather different context) three hundred years ago [11] by Tschirnhausen! Since this curve is too inflexible for general free–form design applications (see [6], [11]) we proceed to higher–order forms.

We can construct the quintic Pythagorean–hodograph curves by choosing quadratic polynomials of the form

$$
\begin{aligned}
u(t) &= u_0(1 - t)^2 + u_1 2(1 - t)t + u_2 t^2\,, \\
v(t) &= v_0(1 - t)^2 + v_1 2(1 - t)t + v_2 t^2\,, \qquad (4.8)
\end{aligned}
$$

to substitute into (4.3). Integration then gives the control points

$$\mathbf{p}_1 = \mathbf{p}_0 + \frac{1}{5}(u_0^2 - v_0^2, 2u_0 v_0),$$

$$\mathbf{p}_2 = \mathbf{p}_1 + \frac{1}{5}(u_0 u_1 - v_0 v_1, u_0 v_1 + u_1 v_0),$$

$$\mathbf{p}_3 = \mathbf{p}_2 + \frac{2}{15}(u_1^2 - v_1^2, 2u_1 v_1) + \frac{1}{15}(u_0 u_2 - v_0 v_2, u_0 v_2 + u_2 v_0),$$

$$\mathbf{p}_4 = \mathbf{p}_3 + \frac{1}{5}(u_1 u_2 - v_1 v_2, u_1 v_2 + u_2 v_1),$$

$$\mathbf{p}_5 = \mathbf{p}_4 + \frac{1}{5}(u_2^2 - v_2^2, 2u_2 v_2), \tag{4.9}$$

for real values of u_0, u_1, u_2 and v_0, v_1, v_2 ($\mathbf{p}_0$ being again arbitrary). In this case the coefficients of the parametric speed are

$$\sigma_0 = u_0^2 + v_0^2,$$

$$\sigma_1 = u_0 u_1 + v_0 v_1,$$

$$\sigma_2 = \frac{2}{3}(u_1^2 + v_1^2) + \frac{1}{3}(u_0 u_2 + v_0 v_2),$$

$$\sigma_3 = u_1 u_2 + v_1 v_2,$$

$$\sigma_4 = u_2^2 + v_2^2. \tag{4.10}$$

The Pythagorean–hodograph quintics prove to be much more malleable than the cubics; they can have inflection points and may be pieced together to form "spline" curves smoothly interpolating sequences of points in the plane (see [6] for further details).

In general, if $\mathbf{r}(t) = \{x(t), y(t)\}$ is a Pythagorean–hodograph curve of odd degree n, defined by choosing polynomials $u(t)$ and $v(t)$ of degree

$$m = \tfrac{1}{2}(n - 1), \tag{4.11}$$

with Bernstein coefficients $\{u_k\}$ and $\{v_k\}$, then by appealing to the arithmetic procedures for polynomials in Bernstein form [10] we find that the parametric speed (4.4) of $\mathbf{r}(t)$ has Bernstein coefficients

$$\sigma_k = \sum_{j=\max(0,k-m)}^{\min(k,m)} \frac{\binom{m}{j}\binom{m}{k-j}}{\binom{n-1}{k}} (u_j u_{k-j} + v_j v_{k-j}) \tag{4.12}$$

for $k = 0, \ldots, n-1$.

A remarkable feature of Pythagorean–hodograph curves is the possibility of integrating the parametric speed (4.4) in closed form. This gives us an expression for the arc length as a *polynomial* function $s(t)$ of the parameter! Making use of the rule

$$\int \binom{n-1}{k}(1-t)^{n-1-k}t^k \, dt = \frac{1}{n}\sum_{j=k+1}^{n}\binom{n}{j}(1-t)^{n-j}t^j, \qquad (4.13)$$

$k = 0, \ldots, n-1$, for the indefinite integrals of the Bernstein basis functions [10] thus gives

$$s(t) = \int_0^t \sigma(\tau)\,d\tau = \sum_{k=0}^{n} s_k \binom{n}{k}(1-t)^{n-k}t^k, \qquad (4.14)$$

where

$$s_0 = 0 \quad \text{and} \quad s_k = \frac{1}{n}\sum_{j=0}^{k-1}\sigma_j \quad \text{for } k = 1, \ldots, n. \qquad (4.15)$$

Computationally, (4.14) is more friendly than the irreducible integral

$$s(t) = \int_0^t \sqrt{x'^2(\tau) + y'^2(\tau)} \, d\tau \qquad (4.16)$$

that we must contend with for a general polynomial curve $\mathbf{r}(t) = \{x(t), y(t)\}$ (see,[2] for example, [15]).

In order to determine the total arc length of any span $t \in [t_1, t_2]$ of a Pythagorean-hodograph curve, we need only evaluate the polynomial (4.14) at the parameter values t_1 and t_2, and compute the difference

$$s(t_2) - s(t_1). \qquad (4.17)$$

This should be compared with the problem of computing the value of the integral

$$\int_{t_1}^{t_2} \sqrt{x'^2(t) + y'^2(t)} \, dt \qquad (4.18)$$

by means of numerical quadrature [15] for a general polynomial curve $\mathbf{r}(t) = \{x(t), y(t)\}$. Note also that the total arc length S of the Pythagorean–hodograph curve $\mathbf{r}(t)$ on $t \in [0, 1]$ is simply the *mean* of the coefficients of its parametric speed,

$$S = \frac{\sigma_0 + \sigma_1 + \cdots + \sigma_{n-1}}{n} = s(1). \qquad (4.19)$$

Fig. 5 illustrates a few Pythagorean–hodograph curves and indicates their exact arc lengths.

---

[2] The problem of computing arc lengths is of interest when parametric curves are used to describe trajectories for computer animation.

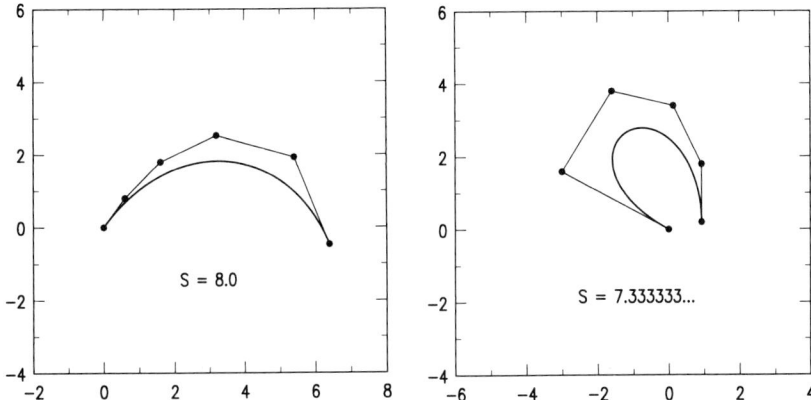

Figure 5. Exact arc lengths of some Pythagorean–hodograph curves

Similarly, it is much easier to determine from (4.14) than from (4.16) the parameter value t_* at which the arc length, measured from $t = 0$, attains a certain prescribed value s_* — *i.e.*, to solve the equation

$$s(t) = s_* \qquad (4.20)$$

with the left–hand side given by (4.14) or (4.16). Indeed, when $s(t)$ is given by (4.14), t_* is just the unique real root of the polynomial $s(t) - s_*$ and we may compute it by the Newton–Raphson iteration

$$t_{k+1} = t_k - \frac{s(t_k) - s_*}{\sigma(t_k)} \quad \text{for } k = 1, 2, \dots, \qquad (4.21)$$

given a starting approximation t_1 (the uniqueness of the desired real root t_* follows from the fact that $ds/dt = \sigma(t) > 0$ and hence $s(t) - s_*$ is strictly monotone–increasing for all real t).

By comparison, to compute the parameter value t_* at which a general polynomial curve $\mathbf{r}(t)$ achieves total arc length s_*, measured from $t = 0$, we must determine when the integral (4.16), *considered as a function of the upper limit of integration*, has the value s_* (see also [28]).

By setting $s_* = k\Delta s$ in (4.21) for $k = 1, \dots, N$, where $\Delta s = S/N$ and the total arc length S is given by (4.19), we may easily compute a uniform rendering of a Pythagorean–hodograph curve, *i.e.*, a sequence of $N + 1$ points on the curve that are distributed *uniformly by arc length* along it (see [6]). By contrast, the usual approach of rendering a polynomial curve $\mathbf{r}(t)$ by evaluating it at a sequence $t_0, \dots, t_N$ of parameter values corresponding to a uniform increment $\Delta t = t_k - t_{k-1}$ gives a distribution that reflects the variation of the speed $\sigma(t)$ along the curve.

5   An Elusive Ideal

The differential geometer's ideal is to parameterize a curve by its arc length s, measured from some fixed point:

$$\mathbf{r}(s) = \{x(s), y(s)\}\,. \tag{5.1}$$

With such a parameterization, the differential characteristics of the curve acquire an exceedingly simple form. Thus, if $\mathbf{T}(s)$ is the unit tangent vector and $\kappa(s)$ the curvature of $\mathbf{r}(s)$, we may write [18]:

$$\mathbf{T}(s) = \frac{d\mathbf{r}}{ds} \quad \text{and} \quad \kappa(s) = \frac{d\theta}{ds}\,, \tag{5.2}$$

where $\theta(s)$ is the angle that $\mathbf{T}(s)$ makes with some fixed reference direction. Thus, $\mathbf{T}(s)$ is simply the curve derivative at each point, while $\kappa(s)$ gives the *rate of rotation* of $\mathbf{T}(s)$ as we traverse $\mathbf{r}(s)$. Indeed, the differential geometer's affinity for the representation (5.1) is so great that he or she frequently refers to it in endearing terms — the "natural" or "intrinsic" parameterization.

It seems reasonable to ask whether there are actually any curves that achieve this ideal using only simple (*i.e.*, polynomial or rational) functions. A trivial example is the straight line

$$x(s) = x_0 + \lambda s\,, \quad y(s) = y_0 + \mu s \tag{5.3}$$

whose direction cosines have the normalization $\lambda^2 + \mu^2 = 1$. As noted above in §1, however, for a polynomial curve of degree $n > 1$ the argument of the radical (1.2) is necessarily a proper polynomial of degree $2(n-1)$ and it is clear that the parametric speed cannot be constant.

For rational curves of the form (1.7) — satisfying the constraints (1.8) — it is not so trivial a matter to establish the impossibility of constant parametric speed (straight lines excepted). It seems remarkable that this problem, being phrased in such simple conceptual terms, has apparently not been resolved until recently [12]. We will provide here only a brief sketch of a proof of the impossibility of unit–speed rational curves, and refer the interested reader to [12] for the full details.

From (1.9) we see that for a rational curve of the form (1.7) to be of unit speed, the polynomials $X(t)$, $Y(t)$, $W(t)$ must satisfy the identity

$$(WX' - W'X)^2 + (WY' - W'Y)^2 \equiv W^4\,. \tag{5.4}$$

Thus, the polynomials $WX' - W'X$, $WY' - W'Y$, and W^2 must have the form of a Pythagorean triple (see §4 above), namely

$$WX' - W'X = (a^2 - b^2)c\,, \quad WY' - W'Y = 2abc\,, \quad W^2 = (a^2 + b^2)c \tag{5.5}$$

for real polynomials $a(t)$, $b(t)$, and $c(t)$. By substituting these expressions into (1.9) and integrating, we see that a unit–speed rational curve would result if we could find a pair of real non–zero polynomials $a(t)$ and $b(t)$ — relatively prime and not both constants — such that the integrals

$$I_x(t) = \int_0^t \frac{a^2(\tau) - b^2(\tau)}{a^2(\tau) + b^2(\tau)} \, d\tau = x(t) - x_0 ,$$

$$I_y(t) = \int_0^t \frac{2\,a(\tau)b(\tau)}{a^2(\tau) + b^2(\tau)} \, d\tau = y(t) - y_0 \qquad (5.6)$$

are *both* rational functions of t.

Now a rational function $f(t)/g(t)$ can be integrated by first computing its *partial fraction decomposition*,

$$\frac{f(t)}{g(t)} = \sum_{r=1}^k \sum_{s=1}^{m_r} \frac{C_{rs}}{(t - z_r)^s} , \qquad (5.7)$$

where $z_1, \ldots, z_k$ are the distinct roots of the denominator, and $m_1, \ldots, m_k$ are their respective multiplicities. The integrals of the individual terms on the right–hand side of (5.7) are well known. In particular, those terms with $s \geq 2$ yield rational contributions on integration, while those with $s = 1$ result in non-rational (logarithmic and arc–tangent) forms; see [13]. Thus, if the integral of (5.7) is to be rational, we must have

$$C_{r1} = 0 \quad \text{for } r = 1, \ldots, k . \qquad (5.8)$$

The coefficients $\{C_{r1}\}$ of the inverse linear terms in the expansion (5.7) are known as the *residues* of $f(t)/g(t)$ at its poles $\{z_r\}$.

The residues also arise in computing the *definite* integral of $f(t)/g(t)$ over the entire real line; from the "calculus of residues" [16] we know that

$$\int_{-\infty}^{+\infty} \frac{f(t)}{g(t)} \, dt = 2\pi i \sum_{\operatorname{Im}(z_r)>0} C_{r1} \qquad (5.9)$$

when the denominator $g(t)$ has no real roots and is of degree two or more greater than the numerator $f(t)$ (the sum on the right–hand side of (5.9) is taken over all poles in the upper half of the complex plane).

Note that the integrands in (5.6) have no real poles, since $\gcd(a, b) = 1$ by assumption. Also, for any relatively prime polynomials $a(t)$ and $b(t)$, we can always choose real numbers α and β — not both zero — such that $\deg(\alpha a + \beta b) < \max(\deg(a), \deg(b))$. Thus, if we were to postulate that the integrals $I_x(t)$ and $I_y(t)$ in (5.6) were *both* rational, then evidently

$$I(t) = \int_0^t \frac{[\alpha a(\tau) + \beta b(\tau)]^2}{a^2(\tau) + b^2(\tau)}\, d\tau$$

$$= \tfrac{1}{2}(\alpha^2 - \beta^2)\, I_x(t) + \alpha\beta\, I_y(t) + \tfrac{1}{2}(\alpha^2 + \beta^2)t \qquad (5.10)$$

would also be rational, *i.e.*, the residues of the integrand $(\alpha a + \beta b)^2/(a^2 + b^2)$ must be zero at each of its poles. But by (5.9), this would imply that

$$\int_{-\infty}^{+\infty} \frac{[\alpha a(t) + \beta b(t)]^2}{a^2(t) + b^2(t)}\, dt = 0, \qquad (5.11)$$

which is an obvious falsity, since the integrand has a finite positive value for all real t when $a(t)$, $b(t)$ are both non–zero and $\mathrm{GCD}(a, b) = 1$!

Hence, the supposition that there exist pairs of non–zero, relatively prime real polynomials $a(t)$, $b(t)$ for which the integrals $I_x(t)$, $I_y(t)$ in (5.6) are both rational must be erroneous. It is *impossible* to parameterize any plane curve, except a straight line, by rational functions of its arc length! The proof may be extended to space curves in three or more dimensions [12].

6   Some Strange Animals

In order to present examples of curves that *do* have an explicit arc–length parameterization, we clearly need to venture beyond the realm of rational functions. There we shall encounter some rather exotic creatures, whose appeal is of a frankly intellectual, rather than practical, nature.

Let us begin, however, with a not–so–strange example. From a well–known identity, it comes as no surprise that the parameterization of the circle in terms of trigonometric functions

$$x(s) = \cos s\,, \quad y(s) = \sin s\,, \qquad (6.1)$$

where $0 \le s < 2\pi$, is of unit speed:

$$\left(\frac{dx}{ds}\right)^2 + \left(\frac{dy}{ds}\right)^2 = \sin^2 s + \cos^2 s \equiv 1\,. \qquad (6.2)$$

Similarly, we might be motivated by the kinship of the trigonometric and hyperbolic functions to concoct the form

$$x(s) = \sin^{-1}(\tanh s)\,, \quad y(s) = \ln(\cosh s) \qquad (6.3)$$

for $-\infty < s < +\infty$, which owes its constancy of parametric speed to an equally well–known identity,

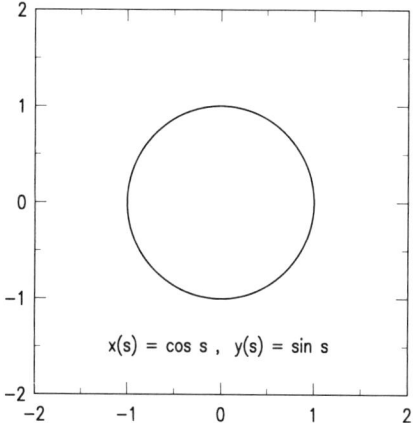

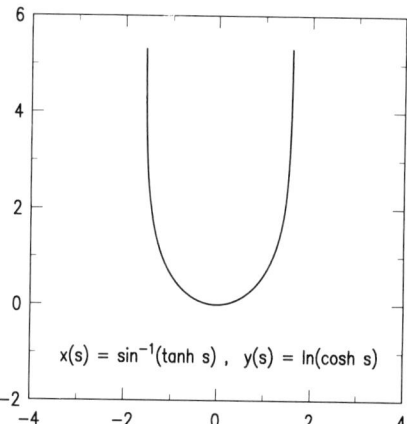

Figure 6. A very familiar unit-speed curve

Figure 7. The hyperbolic counterpart to Figure 6

$$\left(\frac{dx}{ds}\right)^2 + \left(\frac{dy}{ds}\right)^2 = \text{sech}^2 s + \tanh^2 s \equiv 1. \qquad (6.4)$$

The locus of (6.1), shown in Fig. 6, is an old friend. But the torpedo–like trace of (6.3), illustrated in Fig. 7, is doubtless a complete stranger to most readers.

Now although the sum of the squares of two polynomials can never be a constant, there are *pairs of rational functions* whose squares sum to unity. We know already from §5 that the curves defined by taking such pairs to be components of a hodograph can never be rational curves (other than straight lines). It is nevertheless of interest to examine what curiosities might arise from these hodographs. By way of example, we cite

$$x(s) = \sqrt{1 + s^2}, \ y(s) = \sinh^{-1} s \qquad (6.5)$$

with $-\infty < s < +\infty$, for which the reader may care to verify that

$$\left(\frac{dx}{ds}\right)^2 + \left(\frac{dy}{ds}\right)^2 = \frac{s^2}{1 + s^2} + \frac{1}{1 + s^2} \equiv 1. \qquad (6.6)$$

The locus (6.5) is shown in Fig. 8. Other examples of this sort are readily constructed once the technique of integrating rational functions by means of partial–fraction expansions is mastered.

Forging relentlessly ahead in the menagerie of mathematical functions, we encounter the family of unit–speed curves

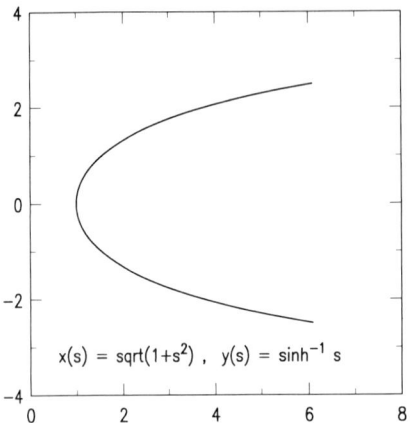

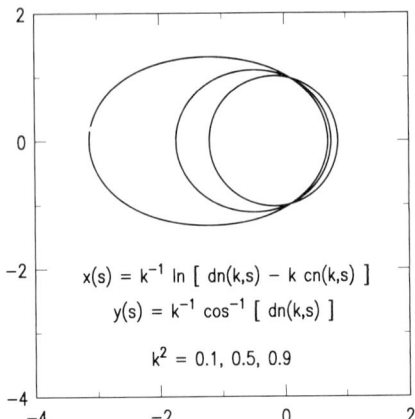

Figure 8. A unit-speed curve with a rational hodograph

Figure 9. Examples of the unit-speed curves (6.7)

$$x(s) = \frac{1}{k}\ln\left[\mathrm{dn}(k,s) - k\,\mathrm{cn}(k,s)\right], \; y(s) = \frac{1}{k}\cos^{-1}\left[\mathrm{dn}(k,s)\right] \qquad (6.7)$$

where $0 \le s < 4K$ and $0 \le k \le 1$, which are parameterized in terms of the *Jacobian elliptic functions* $\mathrm{sn}(k,s)$, $\mathrm{cn}(k,s)$, and $\mathrm{dn}(k,s)$.

(These functions may be defined as follows. For $0 \le k \le 1$, define the angle $\phi(k,s)$ by

$$s = \int_0^{\phi(k,s)} \frac{d\theta}{\sqrt{1 - k^2 \sin^2 \theta}} \qquad (6.8)$$

for each value of s. The expression on the right–hand side is an *incomplete elliptic integral of the first kind*, and the Jacobian elliptic functions are defined implicitly in terms of this integral by the relations

$$\mathrm{sn}(k,s) = \sin \phi(k,s), \; \mathrm{cn}(k,s) = \cos \phi(k,s), \qquad (6.9)$$

and

$$\mathrm{dn}(k,s) = \sqrt{1 - k^2 \sin^2 \phi(k,s)}. \qquad (6.10)$$

The functions $\mathrm{sn}(k,s)$ and $\mathrm{cn}(k,s)$ have period $4K$ (and $\mathrm{dn}(k,s)$ has period $2K$), where

$$K = \int_0^{\pi/2} \frac{d\theta}{\sqrt{1 - k^2 \sin^2 \theta}} \qquad (6.11)$$

is the *complete* elliptic integral of the first kind with modulus k. For further details, see [3], [20], or [30].)

The unit–speed nature of (6.7) follows directly from the identity

$$\left(\frac{\mathrm{d}x}{\mathrm{d}s}\right)^2 + \left(\frac{\mathrm{d}y}{\mathrm{d}s}\right)^2 = \mathrm{sn}^2(k,s) + \mathrm{cn}^2(k,s) \equiv 1 \qquad (6.12)$$

for the elliptic functions; the expressions on the right–hand sides of (6.7) are just the indefinite integrals of $\mathrm{sn}(k,s)$ and $\mathrm{cn}(k,s)$ [20]. Note that in (6.7) the value of $\cos^{-1}[\mathrm{dn}(k,s)]$ is to be taken in $[-\pi/2, 0]$ or $[0, \pi/2]$ according to whether $0 \le s < 2K$ or $2K \le s < 4K$, respectively.

The values of the functions (6.9) and (6.10) have been tabulated (see, for example, [21]) but, unlike their trigonometric and hyperbolic counterparts, they are unfortunately not widely available as function calls in high–level languages such as FORTRAN or C — nor as keys on scientific calculators. However, it is not difficult to write routines that furnish their values to a prescribed tolerance for any k and s. To that end, we appeal to the Fourier series

$$\mathrm{sn}(k,s) = \frac{1}{k}\frac{2\pi}{K}\sum_{r=0}^{\infty}\frac{q^{r+\frac{1}{2}}}{1-q^{2r+1}}\sin\frac{(2r+1)\pi s}{2K},$$

$$\mathrm{cn}(k,s) = \frac{1}{k}\frac{2\pi}{K}\sum_{r=0}^{\infty}\frac{q^{r+\frac{1}{2}}}{1+q^{2r+1}}\cos\frac{(2r+1)\pi s}{2K},$$

$$\mathrm{dn}(k,s) = \frac{2\pi}{K}\left[\frac{1}{4} + \sum_{r=1}^{\infty}\frac{q^r}{1+q^{2r}}\cos\frac{r\pi s}{2K}\right], \qquad (6.13)$$

which converge quite nicely, the quantity

$$q = \mathrm{e}^{-\pi K'/K} \qquad (6.14)$$

being given in terms of (6.11) and of

$$K' = \int_0^{\pi/2}\frac{\mathrm{d}\theta}{\sqrt{1 - k'^2\sin^2\theta}}, \qquad (6.15)$$

where the complementary modulus k' is defined by $k'^2 = 1 - k^2$. (When s is regarded as a *complex* variable, the functions $\mathrm{sn}(k,s)$, $\mathrm{cn}(k,s)$, and $\mathrm{dn}(k,s)$ have the complex periods $2\mathrm{i}K'$, $2(K + \mathrm{i}K')$, and $4\mathrm{i}K'$, respectively.)

Examples of the unit–speed curves (6.7) are shown in Fig. 9 for several values of k. These loci are reminiscent of the progression from a circle to a parabola through a sequence of ellipses of increasing eccentricity.[3] In fact, one motivation for our dwelling upon these curves is that they incorporate

---

[3] But a detailed investigation reveals that the curves (6.7) are actually slightly "fatter" than ellipses of a given axial ratio (the x and y widths of the curves (6.7) are in the ratio $\ln[(1 + k)/(1 - k)] : 2\sin^{-1} k$).

two of our preceding examples — the circle (6.1) and torpedo–curve (6.3) — as special instances, as can be seen from the limiting forms

$$\mathrm{sn}(0,s) = \sin s\,,\ \mathrm{cn}(0,s) = \cos s \tag{6.16}$$

$$\mathrm{sn}(1,s) = \tanh s\,,\ \mathrm{cn}(1,s) = \mathrm{sech}\,s \tag{6.17}$$

of the Jacobian elliptic functions, for $k = 0$ and $k = 1$.

7   Backtracking

We have already encountered in §3 an "improperly parameterized" curve which, on reaching a certain point, exhibits the curious behavior of coming to a halt, selecting reverse gear, and re–tracing its locus up to that point in the opposite sense. This pathological nature was reflected in the vanishing of the parametric speed at the crucial point, and in the symmetry of the hodograph about the origin.

A thorough discussion of how to detect and avoid such peculiar behavior in the case of polynomial and rational curves has been given by Sederberg [23], [24]. Given a polynomial curve of the form

$$x(t) = \sum_{k=0}^{n} a_k t^k\,,\ y(t) = \sum_{k=0}^{n} b_k t^k \tag{7.1}$$

we say it has an *improper parameterization* if there exists a polynomial

$$p(t) = \sum_{j=0}^{m} c_j t^j \tag{7.2}$$

of degree m (a divisor of n) ≥ 2, and a pair of polynomials

$$\tilde{x}(u) = \sum_{k=0}^{n/m} \alpha_k u^k\,,\ \tilde{y}(u) = \sum_{k=0}^{n/m} \beta_k u^k \tag{7.3}$$

such that

$$x(t) \equiv \tilde{x}(p(t)) \quad \text{and} \quad y(t) \equiv \tilde{y}(p(t))\,, \tag{7.4}$$

Under such conditions, it is clear that equations (7.3) describe the same locus as (7.1), but using polynomials of lower degree.

The original polynomials (7.1) are said to be *composite*, and (7.2) is their *common component* — which we assume to be of the highest possible degree, *i.e.*, the polynomials (7.3) have no common component. The parametric speed $\sigma(t)$ of the original (improper) parameterization (7.1) can be related to that $\sigma(u)$ of the new (proper) parameterization (7.3) by the chain rule:

$$\sigma(t) = \left|\frac{dp}{dt}\right| \sigma(u),$$
(7.5)

at corresponding values of u and t. (The equation $u = p(t)$ defines m values of t — real or complex conjugates, counted with multiplicity — corresponding to each real value of u.)

Thus, we see that $\sigma(t)$ vanishes at every root of dp/dt, of which there are $m - 1$ (not necessarily all real or distinct). At the real roots of dp/dt of *odd* multiplicity, the curve (7.1) reverses and begins to re-trace itself backwards; at real roots of *even* multiplicity, it comes to a stop but then proceeds forward in the same sense.

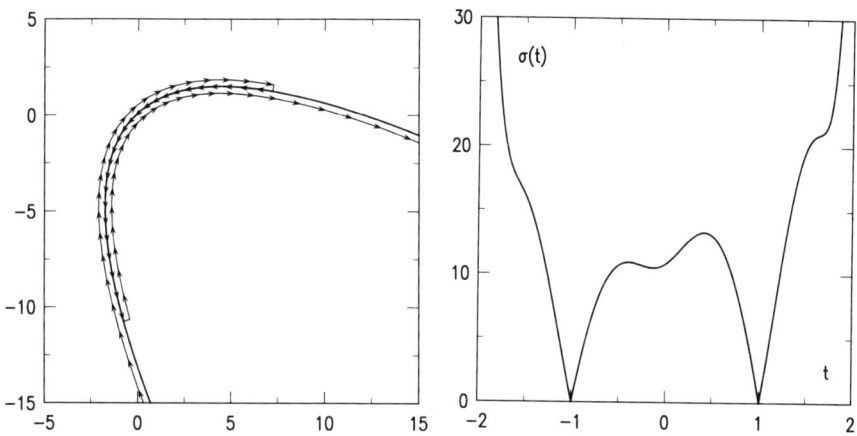

Figure 10. To and fro along an improperly parameterized curve

In Fig. 10 we show the example

$$\begin{aligned}
x(t) &= t^6 - 6t^4 + 2t^3 + 9t^2 - 6t - 2.75, \\
y(t) &= -t^6 + 6t^4 + 3t^3 - 9t^2 - 9t + 1.25,
\end{aligned}$$
(7.6)

for which the common component is $p(t) = t^3 - 3t - 0.5$. On setting $u = p(t)$, we see that (7.6) actually reduces to the parabola $x(u) = u^2 + 3u - 1.5$, $y(u) = -u^2 + 2u + 2.5$. The improper parameterization (7.6) reverses at the roots of $dp/dt = 3(t^2 - 1)$, i.e., at $t = \pm 1$; note that $\sigma(t)$ drops to zero at these points.

Such behaviour is clearly wasteful and annoying in rendering a curve, and will no doubt wreak havoc with curve intersection procedures. Unfortunately, the problem of algorithmically decomposing polynomials is not trivial (see [17]), although we might take some comfort in the fact that curves of *prime* degree (*e.g.*, the ubiquitous cubics and quintics) will never indulge in such mischief. The rational curve case is somewhat more involved (see [24]).

8 Obey the (Rational) Speed Limits!

Having strayed afar, let us now conclude on a somewhat more practical note. Since non–uniform parametric flow is an inescapable fact of life in the world of polynomial and rational parameterizations, we are naturally curious as to how its effects can be minimized. In particular, we note that such curves can be re–parameterized, without affecting their degrees, by appealing to certain special transformations of the parameter. How are such re–parameterizations best employed to minimize the variation in parametric speed over a finite segment?

We note that when the general parameter transformation

$$u = u(t) \tag{8.1}$$

is applied to $\mathbf{r}(t)$, the parametric speed $\sigma(u)$ of the re–parameterized curve $\mathbf{r}(u)$ becomes

$$\sigma(u) = \frac{\sigma(t)}{du/dt} \tag{8.2}$$

for corresponding values of u and t (we assume that the transformation (8.1) satisfies $du/dt > 0$ for all t, so the denominator in (8.2) never vanishes, and we traverse the curve in the same sense for increasing u as for increasing t).

Polynomial curves offer little scope for adjusting their parametric speed without increasing their degree. The most general parameter transformation that preserves the degree of a polynomial curve is the linear map

$$u(t) = at + b \quad \text{where } a > 0, \tag{8.3}$$

which gives $\sigma(u) = \sigma(t)/a$, the constant b being irrelevant. Thus, we can uniformly scale the speed over an interval of interest by an appropriate choice for a, but there is no prospect of remedying *relative* speed variations over that interval (note also that such scalings do not maintain $[0,1]$ as the parameter domain of the curve).

Rational curves are somewhat more flexible, since they admit fractional linear parameter transformations of the form

$$u(t) = \frac{at + b}{ct + d} \quad \text{where } ad - bc > 0 \tag{8.4}$$

without any increase in degree. The transformation (8.4) is also known as a "bilinear" map — or a *homography* in the language of one–dimensional projective geometry [27]; see also [5], [22]. In this case, the parametric speed becomes

$$\sigma(u) = \frac{(ct + d)^2}{ad - bc}\,\sigma(t)\,. \tag{8.5}$$

Note that the multiplying factor in (8.5) has a double root at $-d/c$. We shall generally wish to choose the constants of the transformation (8.4) so as to ensure that this root lies outside the interval of interest. In such cases, the multiplying factor is either monotone–increasing or monotone–decreasing over the given parameter interval.

Now the essential freedoms of the parametric speed transformation (8.5) are illustrated by re–writing it in the form

$$\sigma(u) = k\,(t - z)^2\,\sigma(t)\,, \tag{8.6}$$

where in general the root location $z = -d/c$ and scale factor $k = c^2/(ad - bc)$ may be individually adjusted. Of special interest, however, is the case where we wish the transformation to map $t \in [0,1]$ into $u \in [0,1]$ (*i.e.*, 0 and 1 are *fixed points* of the bilinear map). In that case, setting $u(0) = 0$ and $u(1) = 1$ in (8.4) reveals that $b = 0$ and $a = c + d$, and the scale factor k may be regarded as being determined by the root z through

$$k = \frac{1}{z(z - 1)}\,. \tag{8.7}$$

To avoid having $\sigma(u)$ vanish on the parameter interval $[0,1]$ we need to choose either $z < 0$ or $z > 1$ in (8.6). With k given by (8.7), the factor dt/du that maps $\sigma(t)$ into $\sigma(u)$ is then uniformly increasing or decreasing from $z/(z - 1)$ to $(z - 1)/z$ on $[0,1]$, according to whether $z < 0$ or $z > 1$.

We can use (8.6) with (8.7) to attempt to "balance" the parametric speed over a rational curve segment for which $\sigma(t)$ is monotonically increasing or decreasing,[4] while maintaining $[0,1]$ as the parameter domain. If $\sigma_0 = \sigma(0)$, $\sigma_1 = \sigma(1)$, and we choose

$$z = \frac{\sqrt{\sigma_1}}{\sqrt{\sigma_1} - \sqrt{\sigma_0}} \quad \begin{cases} < 0 & \text{if } \sigma_0 > \sigma_1, \\ > 1 & \text{if } \sigma_0 < \sigma_1, \end{cases} \tag{8.8}$$

then under the transformation defined by (8.6) and (8.7) the two end–point parametric speeds will become equal, having the value

$$\sigma_m = \sqrt{\sigma_0\sigma_1} \tag{8.9}$$

given by the *geometric mean* of the original (disparate) values σ_0 and σ_1. After the transformation, the value of $\sigma(u)$ between $u = 0$ and $u = 1$ is correspondingly more nearly constant than was $\sigma(t)$.

---

[4] Verifying this, however, is not such a straightforward matter.

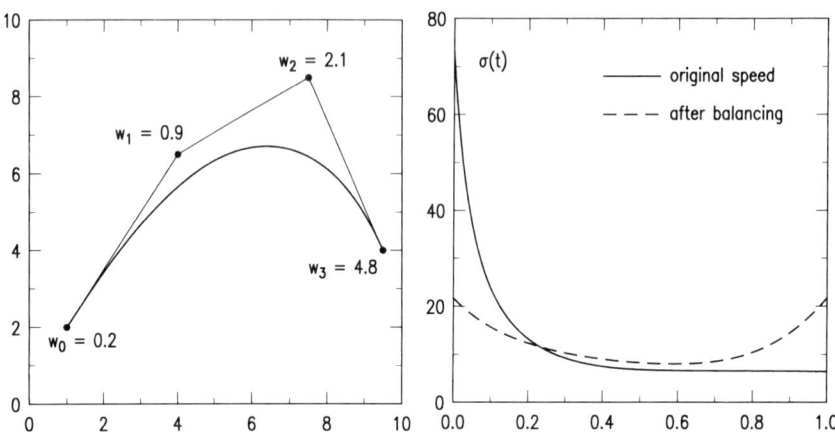

Figure 11. "Balancing" the speed of a rational cubic on $[0, 1]$

By way of example, consider the parameter transformation $t \in [0, 1] \rightarrow u \in [0, 1]$:

$$u(t) = \frac{(1-z)t}{t - z}, \tag{8.10}$$

which may be inverted to give

$$t = \frac{zu}{u + z - 1} \quad \text{and} \quad 1 - t = \frac{(z-1)(1-u)}{u + z - 1}. \tag{8.11}$$

If we take z as in (8.8) and substitute the above expressions into the equation of a rational cubic,

$$\mathbf{r}(t) = \frac{\displaystyle\sum_{k=0}^{3} w_k \mathbf{p}_k \binom{3}{k} (1-t)^{3-k} t^k}{\displaystyle\sum_{k=0}^{3} w_k \binom{3}{k} (1-t)^{3-k} t^k}, \tag{8.12}$$

for example, we find that the transformed curve $\mathbf{r}(u)$ has the same control points $\{\mathbf{p}_k\}$ but different weights,

$$\mathbf{r}(u) = \frac{\displaystyle\sum_{k=0}^{3} w'_k \mathbf{p}_k \binom{3}{k} (1-u)^{3-k} u^k}{\displaystyle\sum_{k=0}^{3} w'_k \binom{3}{k} (1-u)^{3-k} u^k}, \tag{8.13}$$

where

$$w'_k = (z-1)^{3-k} z^k w_k \quad \text{for } k = 0, 1, 2, 3.\tag{8.14}$$

The new parameterization (8.13) has equal speeds at its end points, and if the speed of the original form (8.12) was more–or–less uniformly increasing or decreasing on $[0,1]$ we can expect (8.13) to exhibit less variation in speed over its extent. An example is shown in Fig. 11.

If we relax the requirement of maintaining $[0,1]$ as the parameter domain of our rational curve, we have greater flexibility in using the transformation (8.5) — for example, to minimize the greatest deviation of $\sigma(t)$ from unity, or to ensure that it does not violate prescribed upper and lower bounds, $\sigma_{\max}$ and $\sigma_{\min}$. (We assume that the given segment $\mathbf{r}(t)$ is sufficiently small that $\sigma(t)$ does not undulate wildly along it — if so, we would have to subdivide and apply the appropriate individual transformations to each subsegment.) The details of such schemes lie beyond our present scope, deserving a systematic study in their own right.

References

1. Bézier, P. (1972). *Numerical Control – Mathematics and Applications,* Wiley, London.

2. Bruce, J.W. and Giblin, P.J. (1984). *Curves and Singularities,* Cambridge University Press.

3. Cayley, A. (1895). *An Elementary Treatise on Elliptic Functions,* 2nd edn., Dover, New York (reprint).

4. Farin, G. (1988). *Curves and Surfaces for Computer Aided Geometric Design,* Academic Press, Boston.

5. Farin, G. and Worsey, A. (1990). Reparameterization issues for rational curves, preprint.

6. Farouki, R.T. (1992). Pythagorean–hodograph curves in practical use. In *Geometry Processing for Design and Manufacturing* (R.E. Barnhill, ed.) SIAM, Philadelphia, 3–33.

7. Farouki, R.T. and Neff, C.A. (1990). Analytic properties of plane offset curves. *Comput. Aided Geom. Design* 7, 83–99.

8. Farouki, R.T. (1990). Algebraic properties of plane offset curves. *Comput. Aided Geom. Design* 7, 101–127.

9. Farouki, R.T. and Rajan, V.T. (1987). On the numerical condition of polynomials in Bernstein form. *Comput. Aided Geom. Design* 4, 191–216.

10. Farouki, R.T. (1988). Algorithms for polynomials in Bernstein form. *Comput. Aided Geom. Design 5*, 1–26.

11. Farouki, R.T. and Sakkalis, T. (1990). Pythagorean hodographs, *IBM J. Res. Develop. 34*, 736–752.

12. Farouki, R.T. (1991). Real rational curves are not "unit speed". *Comput. Aided Geom. Design, 8*, 151–157.

13. Fichtenholz, G.M. (1971). *The Indefinite Integral*, Gordon and Breach, New York.

14. Forrest, A.R. (1972). Interactive interpolation and approximation by Bézier polynomials. *Computer J. 15*, 71–79.

15. Guenter, B. and Parent, R. (1990). Computing the arc length of parametric curves. *IEEE Comput. Graph. Applic. 10*, 72–78.

16. Henrici, P. (1974). *Applied and Computational Complex Analysis, Vol. 1*, Wiley, New York.

17. Kozen, D. and Landau, S. (1989). Polynomial decomposition algorithms. *J. Symb. Comput. 7*, 445–456.

18. Kreyszig, E. (1959). *Differential Geometry*, University of Toronto Press.

19. Kubota, K.K. (1972). Pythagorean triples in unique factorization domains. *Amer. Math. Monthly 79*, 503–505.

20. Lawden, D.F. (1989). *Elliptic Functions and Applications*, Springer-Verlag, New York.

21. Milne-Thomson, L.M. (1950). *Jacobian Elliptic Function Tables*, Dover, New York.

22. Patterson, R.R. (1985). Projective transformations of the parameter of a Bernstein-Bézier curve. *ACM Trans. Graph. 4*, 276–290.

23. Sederberg, T.W. (1984). Degenerate parametric curves. *Comput. Aided Geom. Design 1*, 301–307.

24. Sederberg, T.W. (1986). Improperly parameterized rational curves. *Comput. Aided Geom. Design 3*, 67–75.

25. Sederberg, T.W. and Meyers, R.J. (1988). Loop detection in surface patch intersections. *Comput. Aided Geom. Design 5*, 161–171.

26. Sederberg, T.W. and Wang, X. (1987). Rational hodographs, *Comput. Aided Geom. Design 4*, 333–335.

27. Semple, J.G. and Kneebone, G.T. (1952). *Algebraic Projective Geometry*, Oxford University Press.

28. Sharpe, R.J. and Thorpe, R.W. (1982). Numerical method for extracting an arc length parameterization from parametric curves. *Comput. Aided Design* **12**, 79–81.

29. Uspensky, J.V. (1948). *Theory of Equations*, McGraw–Hill, New York.

30. Whittaker, E.T. and Watson, G.N. (1952). *A Course of Modern Analysis*, Cambridge University Press.

Algorithms for Implicitizing Rational Parametric Surfaces

D. Manocha[1] and J. F. Canny[2]

[1]*University of North Carolina, Chapel Hill and* [2]*University of California, Berkeley*

1 Introduction

Currently most geometric modeling systems use parametric form for representing surfaces. For computational reasons, they restrict attention to rational functions. A surface represented parametrically by rational functions is known as a *rational surface*. The parametrization of a rational surface represented in terms of homogeneous coordinates is:

$$(x, y, z, w) = (X(s,t), Y(s,t), Z(s,t), W(s,t)), \qquad (1.1)$$

where $X(s,t)$, $Y(s,t)$, $Z(s,t)$ and $W(s,t)$ are polynomials in the indeterminates s and t. Every rational surface can be represented implicitly as the zero set of an irreducible homogeneous polynomial $F(x,y,z,w) = 0$. Although the parametric formulation is useful for tracing, rendering and surface fitting, many operations like surface intersection desire one of the surfaces to be represented implicitly. Moreover, the implicit representation can be used for testing whether a point lies on the boundary and to represent an object as a semi-algebraic set. The process of converting from parametric to implicit is known as *implicitization*.

There are two known techniques for implicitization. Both these techniques reduce the problem of implicitizing rational surfaces to eliminating two variables from three parametric equations. The first technique involves the use of Elimination theory. In [12, Chapter 5] the two variables are eliminated in succession by using the Sylvester resultant for two equations. The resulting expression does not correspond to the resultant of three parametric equations and contains an extraneous factor. The Dixon formulation for computing the resultant has been used to implicitize tensor product surfaces in [26]. However it is limited to tensor product surfaces and cannot be used for implicitizing other parametrizations like the triangular patches. Bajaj et al. [1] suggest the use of Macaulay's formulation for computing the resultant of three parametric equations, which is used for implicitizing.

It expresses the resultant as a ratio of two determinants. Many a time, both the determinants evaluate to zero, even if the resultant is not zero. To compute the resultant we need to perturb the equations and use limiting arguments. This corresponds to computing the characteristic polynomial of the two matrices and the resultant is expressed as the constant term of the ratio of two characteristic polynomials [5]. Perturbation corresponds to introducing an additional variable and thereby increasing the symbolic complexity of the resulting expression. Later on, we show that this technique is expensive in practice. In general, it is believed that techniques based on elimination theory can result in extraneous factors along with the implicit representation and their separation can be a difficult task [11,7,13]. Furthermore, these algorithms are not able to implicitize parametric surfaces like bicubic patches in a reasonable amount of time and space [13].

The second technique utilizes Gröbner bases. It computes a canonical representation of the ideal generated by the parametric equations, by defining a suitable ordering of the variables [4; 12, Chapter 7]. This technique is fairly expensive in practice and even for low degree parametrizations it may take a lot of time.

We analyze the problem of implicitization. In particular, we formulate the three parametric equations in such a manner that their resultant corresponds exactly to the implicit representation. These parametric equations are of the same degree and their resultant is expressed as the determinant of a matrix. Thus, we do not need to perturb the given equations or compute the characteristic polynomial of the given matrix. We consider two types of parametrizations: total degree bounded[1] and tensor product. In each case the implicit representation corresponds to the determinant of a matrix.

The algorithms mentioned above fail altogether when a given parametrization has base points in the parametric domain. A base point in the domain, say $s = s_0, t = t_0$, corresponds to a common solution of the following four equations

$$X(s,t) = 0, \quad Y(s,t) = 0, \quad Z(s,t) = 0, \quad W(s,t) = 0.$$

It turns out that many rational parametrizations have base points. All quadric surfaces, e.g. spheres, cylinders, cones are rational surfaces with base points. Moreover, all tensor product surfaces have base points at infinity. There is a special formulation of resultants [9], which has been used to implicitize tensor product surfaces [26]. However this technique fails when there are base points in the affine domain or excess base points at infinity. In general, any faithful parametrization of a rational surface, whose algebraic degree is not a perfect square, has base points. These base points blow up to rational curves on the surface (known as seam curves).

---

[1] These parametrizations include the triangular patches.

Parametrizations with base points are analyzed in [18] and techniques for implicitizing them are presented.

In this paper we present algorithms to implicitize parametric surfaces with and without base points. The strength of the algorithms lie in the fact that we do not use multivariate factorization. We also describe an implementation based on interpolation techniques.

The rest of the paper is organized as follows. In Section 2 we specify our notation and present some background material from algebraic geometry. In Section 3 we analyze the problem of implicitization and show how we can use resultants or Gröbner bases on a parametrization without any base points to compute the implicit representation. In Section 4 we highlight many properties of rational surfaces with base points and show why the previous techniques of implicitization using resultants or Gröbner bases fail on such parametrizations. We also consider tensor product parametrizations as a special case of rational surface consisting of base points. We present algorithms for implicitizing parametric surfaces in Section 5 and in Section 6 we discuss its implementation.

2 Background

A rational parametrization is a vector valued function of the form

$$\mathbf{F}(s,t) = (X(s,t), Y(s,t), Z(s,t), W(s,t)). \tag{2.1}$$

We use lower case letters like s, t, x or y to denote scalar variables and upper case letters to represent scalar functions like $W(s,t)$ or $F(x,y,z)$ and homogeneous functions like $\overline{F}(x,y,w)$. Bold face upper case letters, like $\mathbf{F}(s,t)$, are used to represent vector valued functions and lower case bold face letters like $\mathbf{p}$ and $\mathbf{q}$ represent tuples like (s,t,u).

In (2.1), $X(s,t)$, $Y(s,t)$, $Z(s,t)$ and $W(s,t)$ are bivariate polynomials and assumed to have *power basis* representation. All tensor product Bézier, B-spline surfaces can be converted into power basis representation. The degree of parametrization, (2.1) is the maximum of the degrees of $X(s,t)$, $Y(s,t)$, $Z(s,t)$ and $W(s,t)$.

2.1 Rational Surface

In geometric modeling a surface parametrization, (2.1), indicates a mapping of the form

$$\mathbf{F} : \mathbf{R}^2 \to \mathbf{R}^3.$$

In fact the domain is often restricted to a finite interval, of the form $[a_1, b_1] \times [a_2, b_2]$ or a triangle. Since the field of real numbers is not algebraically closed, it is often useful to extend this definition to its algebraic closure, $\mathbb{C}$, the set of complex numbers. Hence we consider the parametrization as a vector valued function

$$\mathbf{F} : \mathbb{C}^2 \to \mathbb{C}^3.$$

Till now we have viewed our surface as a geometric object in *affine* space. However, there are a lot of advantages in considering the object in *projective* space. Projective n-dimensional space consists of the affine n-dimensional space plus the points at *infinity*.

Let $s = s_0, t = t_0$ be a solution of $W(s,t) = 0$. Since (2.1) is a homogeneous representation of the surface, $\mathbf{F}(s_0, t_0)$ is a point at infinity. Furthermore, the parameters s and t can correspond to infinity as well. For example all tensor product surfaces have base points at infinity. Thus, $\mathbf{F}(s,t)$ should be regarded as the following function:

$$\mathbf{F} : \mathbf{P}^2 \to \mathbf{P}^3$$

where $\mathbf{P}$ denotes the complex projective space. A parameter value in the domain, $\mathbf{P}^2$, is represented by the tuple (s,t,u) and $u = 0$ corresponds to the parameter values at infinity. The rational surface $\mathbf{F}(s,t)$ should be interpreted as a representation of the form

$$\overline{\mathbf{F}}(s,t,u) = (\overline{X}(s,t,u), \overline{Y}(s,t,u), \overline{Z}(s,t,u), \overline{W}(s,t,u)) \qquad (2.2)$$

where $\overline{X}(s,t,u), \overline{Y}(s,t,u), \overline{Z}(s,t,u)$ and $\overline{W}(s,t,u)$ are homogeneous polynomials in s, t and u and each polynomial has the same degree. Moreover,

$$\mathrm{GCD}(\overline{X}(s,t,u), \overline{Y}(s,t,u), \overline{Z}(s,t,u), \overline{W}(s,t,u)) = 1.$$

2.2   Algebraic Plane Curves

In this section we present some results on algebraic plane curves. A plane curve of degree n, can be represented by an equation $F(x,y) = 0$, where F is a polynomial of degree n. The corresponding homogeneous representation is $\overline{F}(x,y,w) = 0$. Any point on the curve in general is a *simple point*. A few points on the curve are *multiple points* [25, Chapter 2].

Definition: A *multiple point* of order k (or k-fold point, $k > 1$) of a degree n curve, is a point $\mathbf{p}$ of the curve such that a generic line through $\mathbf{p}$ meets the curve in only $n - k$ further points.

Let us investigate the behavior of a curve at a multiple point. We can assume that the point under consideration is the origin, i.e. $\mathbf{p} = (0, 0, 1)$, else we can bring it to the origin by a suitable linear transformation. The curve can be represented as

$$\overline{F}(x,y,w) = \overline{U}_0(x,y)w^n + \overline{U}_1(x,y)w^{n-1} + \ldots + \overline{U}_{n-1}(x,y)w + \overline{U}_n(x,y) = 0,$$

where $\overline{U}_i(x,y)$ is a homogeneous polynomial of degree i in x and y. A generic line through the origin can be represented in the form $x/a = y/b$. The point of this line whose coordinates are $(ka, kb, 1)$, where k is a scalar, lies on the curve if k is any of the roots of the equation

$$\overline{U}_0(a,b) + k\overline{U}_1(a,b) + k^2\overline{U}_2(a,b) + \ldots + k^i\overline{U}_i(a,b) + \ldots + k^n\overline{U}_n(a,b) = 0. \tag{2.3}$$

To make the curve have a k-fold point at the origin corresponds to making the equation, (2.3), have k nonzero roots for every value of the ratio a/b. This can happen, if and only if $\overline{U}_0(x,y)$, $\overline{U}_1(x,y)$, $\ldots$, $\overline{U}_{k-1}(x,y)$ vanish identically. A line corresponds to a tangent at **p**, if it has $k+1$ of its intersections with the curve at **p** and the $n-k-1$ intersections at other points on the curve. All lines of the form $x/a' = y/b'$, where $\overline{U}_k(a',b') = 0$ are tangent to the curve at **p**. There can be at most k such lines.

A simple version of *Bezout's theorem* is used for determining the number of intersections between a curve of degree m and that of degree n. It is assumed that the curves have no component in common. That is:

Two curves of degree m and n intersect at mn points, counted properly with respect to multiplicity.

2.3 Algebraic Sets

Let us consider an algebraically closed field, $\mathbb{C}$ and define a polynomial ring

$$A = \mathbb{C}[x_1, x_2, \ldots, x_m]$$

of m variables over $\mathbb{C}$. All the polynomials used in this section are assumed to be defined over this ring.

Definition: The set of common zeros of a system of polynomials $F_1, \ldots, F_n$ in $x_1, \ldots, x_m$ is called an *algebraic set* and is denoted $V(F_1, \ldots, F_n) \subset \mathbb{C}^m$. An algebraic set $V(F)$ defined by a single polynomial (which is not identically zero) is called a *hypersurface*. If F is linear, then $V(F)$ is called a *hyperplane*.

The union of two algebraic sets is an algebraic set. The intersection of any family of algebraic sets is an algebraic set. The empty set and the whole space are algebraic sets [10, Chapter 1].

If all the F_i are homogeneous, it is more convenient to work with the projective space $\mathbb{P}^{m-1}$, formed by identifying points in $\mathbb{C}^m$ which are scalar multiples of each other. We use the same notation, $V(\overline{F}_1, \ldots, \overline{F}_n) \subset P^{m-1}$ for an algebraic set defined by homogeneous polynomials $\overline{F}_i$.

2.4 Faithful and Unfaithful Parametrizations

In many cases a rational curve or surface can be identically described by a lower degree rational parametrization. Such curves or surfaces have *unfaithful* parametrizations. In particular, a surface parametrization is *faithful* if there is a one to one relationship between the points on the surface and the parameter values, except for a finite number of points and curves on the surface. Another popular terminology for faithful and unfaithful parametrizations are proper and improper parametrizations, respectively.

Consider the following affine parametrization of the unit sphere

$$(x, y, z, w) = (1 - s^2 - t^2, 2s, 2t, 1 + s^2 + t^2).$$

Since the preimage of $(x, y, z, w) = (0, 0, 1, 1)$ consists of a unique point in the parametric domain $(s = 0, t = 1)$, the given parametrization is faithful. If we reparametrize by substituting $s = uv$ and $t = u^2$, we obtain

$$(x, y, z, w) = (1 - u^2 v^2 - u^2, 2uv, 2u^2, 1 + u^2 v^2 + u^4),$$

which is an unfaithful parametrization. There are two points in the preimage of $(0, 0, 1, 1)$, $(u, v) = (1, 0)$ and $(u, v) = (-1, 0)$.

Every rational surface can be represented by a faithful parametrization [6]. However no algorithms are known at the moment for computing the faithful parametrization of an unfaithfully parametrized rational surface. Resultants and Gröbner bases have been used to decide whether a given parametrization is faithful [1; 12, Chapter 7]. Our implicitization algorithm also determines whether a given parametrization is faithful or not.

3 Implicitization

Given a rational surface

$$\overline{\mathbf{F}}(s, t, u) = (x, y, z, w) = (\overline{X}(s, t, u), \overline{Y}(s, t, u), \overline{Z}(s, t, u), \overline{W}(s, t, u)),$$

where $\overline{X}(s, t, u)$, $\overline{Y}(s, t, u)$, $\overline{Z}(s, t, u)$ and $\overline{W}(s, t, u)$ are homogeneous polynomials of degree n. $\overline{\mathbf{F}}$ is a vector valued function defined as

$$\overline{\mathbf{F}} : \mathbb{P}^2 \to \mathbb{P}^3$$

Let the image of $\overline{\mathbf{F}}$ be $\mathcal{Y}$, where $\mathcal{Y} \subset \mathbb{P}^3$. We assume that $\mathcal{Y}$ has dimension 2. For example, $\overline{\mathbf{F}}$ is not a parametrization of the form

$$\overline{\mathbf{F}}(s, t, u) = (x, y, z, w) = (s + t, s + t, s + t, u),$$

whose image is the 1-dimensional line, $x = y = z$.

A parametrization of the form $\overline{\mathbf{F}}$ corresponds to the plane representation of a rational surface in $\mathbf{P}^3$. To a generic point $\mathbf{p}$ of the plane, there corresponds a unique point of $\mathcal{Y}$. The only exceptions are the base points, which blow up to rational curves on the surface. However, there are a finite number of such base points in the parametric domain. At times, the preimage of a point in $\mathcal{Y}$ consists of a curve in the domain. Let $\mathcal{Z}$ be one such curve of the plane, where $\mathcal{Z} \subset \mathbf{P}^2$. Such curves are known as the *fundamental curves* [25, Chapter 6]. If $\overline{\mathbf{F}}$ is a faithful parametrization, then there are open sets $\mathcal{U} \subseteq \mathbf{P}^2$ and $\mathcal{V} \subseteq \mathcal{Y}$ such that $\mathcal{U}$ is isomorphic to $\mathcal{V}$. Moreover, $\mathcal{U}$ and $\mathcal{V}$ are dense in $\mathbf{P}^2$ and $\mathcal{Y}$, respectively (A subset $\mathcal{A}$ is said to be dense in its superset $\mathcal{B}$, if it has the same dimension as that of $\mathcal{B}$). In fact

$$\overline{\mathbf{F}}(\mathcal{U}) = \mathcal{V}.$$

Thus, $\mathbf{P}^2$ and $\mathcal{Y}$ are birationally equivalent [10, Chapter 1, Corollary 4.5]. For more on birational maps we recommend [10, Chapter 1; 20, Chapter 8]. As a result there exists an inverse rational function (with respect to $\overline{\mathbf{F}}$)

$$\overline{\mathbf{F}}^{-1} : \mathcal{Y} \to \mathbf{P}^2,$$

which can be represented as

$$\overline{\mathbf{F}}^{-1}(x, y, z, w) = (s, t, u) = (\overline{S}(x, y, z, w), \overline{T}(x, y, z, w), \overline{U}(x, y, z, w)),$$
$$(3.1)$$

where $\overline{S}, \overline{T}$ and $\overline{U}$ are homogeneous polynomials in x, y, z and w. Furthermore, each of them has the same degree. Just like $\overline{\mathbf{F}}$ is not defined at the base points, $\overline{\mathbf{F}}^{-1}$ is not defined at those values of (x, y, z, w), which correspond to the images of fundamental curves (like $\overline{\mathbf{F}}(\mathcal{Z})$). To show the birational equivalence, we can remove such points and their images from $\mathbf{P}^2$ and $\mathcal{Y}$ and the birational map is, therefore, defined between the corresponding open sets.

Given $\overline{\mathbf{F}}$, resultants and Gröbner bases have been used to compute $\overline{\mathbf{F}}^{-1}$ in [1] and [12, Chapter 7], respectively. It is possible that there is more than one faithful rational parametrization of a given rational surface. All such parametrizations of a given surface are birationally equivalent.

Let $\overline{\mathbf{F}}$ be an unfaithful parametrization. Then $\overline{\mathbf{F}}$ has a corresponding faithful parametrization of the form

$$\overline{\mathbf{G}} : \mathbf{P}^2 \to \mathbf{P}^3$$

$$\overline{\mathbf{G}}(p, q, r) = (x, y, z, w) = (\overline{X}_1(p, q, r), \overline{Y}_1(p, q, r), \overline{Z}_1(p, q, r), \overline{W}_1(p, q, r)).$$

The image of $\overline{\mathbf{G}}$ is $\mathcal{Y}$. Since $\overline{\mathbf{G}}$ is a faithful parametrization, it has an inverse map

$$\overline{\mathbf{G}}^{-1} : \mathcal{Y} \to \mathbf{P}^3$$

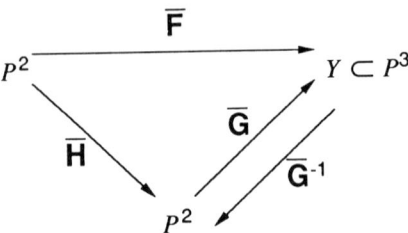

Figure 1. The relationship between the various functions

$$\overline{G}^{-1}(x, y, z, w) = (p, q, r) = (\overline{P}(x, y, z, w), \overline{Q}(x, y, z, w), \overline{R}(x, y, z, w)),$$

where $\overline{P}, \overline{Q}$ and $\overline{R}$ are homogeneous polynomials of same degree. Moreover,

$$(p, q, r) = \overline{G}^{-1}(x, y, z, w) = \overline{G}^{-1}(\mathbf{F}(s, t, u)).$$

Let $\overline{G}^{-1}\overline{F}$ be $\overline{H}$ and

$$(p, q, r) = \overline{H}(s, t, u) = (\overline{P}_1(s, t, u), \overline{Q}_1(s, t, u), \overline{R}_1(s, t, u)),$$

where $\overline{P}_1(s, t, u) = \overline{P}(\overline{X}(s, t, u), \overline{Y}(s, t, u), \overline{Z}(s, t, u), \overline{W}(s, t, u))$. We can similarly define $\overline{Q}_1$ and $\overline{R}_1$. $\overline{H}$ is a rational map of the form

$$\overline{H} : \mathbf{P}^2 \to \mathbf{P}^2.$$

It is possible that $\overline{F}$ has base points, while $\overline{G}$ has no base points. For example, consider the faithful parametrization of a plane

$$\overline{G}(p, q, r) = (x, y, z, w) = (p + r, 2p + q, q - 3r, q + r).$$

$\overline{G}$ has no base points. Substitute

$$(p, q, r) = \overline{H}(s, t, u) = (st + t^2, su + tu, s^2 + su),$$

and the resulting unfaithful parametrization is

$$\begin{aligned}
\mathbf{F}(s, t, u) &= (x, y, z, w) \\
&= (s^2 + t^2 + st + su, 2t^2 + 2st + su + tu, \\
&\quad -3s^2 - 2su + tu, s^2 + tu + 2su).
\end{aligned}$$

$(s, t, u) = (0, 0, 1)$ is a base point of $\overline{F}$.

Let us consider the case when the parametrization, $\overline{F}$, has *no base points* and the map $\overline{F}$, is therefore defined at all points in the domain. Using topological arguments it can be shown that the image of $\overline{F}$ is the zero set of an irreducible and homogeneous polynomial, $\overline{G}(x, y, z, w)$ [18]. The problem of implicitization corresponds to computing $\overline{G}(x, y, z, w)$.

Although a rational surface indicates a map between projective space, for implicitization we restrict ourselves to the affine portion of the image [18]. Consider the following *parametric equations*:

$$\begin{array}{rcl}
\overline{F}_1(s,t,u) & = & x\overline{W}(s,t,u) - \overline{X}(s,t,u) = 0, \\
\overline{F}_2(s,t,u) & = & y\overline{W}(s,t,u) - \overline{Y}(s,t,u) = 0, \\
\overline{F}_3(s,t,u) & = & z\overline{W}(s,t,u) - \overline{Z}(s,t,u) = 0.
\end{array} \qquad (3.2)$$

Let us consider the algebraic set defined as the intersection of the three hypersurfaces $\overline{F}_1, \overline{F}_2$ and $\overline{F}_3$,

$$Q = V(\overline{F}_1, \overline{F}_2, \overline{F}_3) \subset \mathbf{P}^2 \times \mathbf{C}^3,$$

and let Π be the projection function

$$\Pi : \mathbf{P}^2 \times \mathbf{C}^3 \to \mathbf{C}^3$$

such that

$$\Pi(s,t,u,x,y,z) = (x,y,z).$$

Theorem I: *If the given parametrization has no base points then Q consists of a single component. Moreover, that component can be represented as*

$$Q_1 = \{(s,t,u,x,y,z) | x = \frac{\overline{X}(s,t,u)}{\overline{W}(s,t,u)}, y = \frac{\overline{Y}(s,t,u)}{\overline{W}(s,t,u)}, z = \frac{\overline{Z}(s,t,u)}{\overline{W}(s,t,u)}\}.$$

Proof: It is easy to see that $Q_1 \subset Q$. Thus, Q_1 is a component of Q. Let us assume that Q consists of some other component, say P. Since $P \neq Q_1$, $\exists\, \mathbf{p} = (s_1, t_1, u_1, x_1, y_1, z_1) \in P$ and $\mathbf{p} \notin Q_1$. There are two possibilities:

1. $\overline{W}(s_1, t_1, u_1) = 0$.

   We know that $\mathbf{p} \in V(\overline{F}_1, \overline{F}_2, \overline{F}_3)$ and therefore

   $$\overline{F}_1(s_1, t_1, u_1) = 0,$$

   $$\Rightarrow \overline{X}(s_1, t_1, u_1) = x_1\overline{W}(s_1, t_1, u_1) = 0.$$

   Similarly, we can show that $\overline{Y}(s_1, t_1, u_1) = 0$ and $\overline{Z}(s_1, t_1, u_1) = 0$. This implies that (s_1, t_1, u_1) is a base point of $\overline{\mathbf{F}}$, which is contrary to our assumption.

2. $\overline{W}(s_1, t_1, u_1) \neq 0$.

We know that $\mathbf{p} \in Q$ and therefore,

$$\overline{F}_1(s_1, t_1, u_1) = 0$$

$$\Rightarrow x\overline{W}(s_1, t_1, u_1) = \overline{X}(s_1, t_1, u_1)$$

$$\Rightarrow x_1 = \frac{\overline{X}(s_1, t_1, u_1)}{\overline{W}(s_1, t_1, u_1)}.$$

Similarly we can show that

$$y_1 = \frac{\overline{Y}(s_1, t_1, u_1)}{\overline{W}(s_1, t_1, u_1)}, \qquad z_1 = \frac{\overline{Z}(s_1, t_1, u_1)}{\overline{W}(s_1, t_1, u_1)}.$$

This implies that $\mathbf{p} \in Q_1$.

Thus, all points in Q also lie in Q_1 and therefore,

$$Q = Q_1.$$

Thus, Q consist of one component.

Q.E.D.

Since Q is an irreducible algebraic set, each point in $\Pi(Q)$ lies in $\mathcal{Y}$. This follows from the representation of Q or Q_1 in Theorem I. Since Q and $\Pi(Q)$ are two-dimensional algebraic sets, $\Pi(Q)$ correspond to the affine portion of the zero set of the implicit representation of $\overline{F}(s, t, u)$. If the given parametrization is unfaithful, each point in $\Pi(Q)$ has more than one preimages with respect to $\overline{F}$. In this case, $\Pi(Q)$ corresponds to an algebraic set of multiplicity greater than one. Thus,

$$\Pi(Q) = V(H(x, y, z)), \tag{3.3}$$

where $H(x, y, z) = G(x, y, z)^k$, $k \geq 1$. $k = 1$ if and only if $\overline{F}$ is a faithful parametrization. Using Bezout's theorem it can be shown that the algebraic degree of $H(x, y, z)$ is n^2, where n is the degree of the parametrization [24]. The degree of $G(x, y, z)$ is n^2/k. Moreover, k corresponds to the number of points in the (s, t, u) plane, that are the preimages of an arbitrary point in $V(G(x, y, z))$.

3.1 Resultants

Given a set of m homogeneous equations in m variables, it is always possible to combine the equations to obtain from them a single equation $R = 0$, in which these variables do not appear. We are then said to have eliminated the variables and the quantity R is the *resultant* of the system of equations. In general, the resultant of any such system of equations is a function of the coefficients, whose vanishing is the necessary and sufficient condition for the given system to have a non-trivial solution. There are different formulations of computing the resultant of a system of equations. Perhaps the most general one is given by [15]. If all the equations have the same degree, an improved version is given in [16].

Let us consider the system of equations (3.2), as polynomials in s, t, and u with coefficients being functions of x, y, and z. Since $\overline{\mathbf{F}}$ has no base points, the resultant of (3.2), say $R(x, y, z)$, is a nonzero polynomial in x, y and z. Let's consider $V(R(x, y, z))$ and let $(x_1, y_1, z_1) \in V(R(x, y, z))$. The fact that $R(x_1, y_1, z_1) = 0$ implies that $\exists\, s_1, t_1, u_1$ such that $(x_1, y_1, z_1, s_1, t_1, u_1) \in Q$. Thus,

$$\Pi(Q) = V(R(x, y, z)),$$

and from (3.3), it follows that

$$R(x, y, z) = gH(x, y, z),$$

where g is a scalar.

Given $R(x, y, z)$, $G(x, y, z)$ can be expressed as

$$G(x, y, z) = \frac{R(x, y, z)}{\mathrm{GCD}(R(x, y, z), R_z(x, y, z))}, \qquad (3.4)$$

where $R_z(x, y, z)$ is the partial derivative of $R(x, y, z)$ with respect to z.

The resultant of three equal degree homogeneous equations can be expressed as a determinant of a matrix. Such formulations are given in [23;9;21]. In particular the formulation in [9] constructs a matrix of order $2n^2 - n$, where n is the degree of the equations, and its determinant corresponds to the resultant. Some other cases where the resultant can be expressed as the determinant of a single matrix are given in [17]. Thus, we need not perturb the equations for computing the resultant. Later on we show that this formulation of resultants is an efficient and compact representation for the implicit representation.

Example I Let

$$\mathbf{F}(s, t) = (x, y, z) = \left(\frac{st + 1}{t^2}, \frac{s}{t^2}, \frac{s^2}{t^2}\right).$$

After homogenizing we obtain the following system of equations

$$xt^2 - st = 0,$$

$$yt^2 - su = 0,$$

$$zt^2 - s^2 = 0.$$

Considering these equations as polynomials in s, t and u, the resultant is

$$R(x, y, z) = y^4 - 2xy^2z + x^2z^2 - z^3,$$

whose degree is 4, since $n = 2$. Since $\mathbf{F}$ is a faithful parametrization

$$R(x, y, z) = H(x, y, z) = G(x, y, z) = y^4 - 2xy^2z + x^2z^2 - z^3.$$

4   Base Points

A base point is a common solution of

$$\overline{X}(s, t, u) = 0, \quad \overline{Y}(s, t, u) = 0, \quad \overline{Z}(s, t, u) = 0, \quad \overline{W}(s, t, u) = 0.$$

The solution set of any of the polynomials, say $\overline{X}(s, t, u) = 0$, corresponds to an algebraic plane curve in the $\mathbf{P}^2$ plane (denoted by homogeneous coordinates s, t and u). Each curve may have more than one component and the base point corresponds to the intersection of these curves. The *multiplicity* of each base point is equal to the multiplicity of the curves at that point. In other words, a base point has multiplicity k, if it is a k-fold point of $\overline{X}(s, t, u), \overline{Y}(s, t, u), \overline{Z}(s, t, u)$ and $\overline{W}(s, t, u)$. Let

$$\mathcal{S} = V(\overline{X}(s, t, u), \overline{Y}(s, t, u), \overline{Z}(s, t, u), \overline{W}(s, t, u))$$

be the set of base points. Since

$$\mathrm{GCD}(\overline{X}(s, t, u), \overline{Y}(s, t, u), \overline{Z}(s, t, u), \overline{W}(s, t, u)) = 1,$$

$\mathcal{S}$ is therefore a finite set. Let $\mathbf{p} = (s_0, t_0, u_0) \in \mathcal{S}$. Moreover,

$$\overline{\mathbf{F}}(\mathbf{p}) = \overline{\mathbf{F}}(s_0, t_0, u_0) = (0, 0, 0, 0),$$

which does not correspond to any point in the image space. It has been known that base points blow up to rational curves on the surface (known as *seam curves*) [8;25, Chapter 6; 28, page 281; 27]. Furthermore, the degree of the seam curve is bounded by the multiplicity of the corresponding base point. The blow up can be explained in the following manner:

Consider the rational parametrization

$$\mathbf{F}(s, t, u) = (x, y, z, w) = (su + 2tu + s^2, su + 3tu + t^2, su + tu + 2st, su + 4tu).$$

It follows that $\mathbf{p} = (0, 0, 1)$ is a base point of the given parametrization. Let's consider the first neighborhood of $\mathbf{p}$ in the domain. That can be

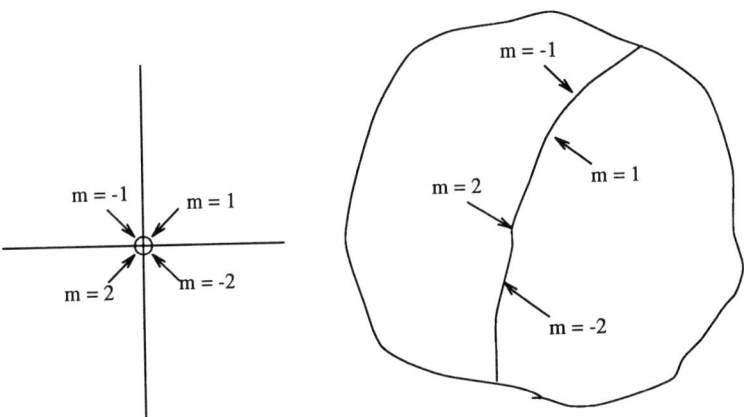

Figure 2. Blowing up of a base point at the origin

obtained by substituting $s = mt$, $u = 1$ and taking the limit $t \to 0$. That is,

$$\mathbf{G}(m) = \lim_{t \to 0} \overline{\mathbf{F}}(mt, t, 1)$$

$$\Rightarrow \mathbf{G}(m) = \lim_{t \to 0} (mt + 2t + m^2t^2, mt + 3t + t^2, mt + t + 2mt^2, mt + 4t)$$

$$\Rightarrow \mathbf{G}(m) = \lim_{t \to 0} (m + 2 + m^2t, m + 3 + t, m + 1 + 2mt, m + 4)$$

$$\Rightarrow \mathbf{G}(m) = (m + 2, m + 3, m + 1, m + 4),$$

which is the rational parametrization of a straight line. Thus, each direction in the first neighborhood of the base point gives rise to a different limit point in the image. For an arbitrary parametrization with a base point at the origin, this has been shown in Fig. 2. For more on blowing up and its applications we recommend [10, Chapter 1; 29, Chapter 4]. If the parametric curves intersect tangentially at a base point, then the first neighborhood cannot be used for computing the seam curve.

Since $\overline{\mathbf{F}}$ is not defined at the base points, we modify its domain and define it as a mapping of the form

$$\overline{\mathbf{F}'} : \mathbf{P}^2 \setminus \mathcal{S} \to \mathbf{P}^3$$

$$\overline{\mathbf{F}'}(s, t, u) = \overline{\mathbf{F}}(s, t, u),$$

where $\mathbf{P}^2 \setminus \mathcal{S}$ represents the difference of two sets. $\mathbf{P}^2 \setminus \mathcal{S}$ is an open and irreducible set of dimension 2. Let $\mathcal{K}$ be the image of $\overline{\mathbf{F}'}$. We know that $\mathcal{K}$ is a 2-dimensional set and $\mathcal{K} \subset \mathbf{P}^3$. In general, $\mathcal{K}$ is a proper subset of an algebraic set $V(\overline{H}(x, y, z, w))$. The problem of implicitization corresponds to computing $\overline{H}(x, y, z, w)$.

The degree of the implicit representation is a function of the number of base points. If a parametrization of degree n has p simple base points, then the implicit representation has degree $n^2 - p$. This is assuming that the parametrization is faithful, else we divide it by a suitable k. This can be explained in the following manner [24]:

The degree of the implicit equation of a parametric surface can be determined by counting the number of times it is intersected by a generic straight line. Let us assume that the parametrization is faithful. Such a line in space can be represented as the intersection of the following 2 planes:

$$a_1 x + a_2 y + a_3 z + a_4 = 0,$$

$$b_1 x + b_2 y + b_3 z + b_4 = 0.$$

To determine the number of intersections of this line with $\overline{\mathbf{F}}$, we substitute for x, y and z and obtain

$$\overline{F}_1(s,t,u) = a_1 \overline{X}(s,t,u) + a_2 \overline{Y}(s,t,u) + a_3 \overline{Z}(s,t,u) + a_4 \overline{W}(s,t,u) = 0,$$

$$\overline{F}_2(s,t,u) = b_1 \overline{X}(s,t,u) + b_2 \overline{Y}(s,t,u) + b_3 \overline{Z}(s,t,u) + b_4 \overline{W}(s,t,u) = 0.$$

These curves are each of degree n in s, t and u. According to Bezout's theorem the two curves intersect in n^2 points. If a parametrization has no base points, each such point of intersection in the (s,t,u) plane is a preimage of a point of intersection between the line and the surface. Thus, the surface has degree n^2. If $\overline{\mathbf{F}}$ has any base points, both $\overline{F}_1(s,t,u)$ and $\overline{F}_2(s,t,u)$ contain the base point. Thus, they intersect at that base point. The base points blow up to 1-dimensional curves on the surface $(V(\overline{H}(x,y,z,w))\backslash \mathcal{K})$. Since there are a finite number of such curves on the surface, it is always possible to find a line which does not intersect the surface along any of these curves. In other words, the line does not intersect with any point in the set $V(\overline{H}(x,y,z,w)) \setminus \mathcal{K}$. For such a line the base point is not the preimage of any point of intersection on the surface, and the line intersects the surface at $n^2 - 1$ or less number of points. If there are p simple base points, then the degree of the implicit equation is $n^2 - p$. Base points of multiplicity k decrease the degree of the surface by at least k^2 [28, page 281; 27]. If $F_1(s,t,u)$ and $F_2(s,t,u)$ intersect tangentially along a k-fold base point, it decreases the degree by at least $k^2 + 1$.

Problem with Resultants

Given $\overline{\mathbf{F}}$, a parametrization with base points, we use resultants to compute the implicit equation. The corresponding parametric equations, (3.2), are

$$
\begin{aligned}
\overline{F}_1(s,t,u) &= x\overline{W}(s,t,u) - \overline{X}(s,t,u) = 0, \\
\overline{F}_2(s,t,u) &= y\overline{W}(s,t,u) - \overline{Y}(s,t,u) = 0, \\
\overline{F}_3(s,t,u) &= z\overline{W}(s,t,u) - \overline{Z}(s,t,u) = 0.
\end{aligned}
$$

The resultant of these equations, by considering s, t and u as variables, is zero. This can be explained in the following manner:

Given $\mathbf{p} = (s_0, t_0, u_0)$, a base point in the parametrization. From the definition of a base point it follows that

$$
\overline{F}_1(s_0, t_0, u_0) = 0 \quad \overline{F}_2(s_0, t_0, u_0) = 0, \quad \overline{F}_3(s_0, t_0, u_0) = 0.
$$

Thus, the given system of equations, (3.2), has a non trivial solution (s_0, t_0, u_0). Moreover, this solution is independent of the coefficients, x, y and z. The resultant is therefore, identically zero.

Problem with Gröbner Bases

The Gröbner bases approach is based on the fact that the implicit equation is contained in the ideal generated by the parametric equations [4; 12, Chapter 7]. Thus, by a suitable ordering of the variables, a polynomial in the Gröbner bases of the ideal corresponds to the implicit equation. However, this method does not work when a given parametrization has base points in its domain. Gröbner bases offer us the flexibility of working in the affine space. If there are base points only at infinity, we can use the affine formulation of the given parametric equations for defining the generating set of the ideal and obtain the implicit equation by computing its Gröbner bases.

Theorem II: *Given a parametrization with base points in the affine domain. Let $\mathcal{I}$ be the ideal generated by the parametric equations. There is no polynomial in $\mathcal{I}$, which is independent of s and t, the variables used for defining the parametric domain.*

Proof: The ideal is of the form

$$
\mathcal{I} = \{xW(s,t) - wX(s,t), yW(s,t) - wY(s,t), zW(s,t) - wZ(s,t)\}.
$$

Let (s_0, t_0) be a base point in the affine domain.

Let us assume that there is a polynomial $F(x, y, z)$, which is independent of s and t in the ideal $\mathcal{I}$. Then $F(x, y, z)$ can be expressed as

$$\begin{aligned}
F(x,y,z) \;=\; & A_1(x,y,z,s,t)(xW(s,t) - wX(s,t)) + \\
& A_2(x,y,z,s,t)(yW(s,t) - wY(s,t)) + \\
& A_3(x,y,z,s,t)(zW(s,t) - wZ(s,t)).
\end{aligned}$$

Let us substitute $s = s_0, t = t_0$ in the above equation. Thus,

$$\begin{aligned}
F(x,y,z) \;=\; & A_1(x,y,z,s_0,t_0)(xW(s_0,t_0) - wX(s_0,t_0)) + \\
& A_2(x,y,z,s_0,t_0)(yW(s_0,t_0) - wY(s_0,t_0)) + \\
& A_3(x,y,z,s_0,t_0)(zW(s_0,t_0) - wZ(s_0,t_0)), \\
\Rightarrow F(x,y,z) \;=\; & 0.
\end{aligned}$$

Hence, the only polynomial independent of s and t, which lies in $\mathcal{I}$ is the zero polynomial. Thus, all other polynomials have a term containing s or t.

Q.E.D.

Given a parametrization with base points in the affine domain, the Gröbner bases of the ideal generated by the parametric equations do not contain the implicit equation. Thus, Gröbner bases cannot be used for implicitizing such parametrizations.

4.1   Tensor Product Surfaces

In computer graphics and geometric modeling, most surfaces are represented as tensor product surfaces. A tensor product surface is a linear combination of bivariate basis functions, where each bivariate function is formed by taking every possible pair, one from one set of univariate functions and the other from another [22]. Typical univariate functions are like the Bernstein polynomials, B-spline basis functions etc. Such a surface is represented as

$$\mathbf{F}(s,t) = (X(s,t), Y(s,t), Z(s,t), W(s,t)) = \sum_j \sum_i \mathbf{A}_{ij} G_i(s) H_j(t),$$

where G_i and H_j are the univariate basis functions and $\mathbf{A}_{ij}$ are 4 dimensional vectors, used to represent the scalars for each component. Any component, say $X(s,t)$, can be represented in the power basis form as:

$$X(s,t) = \sum_{i=0,m} \sum_{j=0,n} X_{ij} s^i t^j, \quad X_{mn} \neq 0,$$

which is of degree $m + n$ in s and t. However, the highest power of s in any monomial is m and that of t is n. After homogenizing we obtain

$$\overline{X}(s,t,u) = \sum_{i=0,m} \sum_{j=0,n} X_{ij} s^i t^j u^{m+n-i-j}, \tag{4.1}$$

which is a homogeneous polynomial of degree $m + n$. Let

$$\overline{\mathbf{F}}(s,t,u) = (\overline{X}(s,t,u), \overline{Y}(s,t,u), \overline{Z}(s,t,u), \overline{W}(s,t,u)).$$

Lemma I: *Every tensor product surface of the form* $\overline{\mathbf{F}}(s,t,u)$ *has a base point of multiplicity* n *at* $(1,0,0)$ *and of multiplicity* m *at* $(0,1,0)$.

Proof Let us consider one of the components, say $\overline{X}(s,t,u)$. It follows from (4.1) that
$$\overline{X}(1,0,0) = 0, \quad \overline{X}(0,1,0) = 0.$$
Similarly $\overline{Y}$, $\overline{Z}$ and $\overline{W}$ vanish at these points, too. Thus, $(1,0,0)$ and $(0,1,0)$ are base points of $\overline{\mathbf{F}}$.

To analyze the multiplicity of $(1,0,0)$, we represent $\overline{X}(s,t,u)$ as

$$\overline{X}(s,t,u) = \overline{X}_0(t,u)s^{m+n} + \ldots + \overline{X}_{n-1}(t,u)s^{m+1} + \overline{X}_n(t,u)s^m +$$

$$\ldots + \overline{X}_{m+n}(t,u),$$

where $\overline{X}_i(t,u)$ is a homogeneous polynomial of degree i in t and u. Since the highest degree of s in any term of $\overline{X}(s,t,u)$ is m, it follows that $\overline{X}_0(t,u), \overline{X}_1(t,u), \ldots, \overline{X}_{n-1}(t,u)$ vanish identically.

Thus, $(1,0,0)$ is a base point of multiplicity n. Similarly, $(0,1,0)$ is a base point of multiplicity m.

Q.E.D.

Since a tensor product surface always has base points at infinity, the resultant of the parametric equations, considering them as total degree bounded polynomials in s, t and u is identically zero. Dixon gave a special formulation of a resultant of three quantics in two independent variables, of the form $X(s,t)$, such that the vanishing of the resultant is a necessary and sufficient condition for the system to have a *non-trivial* solution [9]. The set of trivial solutions consists of

- $(1,0,0)$ of multiplicity n.

- $(0,1,0)$ of multiplicity m.

For total degree bounded polynomials, every base point in $\mathbf{P}^2$ is a *non trivial* base point. We refer to this formulation for tensor product surfaces as *Dixon eliminant*[2]. The term resultant would be used for total degree

---

[2]Dixon gave many other formulations for computing the resultant for a set of equations, including one for total degree bounded polynomials, too. We are only referring to the one used for tensor product surfaces.

bounded parametrizations. The Dixon's eliminant has been used by [26] to implicitize tensor product surfaces.

The degree of the homogeneous formulation of the parametric equations of $\mathbf{F}$ is $m+n$. Hence, the degree of the corresponding implicit equation can be at most $(m+n)^2$. However, base points of multiplicity n and m decrease the degree by n^2 and m^2, respectively, and the implicit representation is of degree $2mn$ or less. If the parametrization has base points in the affine domain or the curves $\overline{X}$, $\overline{Y}$, $\overline{Z}$ and $\overline{W}$ intersect tangentially at $(1,0,0)$ or $(0,1,0)$, the degree of the implicit equation is strictly less than $2mn$ and the Dixon's formulation for such equations is identically zero, too.

Let us see when these curves, $\overline{X}$, $\overline{Y}$, $\overline{Z}$ and $\overline{W}$, intersect tangentially along a base point at infinity. We will later use these constraints for choosing a suitable perturbation. Consider the base point $(1,0,0)$. Let

$$
\begin{aligned}
\overline{X}(s,t,u) &= \overline{X}_n(t,u)s^m + \ldots + \overline{X}_{m+n}(t,u), \\
\overline{Y}(s,t,u) &= \overline{Y}_n(t,u)s^m + \ldots + \overline{Y}_{m+n}(t,u), \\
\overline{Z}(s,t,u) &= \overline{Z}_n(t,u)s^m + \ldots + \overline{Z}_{m+n}(t,u), \\
\overline{W}(s,t,u) &= \overline{W}_n(t,u)s^m + \ldots + \overline{W}_{m+n}(t,u).
\end{aligned}
$$

A line, $t/a = u/b$ is tangent to these curves at $(1,0,0)$ if and only if

$$
\overline{X}_n(a,b) = 0, \quad \overline{Y}_n(a,b) = 0, \quad \overline{Z}_n(a,b) = 0, \quad \overline{W}_n(a,b) = 0.
$$

Since $\overline{X}_n, \overline{Y}_n, \overline{Z}_n, \overline{W}_n$ are homogeneous polynomials in t and u this can happen if and only if

$$
\overline{G}(t,u) = \text{GCD}(\overline{X}_n(t,u), \overline{Y}_n(t,u), \overline{Z}_n(t,u), \overline{W}_n(t,u))
$$

is a homogeneous polynomial of positive degree in t and u. Moreover, (a,b) is one of the roots of $\overline{G}(t,u)$. This constraint is equivalent to saying that

$$
G(t) = \text{GCD}\left(\sum_{j=0,n} X_{mj}t^j, \sum_{j=0,n} Y_{mj}t^j, \sum_{j=0,n} Z_{mj}t^j, \sum_{j=0,n} W_{mj}t^j \right)
$$

is a polynomial of positive degree. $\sum_{j=0,n} X_{mj}t^j$ corresponds to the coefficient of s^m in $X(s,t)$. Similarly the curves intersect tangentially at $(0,1,0)$ if and only if

$$
H(s) = \text{GCD}\left(\sum_{i=0,m} X_{in}s^i, \sum_{i=0,m} Y_{in}s^i, \sum_{i=0,m} Z_{in}s^i, \sum_{j=0,m} W_{in}s^i \right)
$$

is a polynomial of positive degree. In such cases Dixon's formulation can not be directly used for implicitizing. However, Gröbner bases can still be used since the base points are at infinity. Gröbner bases fail when the tensor product surface has base points in the affine domain.

4.2 Implicitizing Parametric Surfaces with Base Points

Parametrizations with base points are analyzed in [18]. The technique for implicitization involves symbolic perturbation, computing the resultant or Gröbner bases of the perturbed system and using GCD of bivariate polynomials to compute the implicit representation.

Given a parametrization with base points, let us perturb one of the three parametric equations, (3.2), say $\overline{F}_3(s,t,u)$ and the resulting perturbed system is

$$
\begin{aligned}
\overline{G}_1(s,t,u) &= x\overline{W}(s,t,u) - \overline{X}(s,t,u) = 0, \\
\overline{G}_2(s,t,u) &= y\overline{W}(s,t,u) - \overline{Y}(s,t,u) = 0, \\
\overline{G}_3(s,t,u) &= z\overline{W}(s,t,u) - \overline{Z}(s,t,u) + \lambda\overline{Z}_1(s,t,u) = 0,
\end{aligned}
\tag{4.2}
$$

where $\overline{Z}_1(s,t,u)$ is a homogeneous polynomial of degree n such that

$$
V(\overline{X}(s,t,u), \overline{Y}(s,t,u), \overline{W}(s,t,u), \overline{Z}_1(s,t,u)) = \phi.
$$

Almost any polynomial of degree n can be chosen for $\overline{Z}_1(s,t,u)$. We will denote this perturbed parametrization as $\overline{\mathbf{G}}$. The resulting parametrization has no base points.

It is still possible that for all choices of $\overline{Z}_1(s,t,u)$ the resultant of $\overline{G}_1(s,t,u)$, $\overline{G}_2(s,t,u)$ and $\overline{G}_3(s,t,u)$ is zero.

Theorem III: *Given a set of three equations of the form, $\overline{G}_1(s,t,u), \overline{G}_2(s,t,u)$ and $\overline{G}_3(s,t,u)$, where $\overline{Z}_1(s,t,u)$ is chosen such that*

$$
V(\overline{X}(s,t,u), \overline{Y}(s,t,u), \overline{W}(s,t,u), \overline{Z}(s,t,u), \overline{Z}_1(s,t,u)) = \phi.
$$

The necessary and sufficient condition that the resultant of $\overline{G}_i$'s does not vanish is that

$$
\overline{P}(s,t,u) = GCD(\overline{X}(s,t,u), \overline{Y}(s,t,u), \overline{W}(s,t,u))
$$

is a constant.

Proof: [18]

Q.E.D.

To circumvent this problem of a vanishing resultant in certain cases we perform a change of coordinates and let the new parametrization be

$$
\overline{\mathbf{F}}'(s,t,u) = (x', y', z', w') = (x, y + kz, z, w)
$$

$$
= (\overline{X}(s,t,u), \overline{Y}(s,t,u) + k\overline{Z}(s,t,u), \overline{Z}(s,t,u), \overline{W}(s,t,u)),
$$

where k is a scalar. The corresponding parametric equations are

$$\overline{G}_1'(s,t,u) \quad = \quad x\overline{W}(s,t,u) - \overline{X}(s,t,u) = 0,$$
$$\overline{G}_2'(s,t,u) \quad = \quad y\overline{W}(s,t,u) - \overline{Y}(s,t,u) - k\overline{Z}(s,t,u) = 0,$$
$$\overline{G}_3(s,t,u) \quad = \quad z\overline{W}(s,t,u) - \overline{Z}(s,t,u) + \lambda\overline{Z}_1(s,t,u) = 0.$$

Since $GCD(\overline{X}(s,t,u), \overline{Y}(s,t,u), \overline{Z}(s,t,u), \overline{W}(s,t,u)) = 1$, for any generic k,

$$GCD(\overline{X}(s,t,u), \overline{Y}(s,t,u) + k\overline{Z}(s,t,u), \overline{W}(s,t,u)) = 1,$$

too. We compute the implicit representation in terms of x', y' and z' and substitute them to obtain an implicit equation in terms of x, y and z.

Thus, it is reasonable to assume that the resultant of $\overline{G}_i$'s is nonzero. Let us express it as a polynomial in λ

$$R(x,y,z,\lambda) \quad = \quad \lambda^i P_i(x,y,z) + \ldots + \lambda^d P_d(x,y,z), \qquad (4.3)$$
$$= \quad \lambda_i S(x,y,z,\lambda)$$

where $i > 0$, $S(x,y,z,0) \neq 0$. In fact i corresponds to the number of base points in the given parametrization (counted properly) [18].

Theorem IV: *$P_i(x,y,z)$ can be expressed as a polynomial of the form*

$$P_i(x,y,z) = H(x,y,z)F(x,y),$$

where $H(x,y,z)$ is the implicit representation and $F(x,y)$ is only a polynomial in x and y and corresponds exactly to the projections of the seam curves on the $X - Y$ plane.

Proof: [18]

Q.E.D.

To compute $H(x,y,z)$ from $P_i(x,y,z)$, we choose 2 generic values of z, say z_1 and z_2, and $GCD(P_i(x,y,z_1), P_i(x,y,z_2))$ would correspond to $F(x,y)$. Thus, the implicit equation can be represented as

$$H(x,y,z) = \frac{P_i(x,y,z)}{GCD(P_i(x,y,z_1), P_i(x,y,z_2))}.$$

A simple consequence of previous theorems is

Theorem V: *Every parametrization has an implicit representation, which has the same coefficient field as the parametric equations.*

Proof: [18]

Q.E.D.

Furthermore, the extraneous factor in the lowest degree term of the resultant, $F(x, y)$, can be used to compute the rational parametrizations of seam curves [18].

5 Algorithm

In the previous sections we have highlighted the technique used for implicitizing rational parametric surfaces. These techniques consist of computing the resultants or Gröbner bases of the parametric equations. If the parametrization has base points, the parametric equations are perturbed and we compute the resultant or Gröbner bases of the perturbed system. We showed that the lowest degree term of the resultant or first polynomial of the Gröbner bases (expressed as polynomials in the perturbing variable) contains the implicit equation. An efficient technique for recovering the implicit equation has been presented, too. Although the Gröbner bases provide us with the additional flexibility of working in the affine space, we prefer resultants for their computational efficiency. In general the running time complexity of the Gröbner bases algorithm can be doubly exponential in the number of variables as compared to the singly exponential complexity of resultants [5]. As a matter of fact, multipolynomial resultant algorithms provide the most efficient methods (as far as asymptotic complexity is concerned) for solving systems of polynomial equations or eliminating variables [1]. In our applications the number of variables is fixed and any argument based on the asymptotic complexity may not have much significance. However we justify our choice with an implementation using resultants and compare its performance with Gröbner bases implementations. Our motivation for choosing resultants stemmed from the fact that most implicit representations are dense polynomials in x, y and z. Gröbner bases seem to perform well on sparse systems and for dense polynomials resultants are considered faster in practice.

Any polynomial parametrization has no base points in the affine domain. Consider the case when such a parametrization has base points at infinity. To implicitize using resultants we need to perturb the system whereas such a problem does not arise while using Gröbner bases. Although perturbation turns out to be relatively expensive in practice, it may be faster to implicitize using resultants of the perturbed system as compared to using the Gröbner bases approach.

5.1 Checking for Base Points

Given a parametrization of degree n,

$$\overline{\mathbf{F}}(s, t, u) = (\overline{X}(s, t, u), \overline{Y}(s, t, u), \overline{Z}(s, t, u), \overline{W}(s, t, u)),$$

the base points correspond to common roots of $\overline{X}(s,t,u), \overline{Y}(s,t,u), \overline{Z}(s,t,u)$ and $\overline{W}(s,t,u)$. In general 4 such polynomials have no roots in common. Consider the intersection of this surface with a line represented as the intersection of the following 2 planes

$$a_1 x + a_2 y + a_3 z + a_4 w = 0$$

$$b_1 x + b_2 y + b_3 z + b_4 w = 0,$$

where a_i's and b_i's can be considered as symbolic variables. To determine the number of intersections of the line with the given surface, we substitute for x, y, z and w and obtain

$$\overline{F}_1(s,t,u) = a_1\overline{X}(s,t,u) + a_2\overline{Y}(s,t,u) + a_3\overline{Z}(s,t,u) + a_4\overline{W}(s,t,u) = 0$$

$$\overline{F}_2(s,t,u) = b_1\overline{X}(s,t,u) + b_2\overline{Y}(s,t,u) + b_3\overline{Z}(s,t,u) + b_4\overline{W}(s,t,u) = 0.$$

The given parametrization has base points, if and only if, $\overline{F}_1(s,t,u)$ and $\overline{F}_2(s,t,u)$ have common roots for all values of a_i's and b_i's. The common roots correspond to the base points. They can be determined by computing the u-resultant of $\overline{F}_1$ and $\overline{F}_2$ [30]. The u-resultant is the resultant of

$$
\begin{aligned}
\overline{F}_1(s,t,u) &= 0 \\
\overline{F}_2(s,t,u) &= 0 \\
\alpha s + \beta t + \gamma u &= 0.
\end{aligned}
\tag{5.1}
$$

The u-resultant is a homogeneous polynomial of degree n^2 in α, β and γ and it decomposes into linear factors of the form $(s_1\alpha + t_1\beta + u_1\gamma)$, where s_1, t_1 and u_1 are functions of a_i's and b_i's. The given parametrization has a base point if and only if there is a factor of the form $(s'\alpha + t'\beta + u'\gamma)$, where s', t' and u' are scalar constants and independent of a_i's and b_i's. In this case (s', t', u') is a base point of the given parametrization. The number of base points (counted properly with respect to multiplicity) is given by the number of factors of the form $(s'\alpha + t'\beta + u'\gamma)$. If there are n^2 base points, the given parametrization is a degenerate parametrization and its image is not a 2-dimensional surface.

The 3 equations (5.1) are of degree n, n and 1. Their resultant can be expressed as the determinant of a matrix [17]. We prefer this formulation of resultant over Macaulay's formulation [15], since it involves computing a single determinant and we do not have to perturb the given equations.

The u-resultant of $\overline{F}_1$ and $\overline{F}_2$ is a homogeneous polynomial of degree $n^2 + 2n$ in 11 variables (a_i's, b_i's, α, β and γ). In practice, this computation can be very time consuming even for low degree parametrizations. Furthermore, we need to factorize the resultant into linear factors and thereby adding to the complexity of the computation. For practical applications we present a probabilistic algorithm.

Consider 4 generic combinations of $\overline{X}(s,t,u), \overline{Y}(s,t,u), \overline{Z}(s,t,u)$ and $\overline{W}(s,t,u)$ of the form

$$\overline{G}_i(s,t,u) = x_i\overline{X}(s,t,u)+y_i\overline{Y}(s,t,u)+z_i\overline{Z}(s,t,u)+w_i\overline{W}(s,t,u), \ 1 \le i \le 4,$$

where x_i's, y_i's, z_i's and w_i's are random numbers. We use these combinations for computing the following u-resultants

$$\overline{R}_1(\alpha,\beta,\gamma) = \text{Resultant}(\overline{G}_1(s,t,u),\overline{G}_2(s,t,u),\alpha s + \beta t + \gamma u),$$

$$\overline{R}_2(\alpha,\beta,\gamma) = \text{Resultant}(\overline{G}_3(s,t,u),\overline{G}_4(s,t,u),\alpha s + \beta t + \gamma u).$$

$\overline{R}_1(\alpha,\beta,\gamma)$ and $\overline{R}_2(\alpha,\beta,\gamma)$ are homogeneous polynomials of degree n^2 in α, β and γ. Let

$$\overline{P}(\alpha,\beta,\gamma) = \text{GCD}(\overline{R}_1(\alpha,\beta,\gamma),\overline{R}_2(\alpha,\beta,\gamma)). \tag{5.2}$$

Let d be the degree of $\overline{P}(\alpha,\beta,\gamma)$. If $d = 0$, the given parametrization has no base points else for each base point $(s^{'},t^{'},u^{'})$, $(s^{'}\alpha+t^{'}\beta+u^{'}\gamma) \mid \overline{P}(\alpha,\beta,\gamma)$. For almost all combinations, $\overline{G}_1,\ldots,\overline{G}_4$, each linear factor of $\overline{P}(\alpha,\beta,\gamma)$ corresponds to a base point. Therefore d corresponds to the number of base points in the given parametrization and the degree of the implicit representation is $n^2 - d$. To reduce the symbolic complexity of the computation, we may specialize α, β or γ (one or two at a time.)

Our implicitization algorithm only needs to know whether the given parametrization has any base points (and not the actual number of base points) and in such cases it perturbs the parametric equations. A simple algorithm to check for the existence of base points is obtained in the following manner. Consider any three generic combinations, $\overline{G}_1(s,t,u)$, $\overline{G}_2(s,t,u)$ and $\overline{G}_3(s,t,u)$, and let R be their resultant. For almost all such combinations, R is zero if and only if the given parametrization has base points. The resultant can be expressed as the determinant of a matrix and all entries of the matrix are numeric. The computation in this case is purely numeric and very fast in practice.

Tensor Product Surfaces

A tensor product surface can either have base points in the affine domain or excess base points at infinity. Given a tensor product surface

$$\mathbf{F}(s,t) = (X(s,t), Y(s,t), Z(s,t), W(s,t)),$$

where any component, say $X(s,t)$, is of the form

$$X(s,t) = \sum_{i=0,m} \sum_{j=0,n} X_{ij} s^i t^j, \quad X_{mn} \neq 0.$$

According to Lemma IV, it has a base point of multiplicity n at $(s,t,u) = (1,0,0)$ and of multiplicity m at $(s,t,u) = (0,1,0)$.

Let us homogenize the given parametrization and take its 4 generic combinations to compute $\overline{P}(\alpha,\beta,\gamma)$, (5.2). Since the parametrization has base points at infinity, $\overline{P}(\alpha,\beta,\gamma)$ can be expressed as

$$\overline{P}(\alpha,\beta,\gamma) = \alpha^{n^2}\,\beta^{m^2}\,\overline{Q}(\alpha,\beta,\gamma),$$

where $\overline{Q}(\alpha,\beta,\gamma)$ is a homogeneous polynomial. Let d be the degree of $\overline{Q}(\alpha,\beta,\gamma)$. The given parametrization has a base point in the affine domain or an excess base point at infinity, if and only if $d > 0$. Only in such cases the Dixon eliminant fails to compute the implicit representation and we need to perturb the parametric equations to compute the implicit representation. The degree of the implicit representation is $2mn - d$.

The technique used to check for the existence of base points is obtained by considering three generic combinations of X, Y, Z and W, say $G_1(s,t)$, $G_2(s,t)$ and $G_3(s,t)$. Let R be the Dixon eliminant of G_1, G_2 and G_3. For almost all combinations, R is zero if and only if the given parametrization has base points in the affine domain or excess base points at infinity.

5.2   Form of Implicit Representation

In this section, we present a simple algorithm to verify whether the implicit representation, $H(x,y,z)$ is independent of z. This algorithm is required before considering the efficient perturbation and thereby perturbing the parametric equation containing the z variable (as shown in (4.2)).

The implicit representation is independent of z, if any line represented as the intersection of the following two planes

$$w_1 x = x_1 w$$

$$w_1 y = y_1 w,$$

where $(x_1, y_1, w_1) = (\overline{X}(s_1,t_1,u_1), \overline{Y}(s_1,t_1,u_1), \overline{W}(s_1,t_1,u_1))$ lies on the surface. In other words the given line intersects the surface at an infinite number of points. This should hold for all (s_1,t_1,u_1), where (s_1,t_1,u_1) correspond to a point in the domain. A simple probabilistic algorithm to check for this property is given below.

Let (s_1,t_1,u_1) correspond to a random point in the domain and consider the equations

$$\overline{F}_1(s,t,u) = \overline{X}(s_1,t_1,u_1)\overline{W}(s,t,u) - \overline{W}(s_1,t_1,u_1)\overline{X}(s,t,u) = 0$$

$$\overline{F}_2(s,t,u) = \overline{Y}(s_1,t_1,u_1)\overline{W}(s,t,u) - \overline{W}(s_1,t_1,u_1)\overline{Y}(s,t,u) = 0$$

and

$$\overline{G}(s,t,u) = \ \text{GCD}(\overline{F}_1(s,t,u), \overline{F}_2(s,t,u)).$$

If $\overline{G}(s,t,u)$ is a constant, then the implicit representation of $\overline{\mathbf{F}}$ is not independent of z. For almost all choices of (s_1,t_1,u_1), the fact that $\overline{G}(s,t,u)$

is a polynomial of positive degree implies that the implicit equation is independent of z.

5.3 Choice of Perturbing Polynomial

If a parametrization has base points, we perturb the polynomials (as shown
in (4.2)) to compute the implicit representation. The only constraint on
the perturbing polynomial $Z_1(s, t, u)$ is imposed by theorem IV, i.e.

$$V(\overline{X}(s, t, u), \overline{Y}(s, t, u), \overline{W}(s, t, u), \overline{Z}_1(s, t, u)) = \phi.$$

To simplify the symbolic complexity of the resulting computation we choose
a perturbing polynomial of the form

$$\overline{Z}_1(s, t, u) = a_1 s^n + a_2 t^n + a_3 u^n,$$

where a_1, a_2 and a_3 are random numbers. For almost all choices of a_1, a_2
and a_3 this polynomial will satisfy the constraint of theorem IV. Otherwise
the resultant of the perturbed system is identically zero.

For tensor product surfaces, $\mathbf{F}(s, t)$, where the highest degree of s in
any monomial is m and the highest degree of t in any monomial is n, we
choose a perturbation polynomial of the form

$$Z_1(s, t) = a_1 s^m t^n + a_2 s^m + a_3 t^n + a_4,$$

where a_1, a_2, a_3 and a_4 are random numbers. In this case we compute the
Dixon eliminant of the perturbed system and extract the implicit equation from its lowest degree term after expressing it as a polynomial in the
perturbing variable.

5.4 Algorithm

Given a parametrization

$$\overline{\mathbf{F}}(s, t, u) = (x, y, z, w) = (\overline{X}(s, t, u), \overline{Y}(s, t, u), \overline{Z}(s, t, u), \overline{W}(s, t, u)),$$

an algorithm for implicitization is

1. Let

$$\overline{P}(s, t, u) = \text{GCD}(\overline{X}(s, t, u), \overline{Y}(s, t, u), \overline{Z}(s, t, u), \overline{W}(s, t, u)).$$

   If $\overline{P}(s, t, u)$, the common factor, is a polynomial of positive degree,
   cancel out the common factor from each component of the parametrization.

2. Check whether the given parametrization has base points.

3. Check whether the implicit representation, $H(x, y, z)$, is independent of z. If it is independent of z, interchange $\overline{Y}(s, t, u)$ and $\overline{Z}(s, t, u)$, such that the new parametrization is equivalent to

$$(x, y, z, w) = (\overline{X}(s, t, u), \overline{Z}(s, t, u), \overline{Y}(s, t, u), \overline{W}(s, t, u)),$$

and its implicit equation will be a function of x and z only. Substitute y for z and the resulting representation would correspond to the implicit representation of $\overline{\mathbf{F}}$.

4. If the parametrization has no base points, compute the resultant of the parametric equations. Let the resultant be $H(x, y, z)$ and the implicit representation, $G(x, y, z)$, can be computed as

$$G(x, y, z) = \frac{H(x, y, z)}{\text{GCD}(H(x, y, z), H_z(x, y, z))}.$$

5. Let $k = \frac{\overline{Z}(s,t,u)}{\overline{W}(s,t,u)}$. If k is a constant, the implicit representation is given by $z = k$.

6. Let

$$\overline{P}(s, t, u) = \text{GCD}(\overline{X}(s, t, u), \overline{Y}(s, t, u), \overline{W}(s, t, u)).$$

If $\overline{P}(s, t, u)$ is a polynomial of positive degree perform a coordinate transformation (follows from Theorem III) and let the new parametrization be

$$\overline{\mathbf{F}}'(s, t, u) = (x', y', z', w') = (x, y + kz, z, w) =$$

$$(\overline{X}(s, t, u), \overline{Y}(s, t, u) + k\overline{Z}(s, t, u), \overline{Z}(s, t, u), \overline{W}(s, t, u)),$$

where k is a random number. The rest of the algorithm is used to implicitize $\overline{\mathbf{F}}'$. Given $G(x', y', z')$, the implicit representation of $\overline{\mathbf{F}}'$, the implicit representation of $\overline{\mathbf{F}}$ corresponds to $G(x, y - kz, z)$.

7. Compute the resultant of the following system of equations

$$
\begin{aligned}
\overline{G}_1(s, t, u) &= x\overline{W}(s, t, u) - \overline{X}(s, t, u) = 0, \\
\overline{G}_2(s, t, u) &= y\overline{W}(s, t, u) - \overline{Y}(s, t, u) = 0, \\
\overline{G}_3(s, t, u) &= z\overline{W}(s, t, u) - \overline{Z}(s, t, u) + \lambda\overline{Z}_1(s, t, u) = 0,
\end{aligned}
$$

where $\overline{Z}_1(s, t, u)$ is a perturbing polynomial. Express the resultant as a polynomial in λ and let $P_i(x, y, z)$ correspond to its lowest degree term.

8. Choose 2 generic values of z, $z = z_1$ and $z = z_2$, and let

$$H(x,y,z) = \frac{P_i(x,y,z)}{\text{GCD}(P_i(x,y,z_1), P_i(x,y,z_2))}.$$

The implicit representation is computed as

$$G(x,y,z) = \frac{H(x,y,z)}{\text{GCD}(H(x,y,z), H_z(x,y,z))}.$$

6 Implementation

The algorithm in the previous section can be easily implemented on top of any computer algebra system. The main operations are computing the matrix entries for resultant formulation, expanding symbolic and numeric determinants, GCD of multivariate polynomials. Most computer algebra systems support these operations. However when it comes to practice, expanding a symbolic determinant becomes a time consuming task. Consider the problem of implicitizing bicubic tensor product Bézier surfaces (with no base points in the affine domain or excess base points at infinity). The implicit representation corresponds to the determinant of a matrix of order 18, where each matrix entry is a linear polynomial in x, y and z. However the computer algebra systems on commonly available workstations (Sun-3's, Sun-4's) are not able to compute such determinants in a reasonable amount of time and space. Some experiments with the implementations of Gröbner bases and resultants in Macsyma 414.62 on a Symbolics lisp machine (with 16MB main memory and 120MB virtual memory) are described in [13]. For many cases of bicubic surfaces, these systems are unable to implicitize such surfaces and fail due to insufficient virtual memory. Only a new algorithm for basis conversion is able to implicitize such surfaces, however it takes about 10^5 seconds, which would be considered impractical for most applications [13].

Let us analyze some properties of the implicit representation. A polynomial equation of degree d in 3 variables can have up to M monomials, where

$$M = \left(\begin{array}{c} d+3 \\ d \end{array} \right).$$

For a polynomial of degree 18 that comes to 1330 terms. In practice the implicit representations are dense polynomials. This is not difficult to see, since a coordinate transformation of the form

$$x = \alpha_1 \overline{x} + \alpha_2 \overline{y} + \alpha_3 \overline{z}$$
$$y = \beta_1 \overline{x} + \beta_2 \overline{y} + \beta_3 \overline{z}$$
$$z = \gamma_1 \overline{x} + \gamma_2 \overline{y} + \gamma_3 \overline{z}$$

would result in a very dense polynomial in $\overline{x}, \overline{y}$ and $\overline{z}$ for almost all choices of α_i, β_i and γ_i. Furthermore the coefficients of the implicit representation are much larger than those of the parametrization. If the absolute magnitude of the coefficients of parametric equations is $|N|$, the coefficients of the implicit representation can have magnitude of the order of $|N|^d$. This follows from the algorithms used for implicitization. Each entry in the matrix has coefficient size equivalent to that of the parametrization and the order of the matrix is equal to the degree of the implicit representation (for tensor product surfaces). For general parametrizations, the order of the matrix is $2n^2 - n$, whereas the implicit equation has degree at most n^2. This property can have significant impact on the numerical stability of the algorithms, when the coefficients of the parametrization and implicit representation are floating point numbers. In particular, any approach based on implicitization can reduce a well-conditioned problem to an ill-conditioned problem [12; 13]. As a result we restrict ourselves to exact arithmetic. In any case, algorithms based on Gröbner bases use rational arithmetic [12].

There are many reasons for the failure and bad performance of implicitization algorithms implemented within the framework of computer algebra systems. Most computer algebra systems use sparse representation for multivariate polynomials and the computations become slow whenever the polynomials generated are dense. Moreover, the algorithms are symbolic in nature and large intermediate expressions may be generated. Their implementations in Lisp-like environments may require a large amount of virtual memory and thereby slow down the computations. Furthermore, these systems use exact arithmetic and represent the coefficients of intermediate expressions as *bignums*. As a result the cost of arithmetic operations is quadratic in the coefficient size. The coefficient size is proportional to the degree of the polynomial expressions being generated and tends to grow exponentially with the degree.

The bottleneck in our algorithm is the symbolic expansion of determinants. The rest of the computations are fast enough in the computer algebra systems. We therefore implemented our algorithm within the framework of a computer algebra system and used a separate implementation for determinant expansion. The main idea is to guess the form of the determinant, say a polynomial of degree d in 3 or 4 variables, corresponding to unperturbed and perturbed parametric equations, respectively, and use Vandermonde interpolation for computing the coefficients. The resulting problem is equivalent to that of interpolating a univariate polynomial which has the same number of coefficients as the multivariate polynomial. As a result, the algorithm only involves numeric computations and no intermediate symbolic expressions are generated. Since the implicit representations are dense polynomials, we use a dense representation for the resultant. To circumvent the problem of coefficient growth and its impact on arithmetic computation, we chose to work in finite fields. The order of the finite fields is about 2^{31} and thereby making the best use of hardware implementations of 32 bit integer arithmetic (available on most workstations). The main problem is in getting a tight bound on the coefficients of the implicit equations. Since the resultant of the parametric equations (perturbed or unperturbed) corresponds to a determinant, we can use Hadamard's bound for computing a bound on the coefficients of the resultant. However, the bound so obtained is rather loose and we use a probabilistic algorithm for computing the coefficients of the implicit equation. The algorithm involves computing the coefficients modulo various primes and use chinese remainder theorem to compute the corresponding bignums. If for two successive prime sequences (the second one contains one more prime than the first sequence) the same bignums are obtained, then those bignums correspond to the coefficients of the implicit equation. More details of interpolation based algorithms to compute resultants and the probabilistic algorithm for computing the coefficients of the resultant are given in [19]. The complexity of the algorithm is output sensitive and is given by $O(|C|M(logM)^2)$, where $|C|$ corresponds to the size of resultant coefficients, and M is the number of terms that can be present in the resultant [19]. If M is small, a simpler algorithm of complexity $O(|C|M^2)$ may be used.

6.1 Interpolation

Let us consider the case when we do not perturb the given equations and each entry of matrix is a linear polynomial of the form $a_1x + a_2y + a_3z + a_4$. Let $D(x_1, y_1, z_1)$ be the determinant of the matrix, when x, y and z are specialized to x_1, y_1 and z_1, respectively. We choose 3 distinct primes, say p_1, p_2 and p_3 and compute M determinants of the form $D(p_1, p_2, p_3)$, $D(p_1^2, p_2^2, p_3^2)$, ..., $D(p_1^M, p_2^M, p_3^M)$, where M corresponds to the number of monomials in the implicit representation. Given M, we use algorithms for Vandermonde interpolation to compute the coefficients of the polynomial $H(x, y, z)$ and thereby reduce the problem to linear interpolation. If we are computing the resultant of perturbed parametric equations, then the resultant is a polynomial of the form $R(x, y, z, \lambda)$. We therefore use 4 distinct primes and their powers for interpolation. In this case each entry of the matrix is of the form

$$a_1x + a_2y + a_3z + \lambda(a_4 + a_5x + a_6y) + a_7.$$

All the computations are performed in a finite field.

Our algorithm requires the value of M. Since it is difficult to compute the actual value, we use an upper bound corresponding to the number of monomials that the polynomials of degree d in 3 or 4 variables can have. Though there are many algorithms available for sparse interpolation, in practice they either require a tight bound on the number of monomials that the polynomials may have or actually figure out the actual monomials first and than compute their coefficients [2; 14]. In the former case, the problem is similar to that of ours and in the latter case, these algorithms seem to take more time in figuring out the actual monomials present in the polynomial and as a result may be slower as compared to the dense interpolation algorithms. Furthermore, the implicit equations and resultants of perturbed systems are generally dense polynomials.

If the resultant or Dixon eliminant of the unperturbed parametric equations does not vanish,

$$M = \binom{d+3}{d},$$

where d is the degree of the implicit representation ($d = 2mn$ for tensor product surfaces of the form $s^m t^n$ and $d = n^2$ for triangular patches of degree n). If the parametrization has base points then the determinant of the perturbed system is a polynomial of degree $2d$ (where $d = 2mn$ or $d = n^2$ depending on the case) in x, y, z and λ. In general such a polynomial can have up to

$$\binom{2d+4}{4}$$

monomials. However, we can use some properties of the resultant of the perturbed parametric equations to improve this bound.

The resultant is a polynomial of degree d in the coefficients of each equation. As a result the sum of the degrees of z and λ in any monomial does not exceed d. According to (4.3), the resultant can be expressed as

$$R(x, y, z, \lambda) = \lambda^i P_i(x, y, z) + \lambda^{i+1} P_{i+1}(x, y, z) + \ldots + \lambda^d P_d(x, y, z),$$

where i corresponds to the total number of base points in the parametrization and $P_j(x, y, z)$ is a polynomial of the form

$$P_j(x, y, z) = Q_d(x, y) + z Q_{d-1}(x, y) + \ldots + z^{d-j} Q_j(x, y),$$

and $Q_k(x, y)$ is a polynomial of degree k in x and y. Thus,

$$M = \sum_{j=i}^{d} \left(\sum_{k=j}^{d} \binom{k+2}{2} \right)$$

$$= \sum_{j=i}^{d} \left(\binom{d+3}{3} - \binom{j+2}{3} \right).$$

Some deterministic and probabilistic algorithms to compute i are presented in Section 5.1. We make use of (4.3) to present a simple and probabilistic algorithm.

Let $(x, y, z) = (x_1, y_1, z_1)$ correspond to a random point in space and compute the determinant $R(x_1, y_1, z_1, \lambda)$, which is a polynomial of degree d in λ. We can use Vandermonde interpolation to compute it. For almost all choices of (x_1, y_1, z_1) the lowest degree of λ in $R(x_1, y_1, z_1, \lambda)$ corresponds to i.

6.2 Performance

The symbolic determinant computation has been implemented in C++ on a Sun-4 (a 10 MIPS machine) and IBM RS/6000 (a 34 MIPS machine). The rest of the algorithm has been implemented on top of Mathematica. The bottleneck of the computation is the determinant computation. In Table 1 and 2 its performance for different parametrizations is given. The timings correspond to a single iteration of the finite field computation. The total number of iterations is a function of the coefficient size of the output. Since the coefficient size is proportional to the degree of the implicit representation, more iterations are needed for higher degree implicit equations. We use finite fields of order 2^{31} and therefore, the number of iterations is bounded by $k + 1$, where k is the minimum integer satisfying the relation

$$|N| < 2^{30k}$$

and $|N|$ is the magnitude of the coefficient of maximum size of the resultant.

Table 1. The performance of implicitization algorithm for parametrizations without base points (a single iteration over a finite field)

Parametrization	Implicit Degree	M	Sun-4	IBM RS/6000
$s^2 + t^2$	4	10	1 sec.	1 sec.
$s^3 + t^3$	9	220	6 sec.	3 sec.
$s^2 t^2$	8	165	4 sec.	2 sec.
$s^3 t^3$	18	1330	100 sec.	23 sec.
$s^3 t^4$	24	2925	430 sec.	118 sec.

Table 2. The performance of implicitization algorithm for parametrizations with base points (a single iteration over a finite field)

Parametrization	Base Points	Implicit Degree	M	Sun-4	IBM RS/6000
$s^2 + t^2$	2	2	74	2 sec.	1 sec.
$s^3 + t^3$	3	6	1064	52 sec.	16 sec.
st^3	2	4	295	4 sec.	2 sec.
$s^2 t^2$	4	4	510	13 sec.	4 sec.
$s^3 t^3$	3	15	15300	4700 sec.	1180 sec.

Thus, we see that the algorithm does not perform well for bicubic parametrizations with base points. In general $3 - 4$ iterations are needed and even on a fast machine like IBM RS/6000 that amounts to 4500 seconds. The main problem is in the perturbation technique, which increases the symbolic complexity of the resultant.

7   Conclusion

In this paper we analyzed the problem of implicitizing rational parametric surfaces. If a parametrization has no base points, the resultant or Dixon eliminant of the parametric equation corresponds exactly to the implicit equation. The base points decrease the degree of the implicit equations and blow up to rational curves on the algebraic surface. We use symbolic perturbation to compute the implicit equations of parametrizations with base points. The strength of our technique lies in the fact that we use GCD of bivariate polynomials to extract the implicit equation from the lowest degree term of the resultant of the perturbed system. Although, similar analysis holds for Gröbner bases, we recommend resultants for efficiency reasons.

Algorithms implemented in the framework of computer algebra systems (available on commonly available workstations) are unable to implicitize parametric surfaces like bicubic patches. We presented an algorithm for computing symbolic determinants and as a result achieved a significant performance improvement compared to the previous implementations of implicitization algorithms. If a parametrization has no base points (or a tensor product surface has no base points in the affine domain or excess base points at infinity), the algorithm is very fast on machines like the IBM RS/6000. Furthermore, the algorithm can be easily parallelized and thereby achieve speedups proportional to the level of parallelism. However, perturbation increases the symbolic complexity of the resultants and the resulting algorithm to implicitize parametric surfaces with base points is slow. We therefore need efficient algorithms to implicitize parametric surfaces with base points.

Acknowledgements

This research was supported in part by David and Lucile Packard Fellowship and in part by a National Science Foundation Presidential Young Investigator Award (# IRI-8958577).

References

1. Bajaj, C., Garrity, T. and Warren, J. (1988) On the applications of multi-equational resultants. Technical report CSD-TR-826, Department of Computer Science, Purdue University.

2. Ben-Or, M. and Tiwari, P. (1988) A deterministic algorithm for sparse multivariate polynomial interpolation. *20th Annual ACM Symp. Theory Comp.*, pp. 301–309.

3. Buchberger, B. (1987) Gröbner bases: An algorithmic method in polynomial ideal theory. In *Recent Trends in Multidimensional Systems Theory*, edited by N.K. Bose, pp. 184–232, D. Reidel Publishing Co.

4. Buchberger, B. (1989) Applications of Gröbner bases in non-linear computational geometry. In *Geometric Reasoning*, eds. D. Kapur and J. Mundy, pp. 415–447, MIT Press.

5. Canny, J.F. (1990) Generalized characteristic polynomials. *Journal of Symbolic Computation*, 9, No. 3.

6. Castelnuovo, G. (1894) *Sulla razionalità delle involuzioni piane. Mathematische Annalen*, **44**, 125–155, (in Italian).

7. Chuang, J.H. and Hoffmann, C.M. (1989) On local implicit approximations and its applications. *ACM Transactions on Graphics*, 8, no. 4, 298–324.

8. Clebsch, A. (1868) *Ueber die abbildung algebraischer flächen insbesondere der vierten and fünften ordnung. Mathematische Annalen*, **1**, 253–316, (in German).

9. Dixon, A.L. (1908) The eliminant of three quantics in two independent variables. *Proceedings of London Mathematical Society*, 6, 49–69, 473–492.

10. Hartshorne, R. (1977) *Algebraic Geometry*, Springer-Verlag.

11. Hoffmann, C. (1988) A dimensionality paradigm for surface interrogations. Tech. report CSD-TR-837, Department of Computer Science, Purdue University.

12. Hoffmann, C. (1989) *Geometric and Solid Modeling: An Introduction*, Morgan Kaufmann Publishers Inc.

13. Hoffmann, C. (1990) Algebraic and numeric techniques for offsets and blends. In *Computation of Curves and Surfaces*, eds. W. Dahmen et. al., pp. 499–529, Kluwer Academic Publishers.

14. Kaltofen, E. and Lakshman, Y. (1988) Improved sparse multivariate polynomial interpolation algorithms. *Lecture Notes in Computer Science*, **358**, 467–474, Springer-Verlag.

15. Macaulay, F.S. (1902) On some formula in elimination. *Proceedings of London Mathematical Society*, 3–27.

16. Macaulay, F. S. (1921) Note on the resultant of a number of polynomials of the same degree. *Proceedings of London Mathematical Society*, 14–21.

17. Morley, F. and Coble, A.B. (1927) New results in elimination. *American Journal of Mathematics*, **49**, 463–488.

18. Manocha, D. and Canny, J.F. (1992) The implicit representation of rational parametric surfaces, *Journal of Symbolic Computation*, **15**, No. 2, 485–510.

19. Manocha, D. and Canny, J. F. (1993) Multipolynomial resultant Algorithms, *Journal of Symbolic Computation*, **13**, 99–122.

20. Mumford, D. (1976) *Algebraic Geometry I: Complex Projective Varieties*, Springer-Verlag.

21. Morley, F. (1925) The eliminant of a net of curves. *American Journal of Mathematics*, **47**, 91–97.

22. Pratt, M. J. (1986) Parametric curves and surfaces as used in computer aided design. In *The Mathematics of Surfaces*, ed. by J.A. Gregory, Clarendon Press, Oxford.

23. Salmon, G. (1885) *Lessons Introductory to the Modern Higher Algebra*, G.E. Stechert & Co., New York.

24. Salmon, G. (1914) *A Treatise on the Analytic Geometry of Three Dimensions*, Longmans, Green, London.

25. Semple, J.G. and Roth, L. (1985) *Introduction to Algebraic Geometry*, Clarendon Press, Oxford, Great Britain.

26. Sederberg, T.W., Anderson, D.C. and Goldman, R.N. (1984) Implicit representation of parametric curves and surfaces. *Computer Vision, Graphics and Image Processing*, **28**, 72–84.

27. Sederberg, T.W. (1990) Techniques for cubic algebraic surfaces. *IEEE CG&A*, 14–25.

28. Snyder, Virgil et. al. (1970) *Selected Topics in Algebraic Geometry*, Chelsea Publishing Company, Bronx, New York

29. Walker, Robert J. (1950) *Algebraic Curves*, Princeton University Press, New Jersey.

30. van der Waerden B. L. (1950) *Modern Algebra*, (third edition) F. Ungar Publishing Co., New York.

The Zero-Eye Subdivision Method for Ray Tracing Parametric Surfaces

Charles Woodward

Helsinki University of Technology

1 Introduction

The problem of computing the intersection point between a line (ray) and a parametric surface is of central importance to many areas of computer-aided geometric design and computer graphics. *Ray tracing* [1] and the two-pass *radiosity* method [2], the most high-quality visualization methods known to date, have to call the ray intersection routine up to millions of times per picture. This makes especially the speed of the intersection routine important. The intersection method should also provide good accuracy and robustness, preferably within reasonable memory space. *Geometric modelling interrogations* [3] often perform intersection computations only a few times before the model is changed, which also makes the preprocessing time an important factor.

The surface intersection method called *adaptive zero-eye subdivision* is evaluated in this article. The method has optimal robustness and memory behaviour and it requires little or no preprocessing. It also provides speed comparable to any numerical solution method. The discussion is directed first of all to ray tracing visualization but the method can as well be applied to any other ray intersection computation task.

2 Bézier Representation

We briefly review the elementary properties of the Bézier representation [4]. A Bézier curve segment is expressed as

$$C(u) = \sum_{i=0}^{k} V_i B_{i,k}(u) \qquad u \in [0,1], \tag{2.1}$$

where k is the order of the curve, the vectors V_i are called *control points* and the Bézier *basis functions* $B_{i,k}()$ are defined by recurrence relations [4].

133

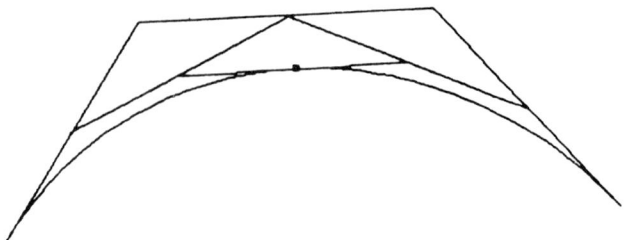

Figure 1. Cubic Bézier curve subdivision

A Bézier surface patch is defined as the tensor product of Bézier curves,

$$S(u,v) = \sum_{i=0}^{k}\sum_{j=0}^{l} V_{i,j} B_{i,k}(u) B_{j,l}(v), \qquad u,v \in [0,1]. \qquad (2.2)$$

Bézier curves and surfaces share basically equal properties:

- *Orthogonal basis.* Any polynomial parametric curve or surface can be converted to the Bézier representation.

- *Boundary conditions.* A Bézier curve segment starts from V_0 and ends at V_k, where the curve tangent directions are towards V_1 and V_{k-1}. Surface patch boundaries are equal to the Bézier curves defined by the control mesh boundary polygons.

- *Convex hull property.* A curve or a surface is located inside the convex hull of its control points. Following from the boundary conditions, this convex hull is relatively tight (compared with B-splines, for instance).

- *Invariance to affine transformations.* A curve or a surface can be translated, rotated and scaled by applying the corresponding transformation matrix to the control points. Affine transformations include orthogonal projection, but not the perspective one.

- *Subdivision.* A Bézier curve can be subdivided at the parametric midpoint into two segments by recursive additions and divisions by two [5] (Fig. 1). Surfaces are divided into four subpatches by curve subdivision in the two directions of the control mesh. The subdivided control points converge fast towards the actual curve or surface.

Taken generally, the presented results apply to any parametric surface representation having the last three of the listed properties, such as rational and even multivariate ones. Due to space restrictions, however, the following discussion primarily considers only the *bicubic* polynomial case $k = l = 4$.

3   Object Space Methods

The problem of computing the intersection point between a ray $R(t)$ and a surface patch $S(u, v)$ can be expressed as

$$S_c(u, v) = R_c(t), \qquad c = x, y, z. \tag{3.1}$$

With bicubic surfaces the polynomial degree of the system of equations is already too high for analytical solutions. Several approximative solutions have been presented based either on numerical methods or surface subdivision.

Bivariate *Newton iteration* [6] [7] is the most popular one among the numerical methods. The method is efficient and accurate with quadratic convergence properties, but to guarantee convergence a dense grid of starting values must be provided. The choice of sampling density is heuristic and it must be exaggerated for robustness, which is experienced as precomputation and memory costs. Further, several iteration procedures may have to be applied with the same patch in order to detect multiple ray intersections.

The *interval arithmetic* Newton method [8] converges more reliably, having advantages also in utilizing image coherence [9]. However, it is doubtful whether possible coherence properties are generally enough to compensate for the basically high computation costs. *Kajiya's method* based on Laguerre's iteration and algebraic geometry is originally reported to be quite costly [10]. Moreover, the method is unstable with flat patches and a robust implementation is again difficult to realize [11].

The *dynamic subdivision* method, as originally reported by Whitted [1], bounds the surface patches first by spheres. If the bounding sphere is hit by the ray the patch is divided into four subpatches which are processed recursively until a given depth or size tolerance. The method is robust and it does not require global memory, but for good accuracy the total amount of computation during the subdivision grows too much.

Static subdivision, as we call it, breaks the surfaces into small polygons already in preprocessing [12]. Careful implementation of static subdivision with suitable space subdivision [13] makes it apparently the fastest available method to produce ray traced images of parametric surfaces. Drawbacks are found in precomputation times, limited accuracy and, obviously, memory costs.

4   Zero-Eye Principle

Rendering parametric surfaces in perspective view is generally (both in vector and raster graphics) considered difficult, as the perspective viewing transformation cannot be applied directly to the control points of the surface. Ray tracing is a method of perspective viewing where even the eye

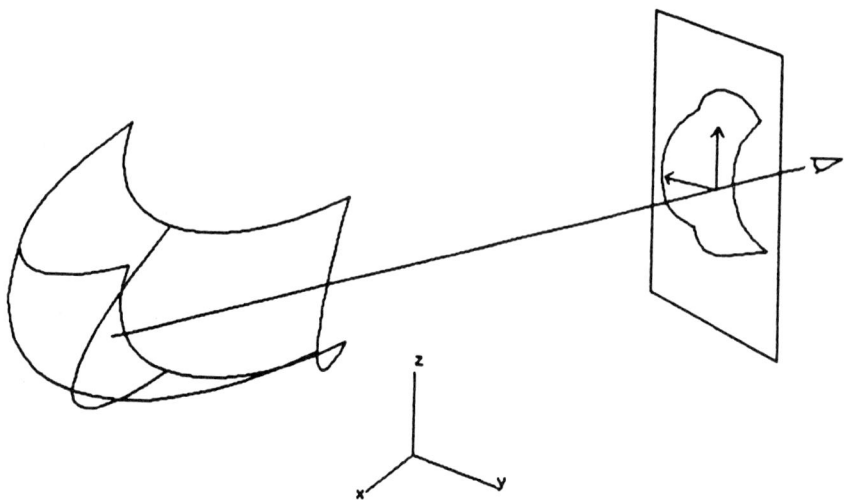

Figure 2. Orthogonal transformation of a surface patch to the ray's viewing plane

point changes with secondary rays from one point to another. To avoid the projection problems, the previously mentioned surface intersection methods perform their computation directly in the 3D object space.

However, the sensation of a perspective view in ray traced images is actually composed of thousands of independent pixel computations where the perspective projection is never needed. Each ray sees just a *point* of the world, and the world around can be taken just as well orthogonal as having perspective. For each ray there also exists a relevant projection system, that whose viewing axis is defined by the ray itself (Fig. 2).

Suppose the intersection between a ray $R(t)$ and a surface patch $S(u,v)$ is to be determined. As the patch is transformed to the ray's orthogonal viewing system $(\mathbf{x},\mathbf{y},\mathbf{z})$, the intersection point becomes located on the depth axis of the viewing system. The patch projection in the viewing plane has correspondingly the intersection point seen at the plane origin (the *zero-eye*):

$$\mathbf{S}_c(u,v) = 0, \quad c = \mathbf{x}, \mathbf{y}. \tag{4.1}$$

The two unknowns (u,v) are obviously easier to solve from (4.1) than the system of three equations for three unknowns (3.1). The ray parameter value (assuming a normalized direction vector) is then obtained directly as

$$t = \mathbf{S}_{\mathbf{z}}(u,v). \tag{4.2}$$

This principle of ray tracing computation was first suggested for surface

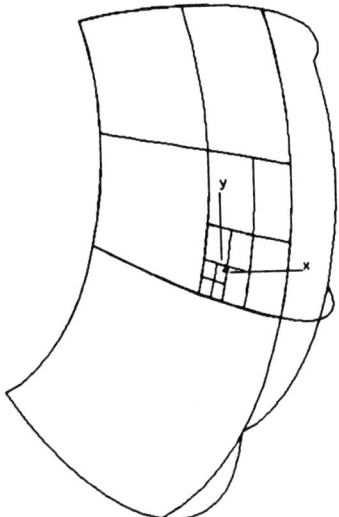

Figure 3. Subdivision in viewing plane

subdivision by Rogers [14]. Improvements leading to a practical implementation were given later in [15].

In the present paper, the method is carried further in two elementary
aspects. First, we show that the subdivision process can be performed
efficiently in integer arithmetic with less than three extra bits needed to
preserve accuracy under any number of subdivisions. Second, a method
to rush through the subdivision tree directly to the intersected subpatch is
presented. The zero-eye method is compared with other surface intersection
methods, and further implications of the results are pointed out in the
hardware implementation context.

5   Subdivision Process

The intersection point computation proceeds basically as follows. First the
plane components **x** and **y** of the surface patch are computed by the viewing
matrix. The surface patch is subdivided if its bounding box contains the
plane origin (Fig. 3):

$$\exists\, \mathbf{V}_{c,i,j} \leq 0, \quad \exists\, \mathbf{V}_{c,i,j} \geq 0, \quad c = \mathbf{x}, \mathbf{y}. \tag{5.1}$$

The sign test (5.1) is applied for the four subpatches and the procedure
is repeated recursively until given depth. At the bottom of the recursion the
intersection point (u, v) values are approximated. The transformation of

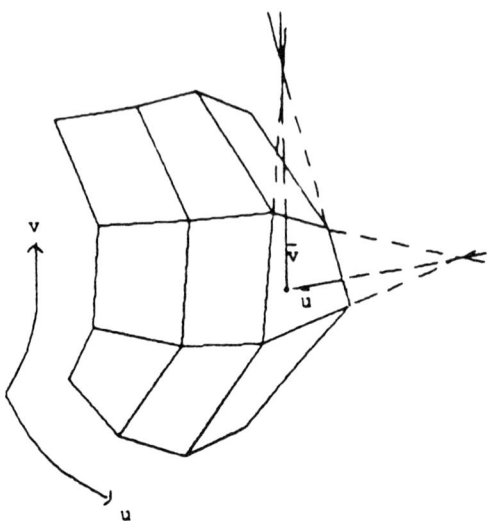

Figure 4. Intersection point parameters

the **z** component of the original surface patch is computed. The ray depth and the surface normal are then available by the $\mathbf{S_z}$ and $\delta S/\delta u \times \delta S/\delta v$ equations.

The basic algorithm is refined in several steps in this and the following sections. The complete algorithm is summarized in Appendix A2.

It is first noted that a slight modification in the *computation order* is useful. Instead of subdividing a patch at once, it is better to compute the **x** subpatch components first. If the sign test fails for some (or all) the subpatches in **x**, subdivision in the **y** component can be omitted.

To derive the *intersection point parameters* (u, v), a refined intersection test at the bottom of the recursion saves a lot of subdivision (c.f., [16]). Each of the rectangular facets of the intersected subpatch control mesh are tested against the origin by point–in–polygon. If the test is successful, the intersection point parameter values inside the subpatch are approximated (Fig. 4). To obtain the global (u, v) values, the local parameter values are added to the known parameter values at the subpatch corners.

For moderate levels of subdivision (under ten or so) it is efficient to approximate also the *depth value and normal vector* by subdivision. The subpatch **z** control points are evaluated by 1D subdivision along the known subdivision path. The depth value and the normal vector are obtained by averaging the corresponding subpatch control mesh vertice normals by the local parameter values. The normal vector is finally transformed to the object space by the inverse viewing matrix.

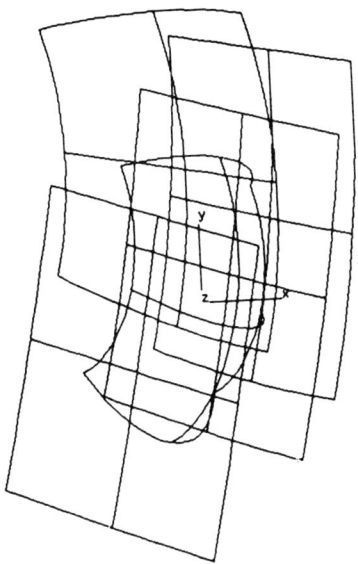

Figure 5. Scaled integer subdivision

6 Integer Transformation

The subdivision computation is speeded up significantly by scaling the control mesh to large *integer numbers* after the viewing transformation. The Bézier subdivision algorithm [5] can then be implemented with just integer addition and bit shift operations.

Generally, each bit shift in the subdivision equations loses accuracy [16]. However, the zero-eye problem (4.1) has the advantage of centering the subdivision at the origin. The extents of the subpatches are approximately halved in each subdivision. Instead of filling the control point numbers with zero bits from the left, the available integer word space can be used by *scaling the numbers by two*, *i.e.*, omitting one bit shift in each subdivision call (Fig. 5).

Appendix A1 proves that the scaled subpatches never grow more than six times larger than the original patch. Thus, 'two and half bits' are enough to prevent overflow and have the remaining bits significant through any number of subdivisions.

Another limit concerning accuracy, however, is set up by the *formulation* of the subdivision equations. For instance, with cubic curves the recursive subdivision equations perform a total of six bit shifts to derive the midpoint of the curve, the open equation $(V_0 + 3(V_1 + V_2) + V_3)/8$ takes three shifts,

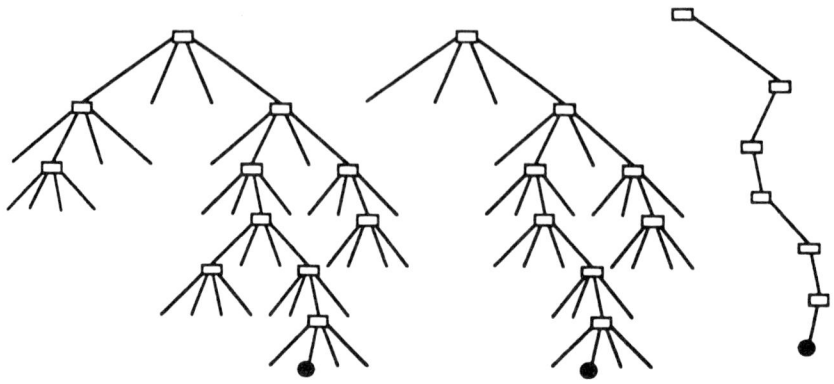

Figure 6. Subdivision tree with (a) complete recursion; (b) termination at first intersection; (c) adaptive branching

and the scaled open equation only two.

The open subdivision equations give the best accuracy under large numbers of subdivisions; bicubic patches on n bit integers can then be scaled up to the maximum absolute size of $2^{(n-5.5)}$, where 3 bits are needed for the subdivision additions and 2.5 bits to prevent overflow.

7 Adaptive Branching

If it is known (see Section 8) that the surface patch does not have an interior silhouette (*i.e.*, it is *visible*) from the ray's viewing direction, there can only be one intersection point with the ray. Once an intersection point has been found, the recursive computation can be terminated immediately. Figs. 6a–b depict how this might affect the subdivision tree structure.

To take even greater advantage of the visibility condition, we introduce a method for *adaptive subdivision branching*. It usually leads directly to the possible intersection point, thus completely avoiding forming the recursion stack and the computation of non-intersected subpatches (Fig. 6c).

Before subdividing a patch, the corner points of its subpatches are computed (Fig. 7). Four of these points are equal to the patch corners and four others are obtained as the boundary curve midpoints. For efficiency reasons, the middle corner is approximated as the mean of the four interior control points.

The four rectangles defined by the subpatch corners are checked by point–in–polygon to contain the origin. If the test is affirmative, the corresponding subpatch is the most likely one to contain the origin and it will

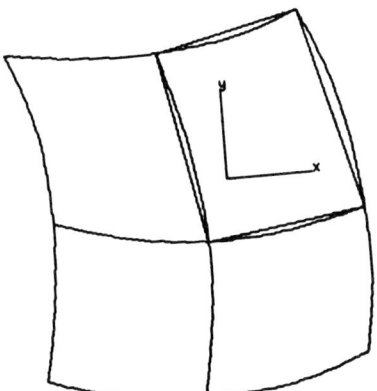

Figure 7. Adaptive branching: upper right subpatch processed first

be processed first.

If an intersection is not found within the first patch, the other sub-patches are evaluated and processed non-adaptively. This is better than trying to pursue adaptation, as the original patch is most likely not intersected at all.

In the test cases (Section 10) the intersection point was found directly in more than 90% of the visible intersected cases. Accounting also for the silhouette cases and those not intersected, less than two subpatches per subdivision call had to be computed on average and only one subpatch processed further. These numbers are practically independent of the chosen subdivision level, as the subpatches become less curved and the adaptive branching criterion stronger the deeper the subdivision proceeds.

8   Visibility Test

A patch does not have a silhouette if none of its normal vectors is orthogonal to the ray. As this is too difficult to test exactly, however, we recommend an approximate method based on the so-called *normals patches* [3]. Their original application in approximating the silhouette curve is:

1. Compute sixteen unit normals of the patch.

2. When transformed to origin, the normals define sixteen points on the unit sphere. Interpolate these with a cubic surface $N(u, v)$.

3. The silhouette curve parameters are obtained as the solution of the plane intersection problem $\mathbf{N_z}(u, v) = 0$.

The normals patch involves obvious approximation risks for actually *representing* the silhouette curve. However, it gives a very reliable answer to possible silhouette *existence*. Thus,

- Precompute the normals patches $N(u, v)$ in Bézier representation.

- With each intersection test, compute the viewing transformation $N_Z(u, v)$. The patch is visible if all the control points of N_Z have equal sign.

Note that interpolating the unit normals with a Bézier patch makes the test more pessimistic and thus more reliable than testing the unit normal signs directly.

9   Gaps and Bugs

We have chosen to use a *fixed subdivision depth* for each composite surface. Depth six is practically always sufficient for visualization [15]. A subpatch size tolerance [16] to terminate recursion might be more efficient, but it involves the danger of producing gaps between neighbouring patches derived from different subdivision levels.

However, we are still aware of one small bug in the presented implementation: a gap between neighbouring patches can result if different integer scaling factors are used for them. With 32 bit integers, such a gap would be of relative magnitude $2^{-26.5} \approx 10^{-8}$ of the patch extents, causing an intersection to be missed once in some hundreds of ray traced pictures. With antialiasing this does not matter, as the error is not systematic.

To reduce the risk of the gaps, the scaling factors can be restricted to powers of two. If the gaps are to be completely avoided, an equal scaling factor must be chosen for all the patches in a surface according to the greatest patch extents.

10   Test Results

Test results are evaluated with two strongly differing ray traced pictures.

The Saab 9000 CS car model in Fig. 8 is published with permission of Oy Saab–Valmet Ab. It was originally defined with the STRIM modeling system with general order Bézier surfaces which we converted to bicubic patches for ray tracing. The picture is strongly shadowed and it contains a very large number of surface patches.

The other test case in Fig. 9 is the 'Geo' glass vase, designed by Mr. Ken Benson for Iittala–Nuutajärvi Oy. The model was constructed with our CubiX modeling system using trimmed cubic B-spline surfaces.

Figure 8. Ray traced image of the Saab 9000 CS car model

The pictures were computed with a Sun 4/260 workstation using the adaptive space subdivision system EXCELL. Test results without antialiasing are shown in Table 1, where '=' means that the numbers are (about) equal to the previous column, and the '-' values do not have meaning.

The results are interpreted as:

- *Number of rays* tells the number of rays which penetrated the bounding box of the scene and *intersection tests* is the number of parametric surface intersection calls. *Actual intersection* was found in about every three test cases out of four.

- The pictures were computed with two or three different *subdivision depths* to analyze the subdivision behaviour. Subdivision depth six was used in both the displayed pictures.

- *Visible patches* tells how many of the tested patches were determined to be visible to the ray.

- *First pass* denotes how often the intersection point of visible patches was found without recursion. Note that the percentage is practically independent of the subdivision depth.

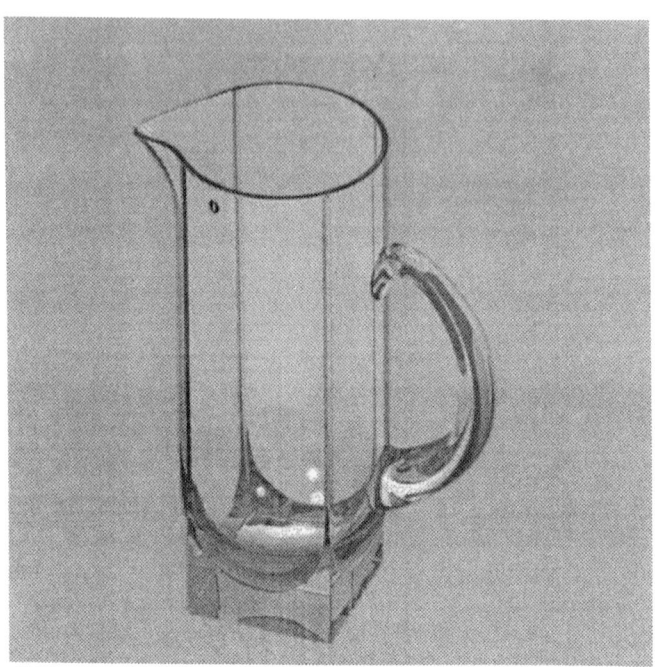

Figure 9. Ray traced image of the 'Geo' glass model

Table 1. Test results

		Saab	=	Geo	=	=
number of patches		9300		630		
pixel resolution		640 x 480		512 x 512		
number of rays		256.000	=	479.000	=	=
intersection tests		62.000		891.000		
actual intersections		45.000		655.000		
subdivision depth		0	6	0	6	12
visible patches		94 %	94 %	91 %	91 %	91 %
first pass		-	99 %	-	94 %	94 %
subpatches/call		-	1.5	-	1.7	1.7
recursion calls		-	5.4	-	6.0	12.2
branches/call		-	0.98	-	1.01	1.02
total cpu time		11 min	13 min	43 min	92 min	141 min
zero-eye time		10 %	23 %	49 %	68 %	81 %

- *Subpatches/call* tells how many subpatches on average were evaluated per recursion call, with subdivision in one coordinate component counted as half. Without sequential treatment of coordinate components and adaptive branching, this number would be exactly four.

- The *recursion calls* number tells how many times the subdivision routine was called on average per intersection test. Due to adaptive branching, the number keeps about equal to the subdivision depth.

- *Branches/call* tells how many subpatches in each call had to be tested further. Also this number keeps about constant (one) in all the cases.

- *Zero-eye time* tells the proportion which the parametric surface intersections took of the total ray tracing time. The zero-eye time includes normal computation by subdivision.

As a general conclusion, we note that the adaptive zero-eye subdivision method leads to statistically non-recursive computation behaviour.

11 Comparisons

The zero-eye subdivision method takes 31 arithmetic floating point operations (*flops*) to compute the viewing matrix, 96 flops for the visibility check, 96-192 flops for the viewing and 16-32 flops for integer scaling transformations, and 245 flops for the parameter and normal computations. Intersected patches thus require a maximum of less than 600 flops and the non-intersected cases about half of this.

All other computation is done in integer arithmetic. According to test results (Section 10), the intersection computation times are doubled from the floating point overhead time (zero subdivisions) after each four or five subdivision levels.

The pointwise Newton method requires at least 1250 flops to compute an intersection point [15]. The zero-eye method thus appears to be about as fast as the Newton method. More precise statements depend on the chosen subdivision depth, implementation, and hardware.

Compared to dynamic object space subdivision, zero-eye subdivision works with only two coordinate components and the integer computation is done without accuracy overhead. The subdivision treats the coordinate components sequentially and branches adaptively, which cannot be implemented efficiently in object space. The expensive bounding object evaluations and intersections in object space are replaced with integer sign checks. Altogether, the zero-eye method performs subdivision about five times faster than the object space method. Accounting for the floating point overhead, it gets faster in total time after one or two levels of subdivision.

Analytic comparisons with static subdivision are difficult as the major part of computation with polygons goes to localizing the intersection detection [1] [15]. Static subdivision is apparently the fastest method available for visualization, but considering memory costs it depends on the application and hardware how to value it against the more accurate methods.

12   Hardware Implementation

The following discussion gives some general remarks regarding possible implementation of the zero-eye subdivision method on hardware. The method is therefore compared with the 'standard' chip architecture [16]. Chip components/solutions of the standard method appear in quotes, whereas new suggestions are written in italics.

Matrix transformation chips are commercially available and they can be used serially with the subdivision chip to perform the zero-eye viewing computations. However, reducing the computation dimensionality from three to two does not alone improve speed, as the coordinate components can be computed in parallel. It is therefore best to transform all the three coordinate components to the viewing system also with the zero-eye method.

After the viewing transformation the 'compute bounding box' and 'test for hit' components can be replaced by just testing *control point signs* (5.1). Applying the *scaled subdivision equations* preserves accuracy during subdivision with the viewing plane coordinate components.

It seems useful to replace the 'mean evaluation' and the 'test accuracy' components by a fixed subdivision depth, and use *point–in–polygon* and *2D line intersection* components to intersect the subdivided control mesh at the bottom of recursion (Fig. 4). This gives better accuracy with much less subdivision and it also avoids the problem of gaps between the subpatches.

The recursion stack is generally considered the greatest bottleneck with the standard chip architecture. For the majority of intersected patches, 'scanning and re-evaluating' the subdivision stack would be avoided with the adaptive branching method. To include a *branching test* in the chip can be done with the point–in–polygon and surface subdivision components which are already included among the previous suggestions.

To actually implement the chip without subdivision stack memory, a *division of labour* between the host computer and the chip must be made. Therefore, the chip immediately returns the patches which are missed with adaptive subdivision to the host. Assuming that 85–95 % of the patches are visible and 90–99 % of visible intersections are found with the chip, it would leave only 5–20 % of the intersected patches to be computed by software. After the hardware viewing transformation, the non-intersected patches are less significant as they are usually discarded at very low subdivision levels.

13   Conclusions

The zero-eye subdivision method for computing ray intersections with parametric surfaces has been described. The method reduces the intersection point computation problem to plane with a transformation of the surface patch to the ray's orthogonal viewing system. The bounding object intersection computation during subdivision is thus replaced with just integer sign comparisons.

New results are the following. The computation can be done efficiently in integer arithmetic with only two and half bits required to preserve accuracy under any number of subdivisions. With surface patches visible to the ray, an adaptive subdivision branching order saves most of the subdivision computation.

Compared to other surface intersection methods, zero-eye subdivision provides optimal memory and robustness and a very good value of speed vs. accuracy. The method offers also new viewpoints for efficient hardware implementation.

14   Epilogue

Parallel to the original presentation of this paper, Nishita, et al. [17] proved that the zero-eye method can be applied to ray tracing rational parametric surfaces, as well.

A ray-surface intersection method called the Bézier clipping algorithm was presented in [17], and compared with an earlier implementation of the zero-eye method [15]. According to the comparisons, the Bézier clipping algorithm was several times faster than zero-eye.

However, in correspondence with one of the authors of [17] it turned out that the test installation of zero-eye [15] had not been done in integer arithemetic, which is crucial for speed. On the other hand, the Bézier clipping algorithm can not be implemented on integer arithmetic. The author of the current paper repeated the tests in [17], concluding that the Bézier clipping algorithm and the one presented in this paper are about as fast.

Acknowledgements

The idea of adaptive subdivision branching was first suggested to the author by Dr. Markku Tamminen. Mr. Panu Rekola who is generally responsible for our ray tracer implementation produced the ray traced pictures and gave other valuable help throughout this work. The surface degree reduction routines were implemented by Mr. André Dolenc. The work has been supported by the Research Support Foundation of Helsinki University of Technology.

References

1. Whitted, T. (1980). An improved illumination model for shaded display. *Comm. ACM, 23(6)*, 96–102.

2. Cohen, M.F., Shenchang, E.C. Wallace, J.R. and Greenberg, D.P. (1988). A progressive refinement approach to fast radiosity image generation. *Computer Graphics (Proc. SIGGRAPH '88), 22(3)*, 75–84.

3. Mortenson, M.E. (1985). *Geometric Modeling*, Wiley.

4. Böhm, W., Farin, G. and Kahmann, J. (1984). A survey of curve and surface methods in CAGD. *Computer Aided Geometric Design, 1(1)*, 1–60.

5. Lane, J.M. and Riesenfeld, R.F. (1980). A theoretical development for the computer generation and display of piecewise polynomial surfaces. *IEEE Trans. on Pattern Analysis and Machine Intelligence, 2(1)*, 35–46.

6. Sweeney, M.J. and Bartels, R.H. (1986). Ray tracing free-form B-spline surfaces. *IEEE Computer Graphics and Applications, 6(2)*, 41–49.

7. Yang, C.-G. (1987). On speeding up ray tracing of B–spline surfaces. *Computer–Aided Design, 19(3)*, 121–130.

8. Toth, D.L. (1985). On ray tracing parametric surfaces. *Computer Graphics (Proc. SIGGRAPH '85), 19(3)*, 171–179.

9. Lischinski, D. and Gonczarowski, J. (1989). Improved techniques for ray tracing parametric surfaces. Submitted to *Visual Computer*.

10. Kajiya, J.T. (1982). Ray tracing parametric patches. *Computer Graphics (Proc. SIGGRAPH '82), 16(3)*, 245–254.

11. van Wijk, J.J. (1988). Correspondence with the author.

12. Fujimoto, A., Tanaka, T. and Iwata, K. (1986). ARTS: accelerated ray tracing system. *IEEE Computer Graphics and Applications, 4(4)*, 16–26.

13. Cleary, J.G. and Wyvill, G. (1988). Analysis of an algorithm for fast ray tracing using uniform space subdivision. *Visual Computer, 4(2)*, 65–83.

14. Rogers, D.F. (1985). *Procedural Elements for Computer Graphics*, McGraw–Hill, pp. 296–305.

15. Woodward, C. (1989). Ray tracing parametric surfaces by subdivision in viewing plane. In *Theory and Practice of Geometric Modeling*, W. Strasser, H.-P. Seidel (eds.), Springer–Verlag, New York, 273–287.

16. Pulleyblank, R. and Kapenga, K. (1987). The feasibility of a VLSI chip for ray tracing bicubic patches. *IEEE Computer Graphics and Applications*, *7(3)*, 33–44.

17. Nishita, T., Sederberg, T.W. and Kakimoto, M. (1990), Ray tracing trimmed rational surface patches, *Computer Graphics*. *Proc. SIGGRAPH 90*, *24(4)*, 337–345.

A1   Subdivision Convergence

It is sufficient first to consider 1D curves within the unit interval:

Lemma 1. *Let $C(u)$ be a 1D cubic Bézier curve with*

$$C(u) = \sum_{i=0}^{4} V_i B_i(u) \qquad 0 \le u \le 1, \quad |V_i| \le 1, \tag{A1}$$

and let $C^k(u)$ be derived with k midpoint subdivisions of $C(u)$. The control points of $C^k(u)$ are restricted by

$$max|V_i^k - V_j^k| \le 6 \times 2^{-k}. \tag{A2}$$

Proof The slowest convergence occurs with $V_0 = -1$ and $V_3 = 1$, otherwise all $V_i^k, k \ge 1$ are closer to each other. After one subdivision the subsegment control polygons are bound to be monotonic, $V_i^k \le V_j^k, i < j$, as in the worst case $V_1 = 1, V_2 = -1$. Thus,

$$\max|V_i^k - V_j^k| \quad = \quad |C^k(1) - C^k(0)| \tag{A3}$$

$$\le \quad |C(u + 2^{-k}) - C(u)|, u \in [0, 1) \tag{A4}$$

$$= \quad 2^{-k}|C(u + 2^{-k}) - C(u)|/2^{-k} \tag{A5}$$

$$\le \quad 2^{-k}\max|C'(u)|, u \in [0, 1], \tag{A6}$$

and a second derivative analysis shows that

$$|C'(u)| = 3 \times | \quad - \quad V_0(1 - u)^2 + V_1(1 - u)(1 - 3u) \tag{A7}$$

$$- \quad V_2 u(3u - 2) + V_3 u^2| \tag{A8}$$

$$\le \quad 6. \tag{A9}$$

The result of the theorem holds similarly for arbitrary dimension curves. Surface subdivision involves curve subdivision twice, so its convergence is at least as fast. Surface patches in zero-eye subdivision additionally obey the condition (5.1), so extending from the notation of Lemma 1 they have

$$\max|\mathbf{V}_{i,j}^k| \le 6 \times 2^{-k}|\mathbf{V}_{i,j}|, \tag{A10}$$

and with the scaled subdivision equations it holds:

$$\max|\mathbf{V}_{i,j}^k| \le 6 \times |\mathbf{V}_{i,j}|. \tag{A11}$$

A2   Summary of the Algorithm

The following pseudo-C description gives the guidelines for an efficient implementation of the zero-eye subdivision method, though necessarily omitting much of the detail. For instance, the normal computation is assumed to be done in the calling routine. Also, it should go without saying that special care must be taken in the programming of the actual subdivision routine.

```
/* GLOBAL VARIABLES */

local int Hit, ClosestHit, Visible;
local double *global_u, *global_v, *global_t;
local double *worldpatch, viewpatch[4][4][3], intpatch[4][4][3], npatch[4][4][3];
static double viewmat[3][4];

/* MAIN ROUTINE */

boolean patch_intersect(ray, t, patch, u, v, depth, normals_patch)
/* when called, t contains the previously found closest intersection depth */
{
    if (first call with this ray) define_viewing_matrix(ray);

    worldpatch = patch;
    for (coord = x, y)
    {
        view_transform(worldpatch, viewpatch, coord);
        int_transform(viewpatch, intpatch, coord);
        if (all_equal_sign(intpatch, coord)) return(FALSE);
    }

    view_transform(normals_patch, npatch, z);
    int_transform(npatch, npatch, z);
    Visible = (all_equal_sign(npatch, z));

    Hit = ClosestHit = FALSE;
    global_t = t; global_u = u; global_v = v;

    sub_intersect(intpatch, 0.0, 0.0, 1.0, depth);

    return(ClosestHit);

}

/* ... continued ... */
```

```
/* RECURSIVE SUBDIVISION */

local sub_intersect(patch, corner_u, corner_v, wid, depth)
{
    if (depth > 0) /* SUBDIVIDE */
    {
        wid /= 2; depth--;
        for (i=0,1 j=0,1) zero_in[i][j] = FALSE;

        if (Visible) /* ADAPTIVE TEST */
        {
            subcorners(patch, corners);
            for (i=0,1 j=0,1)
                if (origin_in_polygon(rectangle defined by the corners of subpatch[i][j]))
                {
                    zero_in[i][j] = TRUE; /* only this subpatch is evaluated */
                    for (coord=x,y) subdivide(patch, subpatches, coord, zero_in);

                    sub_intersect(subpatch[i][j], corner_u+i*wid, corner_v+j*wid,
                                  wid, depth);

                    if (Hit) return;
                    else { Visible = FALSE; /* break adaptive search */ break; }
                }
        }

        /* NON-ADAPTIVE TEST */

        for (i=0,1 j=0,1) zero_in[i][j] = not zero_in[i][j];

        for (coord=x,y) /* subdivide */
        {
            subdivide(patch, subpatches, coord, zero_in);
            for (i=0,1 j=0, 1) zero_in[i][j] = (not all_equal_sign(subpatch[i][j], coord));
        }

        for (i=0,1 j=0,1) if (zero_in[i][j]) /* test further */
            sub_intersect(subpatch[i][j], corner_u+i*wid, corner_v+j*wid, wid, depth);
    }

    else    /* BOTTOM OF RECURSION */
        for (each facet of patch control mesh)
            if (origin_in_polygon(facet)) /* INTERSECTION FOUND */
            {
                local_parameters_of_origin(facet, &local_u, &local_v);
                if (not Hit) view_transform(worldpatch, viewpatch, z);
                compute_distance(viewpatch, local_u, local_v, &t);

                if (t > 0 and t < *global_t ) /* CLOSEST INTERSECTION */
                {
                    *global_t = t;
                    *global_u = corner_u + transformation of local_u;
                    *global_v = corner_v + transformation of local_v;
                    ClosestHit = TRUE;
                }
                Hit = TRUE;
            }
}

/* ... end of pseudocode ... */
```

Fitting Non Regular Parametric Surfaces by Finite Element Methods

J.J. Torrens, F.J. Serón and M.C. López de Silanes

Universidad de Zaragoza

Abstract The aim of this work is to present two methods for fitting a non regular parametric surface on a finite set of points distributed on curvilinear grids. For compensating for the absence of a natural parametrization of the data points, a uniform parametrization is introduced. Then, an approximating surface is constructed, either by interpolation or by smoothing, using rectangular finite elements of Hermite type. The convergence of the methods is studied and some numerical examples are given.

1 Introduction

In some fields, such as Geophysics or Aeronautics, it is common to find the problem of fitting a surface that presents faults or creases from a set of gridded or scattered data points. Such a problem may be modelized as the fitting of a non regular function with values in IR or IR^3 (depending on the surface is given explicity or parametrically). The non regularity is due to the presence of some discontinuity in the function itself (surface with faults) or in some of its derivatives (surface with creases).

In the last years, this problem has been studied by several authors, who have considered both parametric surfaces (cf. [1] [10], [11], [12]) or explicit ones (cf. [2], [4], [5], [6], [7], [8]).

In this paper, we restrict ourselves to the case of a surface defined by a discontinuous parametrization. Two methods are proposed for fitting such a surface from data points placed on curvilinear grids, and they both give an approximating surface defined by a parametrization that belongs to a finite element space constructed from the Bogner–Fox–Schmit generic finite element.

The first method is a Lagrange interpolation in which the coefficients of the basis functions are directly computed using finite difference formulae for approximating the derivatives to which they are related. In the second method, the surface is fitted with a discrete smoothing D^m-spline, which is the solution of a certain minimization problem.

2 Notations and Preliminaries

Given a subset E in IR^2, the closure, the interior and the boundary of E will be denoted, respectively, by $\overline{E}$, $\overset{\circ}{E}$ and ∂E .

For all $\alpha = (\alpha_1, \alpha_2) \in \mathrm{IN}^2$, we shall use the notations $|\alpha| = \alpha_1 + \alpha_2$, $|\alpha|_\infty = \max(\alpha_1, \alpha_2)$ and $\partial^\alpha = \dfrac{\partial^{|\alpha|}}{\partial x_1^{\alpha_1} \partial x_2^{\alpha_2}}$.

For all $m, n \in \mathrm{IN}^*$, we shall write $\mathcal{M}_{m \times n}$ for the space of real matrices with m rows and n columns.

Let p_i, $i = 1, \ldots, n$, denote the canonical projections from IR^n to IR.

Given a connected set E in IR^2 and $m \in \mathrm{IN}$, we shall represent by $P_m(E)$ and $Q_m(E)$ the spaces of restrictions to E of the polynomials whose degrees and less than or equal to m with respect to the set of all variables and with respect to each one of them. Similarly, $P_m(E, \mathrm{IR}^3)$ and $Q_m(E, \mathrm{IR}^3)$ will denote the spaces of functions q with values in IR^3 for which $p_i \circ q \in P_m(E)$ and $p_i \circ q \in Q_m(E)$, $i = 1, 2, 3$, respectively.

In the sequel, ω will be an open set in IR^2.

For all $m \in \mathrm{IN}$, we shall write $H^m(\omega, \mathrm{IR}^3)$ for the Sobolev space of (classes of) all functions φ with values in IR^3 for which $p_i \circ \varphi$, $i = 1, 2, 3$, together with all their partial derivatives $\partial^\alpha(p_i \circ \varphi)$ (in the distribution sense), with $|\alpha| \leq m$, belong to the space $L^2(\omega)$. This Sobolev space is equipped with the norm:

$$\| \varphi \|_{m, \omega, \mathrm{IR}^3} = \left(\sum_{i=1}^{3} \sum_{|\alpha| \leq m} \int_\omega |\partial^\alpha(p_i \circ \varphi)|^2 \mathrm{d}x \right)^{\frac{1}{2}} .$$

Moreover, for $l = 0, \ldots, m$, we shall write

$$(\varphi, \eta)_{l, \omega, \mathrm{IR}^3} = \sum_{i=1}^{3} \sum_{|\alpha| = l} \int_\omega \partial^\alpha(p_i \circ \varphi) \, \partial^\alpha(p_i \circ \eta) \, \mathrm{d}x \quad \forall \varphi, \eta \in H^m(\omega, \mathrm{IR}^3)$$

and

$$|\varphi|_{l, \omega, \mathrm{IR}^3} = \left((\varphi, \varphi)_{l, \omega, \mathrm{IR}^3} \right)^{\frac{1}{2}} \quad \forall \varphi \in H^m(\omega, \mathrm{IR}^3).$$

For all $k \in \mathrm{IN}$ and for $n = 2, 3$, $C^k(\omega, \mathrm{IR}^n)$ will denote the space of functions φ with values in IR^n for which $\partial^\alpha(p_i \circ \varphi)$ is continuous on ω, for $i = 1, \ldots, n$ and for all $\alpha \in \mathrm{IN}^2$ with $|\alpha| \leq k$. Analogously, we shall write $C^k(\overline{\omega}, \mathrm{IR}^n)$ for the space of functions $\varphi \in C^k(\omega, \mathrm{IR}^n)$ such that, for $i = 1, \ldots, n$ and for all $\alpha \in \mathrm{IN}^2$ with $|\alpha| \leq k$, $\partial^\alpha(p_i \circ \varphi)$ admits a continuous extension to $\overline{\omega}$ (which will be denoted in the same way). This space is equipped with the norm:

$$\| \varphi \|_{C^k(\overline{\omega}, \mathrm{IR}^n)} = \max_{1 \leq i \leq n} \ \max_{0 \leq |\alpha| \leq k} \ \max_{x \in \overline{\omega}} \ |\partial^\alpha(p_i \circ \varphi)(x)|.$$

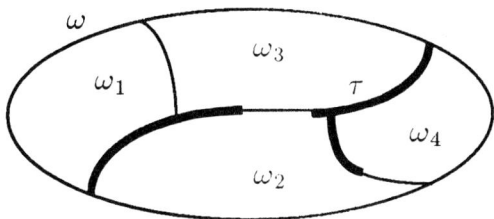

Figure 1.

Let φ be a function defined on a set A and let $B \subset A$. Then, $\varphi|_B$ will denote the restriction of φ to the set B as well as its continuous extension to $\overline{B}$, if this exists.

Suppose that ω is non empty, connected, bounded and it has a Lipschitz-continuous boundary. Let τ be a closed, non empty subset of $\overline{\omega}$. Let $\omega' = \omega \setminus \tau$ and let $\{\omega_1, \ldots, \omega_l\}$ be a family which represents τ in ω, that is, a finite family of pairwise disjoint, connected, non empty, open subsets of ω, such that $\partial\omega_i$, $i = 1, \ldots, l$, is Lipschitz-continuous, $\overline{\omega} = \bigcup_{i=1}^{l} \overline{\omega}_i$ and $\tau \subset \bigcup_{i=1}^{l} \partial\omega_i$. This situation is exemplified in Fig. 1.

Then, given $k \in \mathbb{N}$, we shall write $C_\tau^k(\omega', \mathbb{R}^3)$ for the space of functions $\varphi \in C^k(\omega', \mathbb{R}^3)$ such that, for $i = 1, \ldots, l$, $\varphi|_{\omega_i} \in C^k(\overline{\omega}_i, \mathbb{R}^3)$.

If P is a subset of $\mathbb{R}^3$ and $\varphi \in C_\tau^k(\omega', \mathbb{R}^3)$, then $\varphi^{-1}(P)$ will denote the set of points $d \in \overline{\omega}$ for which there exists a set $B \subset \omega'$ such that $d \in \overline{B}$, $\varphi|_B \in C^0(\overline{B}, \mathbb{R}^3)$ and $\varphi|_B(d) \in P$.

Theorem 1. *$C_\tau^k(\omega', \mathbb{R}^3)$ is a Banach space with norm:*

$$\| \varphi \|_{C_\tau^k(\omega', \mathbb{R}^3)} = \max_{1 \leq i \leq l} \|\varphi|_{\omega_i}\|_{C^k(\overline{\omega}_i, \mathbb{R}^3)}.$$

Moreover, the space $C_\tau^k(\omega', \mathbb{R}^3)$ and the norm $\| \cdot \|_{C_\tau^k(\omega', \mathbb{R}^3)}$ are independent of the choice of the family $\{\omega_1, \ldots, \omega_l\}$ which represents τ in ω.

Theorem 2. *If $m > k + 1$, then $H^m(\omega', \mathbb{R}^3)$ is a subset of $C_\tau^k(\omega', \mathbb{R}^3)$ with continuous injection.*

The proofs of these theorems may be found in [10] or [11]. See also [8], where similar results are stated for real-valued functions.

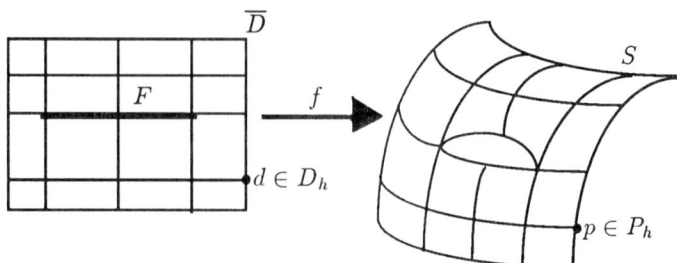

Figure 2.

3 Modelization of the Fitting Problem

Let S be a surface defined by a parametrization $f \in H^m(D', \mathrm{IR}^3)$, where m is (and it will be throughout this paper) an integer greater than or equal to 2, D is a bounded, connected, non empty open set in IR^2 with Lipschitz–continuous boundary, F is a non empty, closed subset of $\overline{D}$ such that there exists a family $\{D_1, \ldots, D_l\}$ that represents it in D, and $D' = D \setminus F$.

By Theorem 2, f may be considered as a continuous function on D'. We assume that f is discontinuous on F, except, at most, on a finite number of points.

Let H be a bounded set of positive real numbers of which 0 is an accumulation point. Then, for any $h \in H$, we consider that the surface S is a finite union of skew quadrilaterals (occasionally triangles) of which a subset of the set of vertices is known and it makes up a system P_h of data points.

A priori, the form of the set F of discontinuity may be very complex, but, in order to simplify the modelization, we assume that F follows a rectangular geometry. More precisely, we suppose that:

- the boundary of D is formed by a finite number of segments which are parallel to the coordinate axis;

- there exists a family $(Q_h)_{h \in H}$ of triangulations of $\overline{D}$ made with rectangles such that, for all $h \in H$, F is a union of sides of rectangles belonging to Q_h and $f^{-1}(P_h) \subset D_h$, D_h being the set of vertices of Q_h.

This hypothesis is illustrated in Fig. 2.

The first question that must be solved is to find a parametrization for the data points, since the set D and the family $(D_h)_{h \in H}$ are unknown. In order to get a uniform parametrization, we assume the existence of:

- a bounded, connected, non empty, open set $\Omega \subset \mathbb{R}^2$, with Lipschitz-continuous boundary formed by segments which are parallel to the coordinate axis;

- a family $(\mathcal{T}_h)_{h \in H}$ of triangulations of $\overline{\Omega}$ made with equal squares with sides of length h;

- a set $\Phi \subset \overline{\Omega}$ which is, for all $h \in H$, a union of sides of squares belonging to $\mathcal{T}_h$.

For every $h \in H$, let B_h be the set of vertices of $\mathcal{T}_h$. Now, we suppose :

For all $h \in H$, there exists a bijection $\psi_h : \overline{\Omega} \to \overline{D}$ such that

$$\psi_h \in C^m(\overline{\Omega}, \mathbb{R}^2) \ , \ \psi_h^{-1} \in C^1(\overline{D}, \mathbb{R}^2), \tag{3.1}$$

$$\psi_h(B_h) = D_h \ , \ \Phi = \psi_h^{-1}(F). \tag{3.2}$$

For every $h \in H$, we set

$$g_h = f \circ \psi_h.$$

Note that the function g_h is a parametrization of S that belongs to $H^m(\Omega', \mathbb{R}^3)$, where $\Omega' = \Omega \setminus \Phi$. Besides, the data points have been uniformly parametrized, since $g_h^{-1}(P_h) \subset B_h$.

In the next two sections, we shall construct, for any $h \in H$, an approximating surface of S fitting its parametrization g_h, either by interpolation or by smoothing.

4 The Interpolation Method

Let h be a fixed element of H.

For all $b \in B_h$, let γ_b be the number of connected components of $S_b \setminus \Phi$, where $S_b = \cup\{ K \in \mathcal{T}_h \mid b \in K \}$. Let us denote every one of them by S_b^j, $j = 1, \ldots, \gamma_b$ (see Fig. 3).

For defining an interpolant of the surface S, we need that all the vertices of the skew quadrilaterals that compound S belong to P_h. Hence, there are γ_b points of P_h attached to every node $b \in B_h$.

Let V_h be the finite element space constructed on the triangulation $\mathcal{T}_h$ from the Bogner–Fox–Schmit generic finite element of class C^k, where $k \in \mathbb{N}, k \geq 1$. Let us remember that such element may be defined as a triple (K, P_K, Σ_K) (cf. [3]), where K denotes any square in $\mathcal{T}_h$, $P_K = Q_{2k+1}(K)$ and $\Sigma_K = \{v \to \partial^\alpha v(b) \mid b \in K \cap B_h, \alpha \in J_k\}$, being $J_k = \{\alpha \in \mathbb{N}^2 \mid |\alpha|_\infty = \leq k\}$.

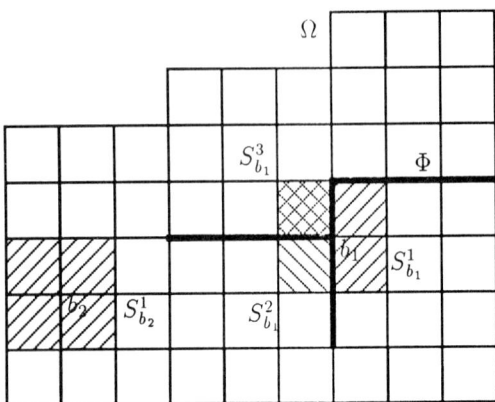

Figure 3. Examples of sets S_b^j

For all $b \in B_h$ and $\alpha \in J_k$, let w_b^α denote the basis function of V_h attached to the node b and the degree of freedom $v \to \partial^\alpha v(b)$.

For all $b \in B_h$, $j = 1, \dots, \gamma_b$ and $\alpha \in J_k$, let

$$w_b^{\alpha,j} = w_b^\alpha \cdot \chi_{\overline{S}_b^j},$$

where $\chi_{\overline{S}_b^j}$ is the characteristic function of $\overline{S}_b^j$, and let V_h' be the space generated by the family $(w_b^{\alpha,j})$. Note that $(V_h')^3 \subset H^{k+1}(\Omega', \mathbb{R}^3) \cap C_\Phi^k(\Omega', \mathbb{R}^3)$.

In this context, we search a Lagrange interpolant $A_h g_h$ of the parametrization g_h of the surface S belonging to $(V_h')^3$. So we set

$$A_h g_h = \sum_{b \in B_h} \sum_{j=1}^{\gamma_b} \sum_{\alpha \in J_k} g_{h,b}^{\alpha,j} \, w_b^{\alpha,j}. \tag{4.1}$$

To calculate the coefficients $g_{h,b}^{\alpha,j} \in \mathbb{R}^3$ of the basis functions, we associate to every node $b \in B_h$ a set of squares K_b^j, $j = 1, \dots, \gamma_b$, with sides of length $(2k+1)h$ and vertices in B_h, such that $\overset{\circ}{K}_b^j \cap \Phi = \emptyset$ and $\overset{\circ}{K}_b^j \cap \overset{\circ}{S}_b^j \neq \emptyset$ (see Fig. 4). Then, we define

$$\forall b \in B_h, \ \forall j = 1, \dots, \gamma_b, \ \forall \alpha \in J_k, \quad g_{h,b}^{\alpha,j} = \partial^\alpha \Pi_{K_b^j} g_h(b), \tag{4.2}$$

where $\Pi_{K_b^j} g_h \in Q_{2k+1}(K_b^j, \mathbb{R}^3)$ is the (vectorial) Lagrange interpolant of $g_h|_{\overset{\circ}{K}_b^j}$ on $B_h \cap K_b^j$.

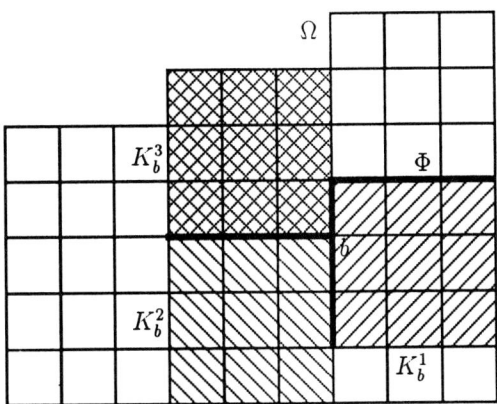

Figure 4. Squares K_b^j for $b \in \Phi$ and $k = 1$

Notice that, given $b \in B_h$, the coefficients $g_{h,b}^{(0,0),j}$ are, in fact, the coordinates of some data points belonging to P_h. Likewise, the expressions of $g_{h,b}^{\alpha,j}$, for $\alpha \in J_k \setminus \{(0,0)\}$, are finite difference formulæ to approximate the corresponding partial derivatives of g_h at the node b. For a full description of the formulæ needed for performing a C^1 or C^2 interpolation, see [10] or [12].

5 The Smoothing Method

Let h be again a fixed element of H.

Let $\gamma = \max_{b \in B_h} \gamma_b$. For $j = 1, \ldots, \gamma$, let $B_h^j = \{ b \in B_h \mid g_h|_{S_b^j}(b) \in P_h \}$ and set $\hat{B}_h = \{ b \in B_h^j \mid j = 1, \ldots, \gamma \}$, which may contain repeated points (belonging to Φ). For all $b \in \hat{B}_h$ and $v \in H^m(\Omega', \mathbb{R}^3)$, we define

$$v(b) = v|_{S_b^j}(b) \ , \ \text{if } b \in B_h^j.$$

Let $N = \operatorname{card} \hat{B}_h$ and (assuming that $\hat{B}_h$ is ordered) set

$$\rho^h v = (v(b))_{b \in \hat{B}_h} \in \mathcal{M}_{N \times 3}.$$

We further suppose that

$$(\rho^h v = 0) \implies (v = 0) \quad \forall v \in P_{m-1}(\Omega', \mathbb{R}^3). \tag{5.1}$$

Let $\tilde{T}_{\tilde{h}}$ be a 'sub-triangulation' of T_h (i.e., a triangulation of $\overline{\Omega}$ whose vertices are vertices of T_h) made with rectangles of diameter less than or equal to $\tilde{h}$ and such that Φ is union of sides of rectangles of $\tilde{T}_{\tilde{h}}$.

Let $\tilde{V}_{\tilde{h}}$ be the finite element space constructed on the triangulation $\tilde{\mathcal{T}}_{\tilde{h}}$ from the Bogner–Fox–Schmit generic finite element of class C^k and let $\tilde{V}'_{\tilde{h}}$ be the space obtained from $\tilde{V}_{\tilde{h}}$ by decomposing its basis functions as we did in section 4. Now, we assume

$$m \leq k + 1. \tag{5.2}$$

Therefore, $(\tilde{V}'_{\tilde{h}})^3 \subset H^m(\Omega', \mathrm{IR}^3) \cap C^k_{\Phi}(\Omega', \mathrm{IR}^3)$.
For any $\varepsilon > 0$, we set

$$J^h_\varepsilon(v) = \|\rho^h v - \rho^h g_h\|^2 + \varepsilon |v|^2_{m,\Omega',\mathrm{IR}^3} \quad \forall v \in H^m(\Omega', \mathrm{IR}^3),$$

where $\| \ . \ \|$ is the Euclidean norm in $\mathcal{M}_{N \times 3}$.

Then, a smoothing approximant of the parametrization g_h is obtained as the solution of the following minimization problem:

$$Find \ \sigma^h_{\varepsilon \tilde{h}} \in (\tilde{V}'_{\tilde{h}})^3 \ such \ that$$

$$J^h_\varepsilon(\sigma^h_{\varepsilon \tilde{h}}) \leq J^h_\varepsilon(v) \ \ \forall v \in (\tilde{V}'_{\tilde{h}})^3. \tag{5.3}$$

Theorem 3. *Problem* (5.3) *has a unique solution* $\sigma^h_{\varepsilon \tilde{h}}$, *the* $(\tilde{V}'_{\tilde{h}})^3$-*discrete smoothing* D^m-*spline relative to* $\hat{B}_h$, $\rho^h g_h$ *and* ε, *which is also the unique solution of the following problem : find* $\sigma^h_{\varepsilon \tilde{h}} \in (\tilde{V}'_{\tilde{h}})^3$ *satisfying*

$$< \rho^h \sigma^h_{\varepsilon \tilde{h}}, \rho^h v > + \varepsilon (\sigma^h_{\varepsilon \tilde{h}}, v)_{m,\Omega',\mathrm{IR}^3} = < \rho^h g_h, \rho^h v > \quad \forall v \in (\tilde{V}'_{\tilde{h}})^3, \tag{5.4}$$

where

$$\forall \xi = (\xi^j_i), \eta = (\eta^j_i) \in \mathcal{M}_{N \times 3}, \quad < \xi, \eta > = \sum_{i=1}^{N} \sum_{j=1}^{3} \xi^j_i \ \eta^j_i.$$

Proof We use the Lax–Milgram lemma and the equivalent norm $(\|\rho^h v\|^2 + |v|^2_{m,\Omega',\mathrm{IR}^3})^{1/2}$ (cf. [9]).

Now, let $M = M(\tilde{h})$ denote the dimension of $\tilde{V}'_{\tilde{h}}$ and let $(w_j)_{1 \leq j \leq M}$ be a basis of $\tilde{V}'_{\tilde{h}}$. Set

$$\sigma^h_{\varepsilon \tilde{h}} = \sum_{j=1}^{M} \alpha_j \ w_j,$$

with $\alpha_j \in \mathrm{IR}^3$, $j = 1, \ldots, M$. Let $b_1, \ldots, b_N$ be the points of $\hat{B}_h$ and consider the matrices:

$$A = (w_j(b_i)) \in \mathcal{M}_{N \times M} \ , \quad R = ((w_j, w_i)_{m,\Omega'}) \in \mathcal{M}_{M \times M}$$

where $(\ . \ , \ . \ )_{m,\Omega'}$ is the usual semi-norm of order m of the Sobolev space $H^m(\Omega')$. Then, we see that (5.4) is equivalent to the problem:

Find $\alpha = (\alpha_j)_{1 \leq j \leq M} \in \mathcal{M}_{M \times 3}$ *such that*

$$(A^T A + \varepsilon R)\alpha = A^T(\rho^h g_h). \tag{5.5}$$

Thus, the initial problem (5.3) has been reduced to solve three linear systems, each one of which has the same symmetric, positive definite matrix $A^T A + \varepsilon R$. Notice that $A^T A$ is the least-squares matrix associated to the basis (w_j).

6 Study of the Convergence

Under some complementary hypotheses about the family $(\psi_h)_{h \in H}$ of parameter changes introduced in (3.1) and (3.2), the convergence of the interpolation method described in section 4 may be established.

Theorem 4. *Assume that (3.1) and (3.2) are satisfied and suppose that*

$$\exists C > 0, \ \forall h \in H, \ \|\psi_h\|_{C^m(\overline{\Omega}, \mathbb{R}^2)} \leq C, \ \|\psi_h^{-1}\|_{C^1(\overline{D}, \mathbb{R}^2)} \leq C, \tag{6.1}$$

$$\forall x \in \overline{\Omega}, \ \exists \psi(x) = \lim_{h \to 0} \psi_h(x). \tag{6.2}$$

For every $h \in H$, let $A_h g_h$ the interpolant of g_h given by (4.1) and (4.2). Then, there exists a parametrization $g \in H^m(\Omega', \mathbb{R}^3)$ of the surface S such that

$$\lim_{h \to 0} \| g - A_h g_h \|_{\hat{m}, \Omega', \mathbb{R}^3} = 0,$$

where $\hat{m} = \min(m - 1, k + 1)$.

Proof By compactness arguments, we first show the existence of g as the (strong) limit of (g_h) in $H^{m-1}(\Omega', \mathbb{R}^3)$ when $h \to 0$. Then, using finite element results and standard techniques, we prove that there exists $C > 0$ such that, for all $h \in H$ and for $j = 0, \ldots, m'$,

$$|g_h - A_h g_h|_{j, \Omega', \mathbb{R}^3} \leq Ch^{m^* - j} \| f \|_{m^*, D', \mathbb{R}^3}$$

where $m' = \min(m, k + 1)$ and $m^* = \min(m, 2k + 2)$. The theorem follows from this inequality and the convergence of (g_h) to g.

With respect to the smoothing method, the reader is referred to [11], where the following result is proved.

Theorem 5. *Let m'' be an integer such that $m + 2 \leq m''$. Suppose that $f \in H^{m''}(D', \mathbb{R}^3)$ and that (3.1) and (6.1) are verified with m'' instead of* *m. Assume that the hypotheses (3.2), (5.1), (5.2) and (6.2) are satisfied and, lastly, suppose that*

$$\exists C > 0, \ \forall h \in H, \ \tilde{h} \le C \, h. \tag{6.3}$$

$$\lim_{h \to 0} \ \sup_{x \in \Omega'} \ \max_{b \in \hat{B}_h} ||x - b||_{\mathbb{R}^2} = 0 \tag{6.4}$$

Then, there exists a parametrization $g \in H^{m''}(\Omega', \mathbb{R}^3)$ of the surface S such that, for any $\varepsilon_0 > 0$, the solution $\sigma^h_{\varepsilon \tilde{h}}$ of (5.3) verifies

$$\lim_{\substack{0 < \varepsilon \le \varepsilon_0 \\ h \to 0, h^{l-m-1}/\varepsilon^{1/2} \to 0}} || \, g - \sigma^h_{\varepsilon \tilde{h}} \, ||_{m, \Omega', \mathbb{R}^3} = 0,$$

where $l = \min(m'', 2k + 2)$.

Remark *The hypotheses (3.1), (3.2), (6.1) and (6.2) are verified (cf. [10], [11]) under reasonable conditions that involve the monotonocity of the family (P_h) of data points and a certain kind of regularity of the triangulations (Q_h) of $\overline{D}$.*

7   Numerical Example

In order to test the interpolation and the smoothing methods, we have considered a surface S given by a parametrization $f = (f_1, f_2, f_3) : D \longrightarrow \mathbb{R}^3$, where $D = (-6, 6) \times (-6, 6)$ and

$$f_1(u, v) \ = \ \text{sgn}(u) \, g_1(|u|, |v|)$$

$$f_2(u, v) \ = \ \text{sgn}(v) \, g_2(|u|, |v|)$$

$$f_3(u, v) \ = \ \begin{cases} g_3(|u|, |v|) & \text{if } |u| + |v| < 7 \\[2mm] \text{sgn}(u) \, \text{sgn}(v) \, g_3(|u|, |v|) & \text{if } |u| + |v| \ge 7 \end{cases}$$

with

$$g_1(u, v) \ = \ \begin{cases} \frac{3}{2} + \frac{3}{2\pi} \cos(\pi(\frac{u+v}{3} + \frac{1}{2})) + \frac{u-v}{2} & \text{if } 0 \le u + v \le 3 \\[2mm] u & \text{if } 3 < u + v \le 7 \\[2mm] \frac{7}{2} + \frac{5}{2\pi} \cos(\pi(\frac{19}{10} - \frac{u+v}{5})) + \frac{u-v}{2} & \text{if } 7 < u + v \le 12 \end{cases}$$

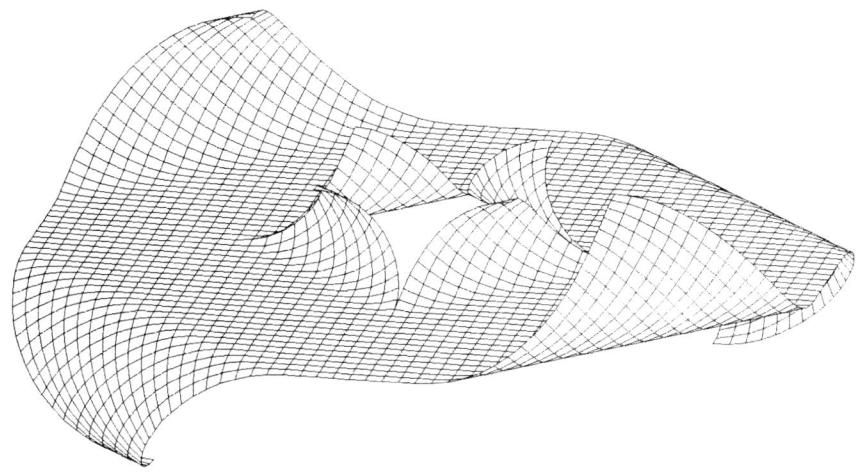

Figure 5. Sample surface S

$$
g_2(u,v) = \begin{cases}
\frac{3}{2} + \frac{3}{2\pi}\cos(\pi(\frac{u+v}{3} + \frac{1}{2})) - \frac{u-v}{2} & \text{if } 0 \le u+v \le 3 \\[2ex]
v & \text{if } 3 < u+v \le 7 \\[2ex]
\frac{7}{2} + \frac{5}{2\pi}\cos(\pi(\frac{19}{10} - \frac{u+v}{5})) - \frac{u-v}{2} & \text{if } 7 < u+v \le 12
\end{cases}
$$

$$
g_3(u,v) = \begin{cases}
\frac{3\sqrt{2}}{2\pi}(1 + \sin(\pi(\frac{u+v}{3} + \frac{1}{2}))) & \text{if } 0 \le u+v \le 3 \\[2ex]
0 & \text{if } 3 < u+v \le 7 \\[2ex]
\frac{-5\sqrt{2}}{2\pi}(1 - \sin(\pi(\frac{19}{10} - \frac{u+v}{5}))) & \text{if } 7 < u+v \le 12
\end{cases} \quad .
$$

This surface has been represented in Fig. 5.

Let us note that $f \in C_F^1(D', \mathrm{IR}^3) \cap H^2(D', \mathrm{IR}^3)$, where

$$F = \{(u,0) \in \mathrm{IR}^2 \mid |u| \le 3\} \cup \{(0,v) \in \mathrm{IR}^2 \mid |v| \le 3\}$$

and $D' = D \setminus F$. Notice that, in this case, $m = 2$.

We have set $H = \{\frac{1}{2^n} \mid n \in \mathrm{IN}\}$ and each triangulation $\mathcal{Q}_h$ of $\overline{D}$ is made with equal squares with sides of length h. So, the data points are regularly distributed on the surface S. We have also taken $\Omega = D$, $T_h = \mathcal{Q}_h$ and $g_h = f$.

In this situation, we have applied the two fitting methods for obtaining different approximating surfaces of class C^1 and C^2.

Three sets P_h of data points have been used, corresponding to $h = \frac{1}{2}, \frac{1}{4}$ and $\frac{1}{8}$. They contain, respectively, 648, 2448 and 9504 points.

For the smoothing method, we have considered three triangulations $\tilde{T}_{\tilde{h}}$ of $\overline{\Omega}$ made with 4×4, 8×8 and 12×12 equal–size squares. In the sequel, we refer to them as Grid 4, Grid 8 and Grid 12. Table 1 summarizes the dimension M of the corresponding finite element spaces $\tilde{V}'_{\tilde{h}}$. Likewise, the smoothing parameter ε has been fixed to $\varepsilon = 10^{-6}$.

Table 1. Dimension M of the space $\tilde{V}'_{\tilde{h}}$

k	Grid 4	Grid 8	Grid 12
1	112	352	720
2	252	792	1620

Table 2. Estimations of the error E^i_{rel} for the sample surface S

h	k	i	Smoothing D²-spline Grid 4	Grid 8	Grid 12	Interpolant
$\frac{1}{2}$	1	1	0.217×10^{-2}	0.223×10^{-3}	0.203×10^{-3}	0.700×10^{-4}
		2	0.223×10^{-2}	0.227×10^{-3}	0.207×10^{-3}	0.709×10^{-4}
		3	0.254×10^{-1}	0.486×10^{-2}	0.228×10^{-2}	0.255×10^{-2}
	2	1	0.322×10^{-3}	0.333×10^{-3}	0.365×10^{-3}	0.997×10^{-4}
		2	0.323×10^{-3}	0.336×10^{-3}	0.369×10^{-3}	0.101×10^{-3}
		3	0.115×10^{-1}	0.361×10^{-2}	0.381×10^{-2}	0.293×10^{-2}
$\frac{1}{4}$	1	1	0.196×10^{-2}	0.211×10^{-3}	0.559×10^{-4}	0.516×10^{-5}
		2	0.201×10^{-2}	0.216×10^{-3}	0.567×10^{-4}	0.522×10^{-5}
		3	0.241×10^{-1}	0.459×10^{-2}	0.162×10^{-2}	0.408×10^{-3}
	2	1	0.302×10^{-3}	0.317×10^{-4}	0.811×10^{-5}	0.583×10^{-5}
		2	0.301×10^{-3}	0.322×10^{-4}	0.780×10^{-5}	0.590×10^{-5}
		3	0.102×10^{-1}	0.201×10^{-2}	0.710×10^{-3}	0.395×10^{-3}
$\frac{1}{8}$	1	1	0.183×10^{-2}	0.206×10^{-3}	0.546×10^{-4}	0.314×10^{-6}
		2	0.187×10^{-2}	0.211×10^{-3}	0.554×10^{-4}	0.318×10^{-6}
		3	0.236×10^{-1}	0.449×10^{-2}	0.159×10^{-2}	0.674×10^{-4}
	2	1	0.292×10^{-3}	0.310×10^{-4}	0.777×10^{-5}	0.334×10^{-6}
		2	0.290×10^{-3}	0.319×10^{-4}	0.725×10^{-5}	0.338×10^{-6}
		3	0.956×10^{-2}	0.184×10^{-2}	0.613×10^{-3}	0.630×10^{-4}

In Table 2, we have detailed the relative error estimations that we have computed for each component of the corresponding parametrization g_h,

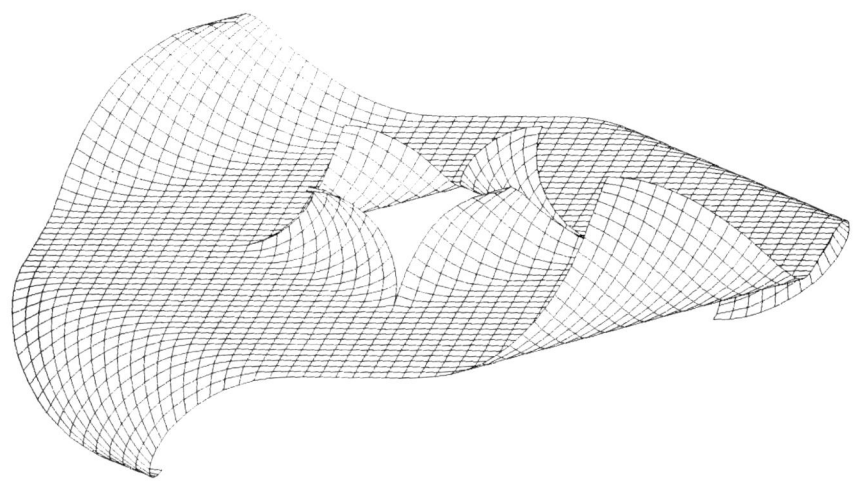

Figure 6. C^2 interpolant of S

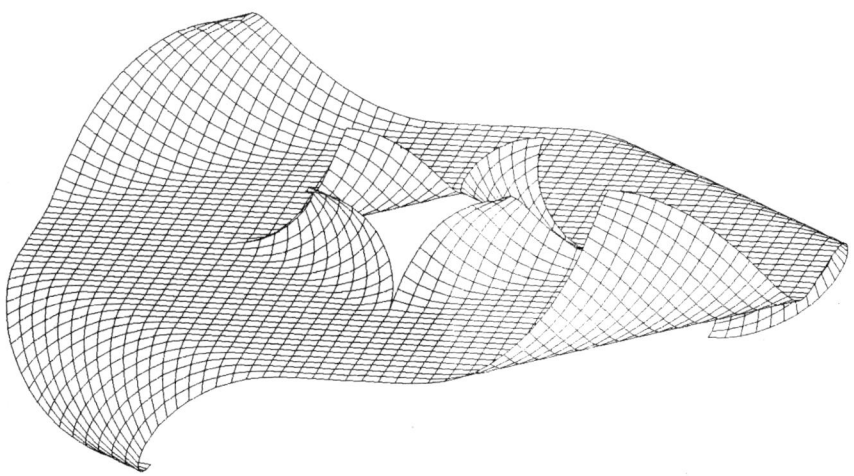

Figure 7. C^2 smoothing D^2-spline of S (for Grid 8 and $\varepsilon = 10^{-6}$)

using the following formula:

$$E_{rel}^i = \left(\frac{\sum\limits_{d \in B} |(p_i \circ g_h)(d) - (p_i \circ a_h)(d)|^2}{\sum\limits_{d \in B} |(p_i \circ g_h)(d)|^2} \right)^{\frac{1}{2}} , \quad i = 1, 2, 3,$$

where B is a set of 1000 points randomly distributed on $\overline{\Omega}$ and a_h is the fitting function of g_h ($A_h g_h$ for interpolation and $\sigma_{\varepsilon \tilde{h}}^h$ for smoothing).

Finally, Fig. 6 and Fig. 7 show an interpolant and a D^2-spline of the surface S, both ones calculated from the set $P_{\frac{1}{2}}$ of data points.

Acknowledgements

This work was supported in part by the Diputación General de Aragón (Spain) under project No. CB-5/88. Also, the first author was assisted by a grant from the Ministerio de Educación y Ciencia of Spain.

References

1. Apprato, D. (1987). Approximation de surfaces paramétrées par éléments finis. Thèse dÉtat, Université de Pauet des Pays de l'Adour.

2. Arcangéli, R. (1989). Some applications of discrete D^m splines. In *Mathematical Methods in Computer Aided Geometric Design*, Tom Lyche and Larry L. Schumaker, editors, Academic Press, 35–44.

3. Ciarlet, P.G. (1978). *The Finite Element Method for Elliptic Problems*, North–Holland.

4. Franke, R. and Nielson, G. (1983). Surface approximation with imposed conditions. In *Surfaces in CAGD*, R.E. Barnhill and W. Boehm, editors, North–Holland, 135–146.

5. Klein, P. (1987). Sur l'approximation et la représentation de surfaces explicites en présence de singularités. Thèse de 3e cycle, Université de Grenoble.

6. Laurent, P.J. (1986). Inf-convolution spline pour l'approximation de données discontinues. *RAIRO*, *20*, 89–111.

7. Mallet, J.L. (1989). Discrete smooth interpolation. *ACM Transactions on Graphics*, *8*, 121–144.

8. Manzanilla, R. (1986). Sur l'approximation de surfaces définies par une équation explicite. Thèse, Université de Pauet des Pays de l'Adour.

9. Necas, J. (1967). *Les Méthodes Directes en Théorie des Équations El-liptiques*, Masson.

10. Torrens, J.J. (1992), Intepolación de superficies paramétricas con discontinuidades mediante elementos finitos. Aplicaciones. Tesis doctoral, Publicaciones del Seminario Matemático Garcia de Galdeano, Serie II, Sección 2, No. 24, Universidad de Zaragoza.

11. Torrens, J.J. (1993), Approximation de surfaces parametrées présentant des discontinuités par D^m-splines d'ajustement discrètes. Publication UA 1204 CNRS No. 93/9.

12. Torrens, J.J., Serón, F.J., López de Silanes, M.C., Arcangéli, R. and Apprato, D. 1990. Interpolaciòn de superficies paramétricas no regulares mediante elementos finitos de Bogner–Fox–Schmit. In *Actas de las I Jornadas Zaragoza–Pau de Matemática Aplicada*, M.C. López de Silanes and B. Ycart, editors, Publicaciones del Seminario Matemático Garcia de Galdeano, Universidad de Zaragoza, 117-193..

Optimized Triangulation of Parametric Surfaces

A. Dolenc and I. Mäkelä

Helsinki University of Technology

1 Introduction

The triangulation of parametrically defined surfaces is, essentially, a problem of facetting which is well-known in CAD and approximation theory. Given a tolerance which determines the upper bound on the maximum Euclidian distance between the original surface and the facets (in our case, triangles) one can choose from a couple of methods.

The fastest method is to determine *a priori* where to split the patch so that tolerances are respected and is, therefore, the most suitable method for visualization purposes. The best algorithms to our knowledge are [12, 13, 14]. The algorithm described in [14] is the fastest since it only requires the control vertices whereas in [12, 13] it is necessary to compute bounds on higher order derivatives. On the other hand, using the latter methods might result in less triangles.

Another alternative is to use recursive subdivision associated with a flatness test (adaptive subdivision) [8]. Although it is slower than the ones mentioned above it has the advantage that it is much easier to implement and is more likely to minimize the number of triangles. We will present in Section 3 a variation of this method.

Finally, one can treat it as a degree reduction problem. However, this approach introduces unnecessary complications. Furthermore, they are relatively too slow, can generate an excessive number of patches and all the methods we know of, except [4], need to be extended to handle trimmed curves.

Regardless of the method used, one must handle the problem of cracks between adjacent patches. In our case, since the input is a collection of trimmed surfaces we must also deal with the problem of correctly triangulating the intersections between surfaces.

Our particular problem is to build an interface between CAD systems and the Stereolithography Apparatus (SLA). A SLA is a machine which permits rapid prototyping, and currently the interface requires that the parts to be machined be triangulated and that the resulting triangulation represent a solid. A broad overview of the manufacturing process can be found in [6] and its use in a specific application (injection moulding) in [5].

Besides representing a solid, the essential requirements on the triangulation are now presented.

Optimization: The number of triangles must be minimized, not that it affects significantly the manufacturing time, but because it reduces considerably the file sizes and because the preprocessor to the SLA machine has a low upper bound on the number of triangles it can handle (e.g. as low as $15000 \sim 17000$ triangles). This requirement induced us to use adaptive subdivision.

Robustness and Reliability: It is essential that the triangulation be *correct*. The benefits of manufacturing processes such as SLA are lost if the data is rejected by the receiving system or if it must be meddled with in order to be processed. There should be no holes between adjacent surfaces and patches. Furthermore, triangles must not intersect except at common vertices and edges (so-called vertex-to-vertex rule).

This requirement posed several problems because the boundaries between adjacent surfaces are described using trimmed curves which are normally the result of the (approximated) intersection between the given surfaces and may not even contain common points. Another problem is that a collection of surfaces may contain holes which are impossible to avoid. Finally, rounding errors can cause triangles to intersect each other. In spite of these problems, the algorithm should yield correct and acceptable results.

1.1   Some definitions

The term *surface* will be used meaning a rectangular array of four-sided *patches* expressed as tensor product polynomials. The implementation requires (i) that the patches be subdivided in a given direction, and (ii) that there exists a flatness test. Our current implementation uses the Bézier representation, and the input surfaces are described using the data exchange standards VDA-FS or IGES.

A *trimmed curve* $\vec{c}(t)$ on a surface $\vec{S}(u, v)$ is defined by two piecewise polynomials $\vec{u}(t)$ and $\vec{v}(t)$ such that $\vec{c}(t) = \vec{S}\left(\vec{u}(t), \vec{v}(t)\right)$. Two intersecting surfaces and the corresponding trimmed curves are illustrated in Figs. 1 and 2, respectively.

2   An Overview of the Algorithm

The *first stage* consists of gathering topological information, which may not be present in the input, and of discretizing the trimmed curves. This

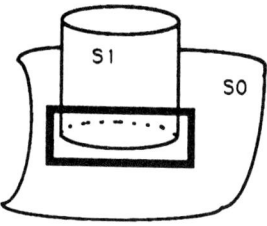

Figure 1. Two intersecting surfaces S_0 and S_1

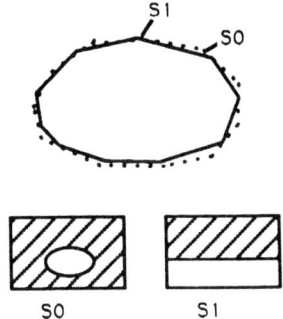

Figure 2. The trimmed curves in 3D and parameter space

preprocessing guarantees that associated to each surface-surface intersection there will be a unique piecewise linear curve in 3D space. We delay the description of this stage to Section 6.

The *second stage* uses adaptive subdivision on each patch in turn. Currently, the flatness test is the one described in [8], but plans are to use the method described in Section 3 since better results are expected. Patches are always split at the midpoint but the direction in which this takes place depends on the shape of the patch, and this information is currently derived from the control mesh. One could argue that midpoint subdivision is not a very good strategy for some patches notably degenerate ones (e.g.triangular patches) but we believe that the gains obtained by such optimizations do not justify the complexity added to the implementation.

Once patches become flat, the triangulation is performed in parameter space. When no trimmed curves are present in a flat domain an attempt is made to merge adjacent triangles. The triangulation algorithm is explained in Section 4 and the merging in Section 5.

The *third* and *final stage* consists of filling the holes which might exist between neighboring surfaces. Evidently, holes larger than a certain tolerance are errors. This stage is described in Section 6.4.

3 Adaptive Subdivision

Let $d(P_{uv}, T_{uv})$ be the distance between the patch P_{uv} and one of the two possible triangulations T_{uv} of P_{uv}. It is shown in [13] that $d(P_{uv}, T_{uv}) \leq \sigma M$, where σ is a measure of the domain of P_{uv} and M is an upper bound on the second order derivatives of the components of P_{uv}. In [12] a sharper bound is given but it requires that the domain be triangular.

Essentially, σ is very easy to compute but sharp bounds for M are not, especially in the case of rationals. For instance, the method suggested in [12] for integral polynomials requires the computation of polynomial roots. Therefore, an upper bound for M is normally computed for each patch and remains fixed thereafter. At each stage of the subdivision only σ is updated. If the intervals are always subdivided at, say, the midpoint then one obtains a regular subdivision of the domain of P_{uv}. Actually, it is then possible to decide how many times to subdivide the domain in order that tolerances are respected, avoiding the need of subdivision altogether.

On the other hand, in [8] at each stage of the subdivision a flatness test is performed which requires computing the distance between a plane and all the control vertices. Clearly this is much more expensive in terms of time and storage requirements.

We would like to combine both methods by computing good upper bounds for M at each stage of the subdivision which is not more expensive computationally than the flatness test described in [8]. The computation of M requires an upper bound for $\|\partial^2 P_{uv}/\partial u^2\|$, $\|\partial^2 P_{uv}/\partial v^2\|$, and $\|\partial^2 P_{uv}/\partial u \partial v\|$, where $\|\cdot\|$ denotes the uniform norm.

The easiest approach is to compute the bounds from the corresponding Bézier representation, but at each subdivision one must handle three surfaces for each component. There is, however, a cheaper route for integral polynomials which gives comparable bounds.

Instead, express the original patch P_{uv} in terms of the Chebyshev polynomials as shown in [9] (a comprehensive study of the properties of the Chebyshev polynomials can be found in [18]). P_{uv} becomes $P_{uv} = \sum_{ij} \lambda_{ij} T_i^u T_j^v$ where T_k is the Chebyshev polynomial of degree k defined over a suitable interval which we will assume here to be the unit interval, $[0, 1]$. Thus, $T_{[0,1]} = T_{[-1,1]}(2x - 1)$.

Next, using the fact that $\|T_k\| = 1$ and the bounds on derivatives discovered by the brothers Markoff (see [18] Section 2.7 or [10]) one can obtain good bounds for the second order derivatives of components P_{uv}^k of P_{uv}:

$$\|\partial^2 P_{uv}^k / \partial u^2\| \leq 2/3 \sum_{i=2,j=0}^{n,m} \left|\lambda_{ij}^k\right| i^2(i^2 - 1)$$
$$\|\partial^2 P_{uv}^k / \partial v^2\| \leq 2/3 \sum_{i=0,j=2}^{n,m} \left|\lambda_{ij}^k\right| j^2(j^2 - 1)$$
$$\|\partial^2 P_{uv}^k / \partial u \partial v\| \leq 4 \sum_{i=1,j=1}^{n,m} \left|\lambda_{ij}^k\right| i^2 j^2$$

where n and m are the degrees of P_{uv}^k in the u and v variable, respectively. (The constant 2 is due to the shifting of the Chebyshev polynomials onto the unit interval.)

At each stage of the subdivision it is possible to update the coefficients λ_{ij} so that the bounds on the derivatives can be re-evaluated. For simplicity, we demonstrate how this can be done using univariate polynomials (we were inspired by [9]). Let $p_n(u) = \sum_{i=0}^{n} \alpha_i T_i(u)$ be a polynomial of degree n defined on $[0,1]$, and let

$$p_n^+(u) = p_n(u/2)$$
$$p_n^-(u) = p_n^+(u + 1) = p_n((u + 1)/2).$$

It follows that $p_n^+(2u) = p_n(u)$ and expanding in terms of the Chebyshev polynomials one obtains $\sum_{i=0}^{n} \beta_i^+ T_i(2u) = \sum_{i=0}^{n} \alpha_i T_i(u)$, where $T_i = \sum_{j=0}^{i} t_{i,j} u^j$. Rewriting in terms of the power series one obtains

$$\sum_{j=0}^{n} u^j \left(\sum_{i=j}^{n} \beta_i^+ t_{i,j} 2^j \right) = \sum_{j=0}^{n} u^j \left(\sum_{i=j}^{n} \alpha_i t_{i,j} \right).$$

Finally, equating the coefficients of the monomials one obtains:

$$\beta_n^+ t_{n,n} 2^n = t_{n,n} \alpha_n$$
$$\beta_j^+ t_{j,j} 2^j = t_{j,j} \alpha_j + \sum_{i=j+1}^{n} (\alpha_i - \beta_i^+ 2^j) t_{i,j}$$

Fortunately, $t_{j,j}$ is of the form 2^e so the term $t_{j,j} 2^j$ on the left-hand side can be reduced to a shifting operation. The coefficients β_j^- can be computed similarly from the coefficients β_j^+ but it requires a little more work (the interested reader can consult the Appendix).

The Chebyshev coefficients t_{ij} can be precomputed and stored for later use. The three-term recurrence formula below computes the Chebyshev polynomials shifted onto the interval $[0,1]$:

$$T_0 = 1$$
$$T_1 = 2x - 1$$
$$T_n = 4x T_{n-1}(x) - 2T_{n-1}(x) - T_{n-2}(x), n > 1.$$

Notice that for practical values of n all computations for t_{ij} can be carried out in integer arithmetic.

This method gives us a flatness test which has the same order of time complexity as [8] ($O(nm)$) if one ignores the initial cost of the conversion to the Chebyshev basis, and it has similar storage requirements. The subdivision cost is also the same, $O(n^2m)$. The advantages are that, besides having good bounds for the derivatives at each stage of the subdivision (the Chebyshev have some nice extremal properties), it is also possible (i) to control in which direction the subdivision should take place, and (ii) which component should be subdivided. The latter is due to the fact that $M = \sum_{k=0}^{2} M^k$, where $M^k = \max\{\|\partial^2 P_{uv}^k/\partial u^2\|, \|\partial^2 P_{uv}^k/\partial v^2\|, \|\partial^2 P_{uv}^k/\partial u\partial v\|\}$.

The above method can be extended to rational polynomials but with the help of other representations. Let $r_n = p_n/q_m$, $n > m$, be a rational polynomial of degree n. In order to find an upper bound for $\|r_n\|$, write q_m in terms of, say, the Bézier basis. So $q_m = \sum_{j=0}^{m} w_j B_{m,j}$ and take $\xi = \min\{|w_j|\}$. If $\xi > 0$ then

$$\|r_n\| = \|p_n q_m^{-1}\| \leq \|p_n\|\|q_m^{-1}\| \leq \xi^{-1}\|p_n\|$$

where $\|p_n\|$ can be computed as explained above. Bounds for derivatives can be computed similarly. If $\xi = 0$ then more accurate methods for bounding $\|q_m\|$ from below must be used instead [12].

4   The Triangulation

It is not practical to optimize the triangulation across the entire surface. As a compromise, the current implementation triangulates rows or columns of patches. The direction is chosen initially depending on the 'look-and-feel' of the surface. For the sake of discussion lets assume one row r is taken at a time. Since the subdivision pattern of all patches $P \in r$ is stored, we have the opportunity to attempt to merge triangles across patch boundaries.

The intersections between the subdivision lines and the trimmed curves is computed and the trimmed curves of the other surfaces to which it is associated must be updated. This is necessary because if a triangle in a surface S_0 with an edge along a trimmed curve is split, then the adjacent triangle in surface S_1 must also be split in order to avoid cracks. The corresponding parameter value in surface S_1 is approximated using the procedure described in Section 6.2.

The parametric domain may take the shape of a polygon with holes when (discretized) trimmed curves are present. Several algorithms exist in the literature which triangulate polygons of all kinds. The one we prefer due to its simplicity is the one described in [17] Section VIII 4.2. It is as slow as the sorting algorithm at hand. The triangulation procedure must partition the row into regions which will be triangulated using such an algorithm and into regions in which the merging procedure will be applied.

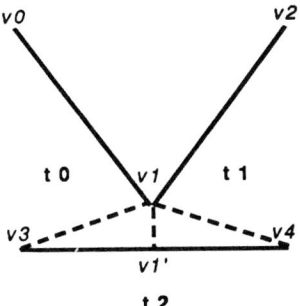

Figure 3. $v_3 v_1 v_4$ is a thin triangle in parameter space

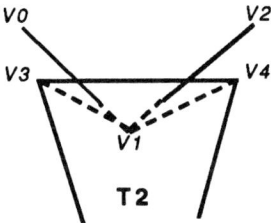

Figure 4. Intersecting triangles in 3D space

The merging procedure is described in the next section and it can be used whenever no trimmed curves are present in the domain.

Last, but not least, we would like to mention a problem caused by rounding errors and the approximation error, and which seems to occur to *any* thin triangle which contains a vertex on a trimmed curve. Let lower case characters denote points in parameter space and upper case characters points in 3D space. Consider, in parameter space, the trimmed curve and the subdivision line (or patch boundary) as shown in Fig. 3. Depending on the tolerances and how close v_1 is to the edge $v_3 v_4$, the edges $V_0 V_1$ and/or $V_1 V_2$ of triangles T_0 and T_1, respectively, will intersect the triangle T_2 (in 3D).

The solution we have adopted is to introduce a new point v_1' which is the projection of v_1 onto the segment $v_3 v_4$ as shown in Fig. 4. Another solution, which is used when v_1 is even closer and requires only one additional triangle but more work, is to delete the point v_1 and replace it by v_1'. In order to avoid inconsistencies with respect to the other surfaces which intersect this one along the trimmed curve, the point in 3D associated to the new vertex v_1' must be adjusted (we use the arithmetic average).

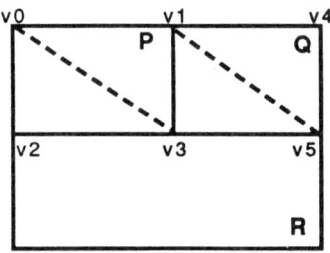

Figure 5.

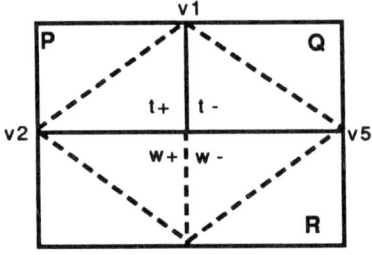

Figure 6.

5   Merging Triangles

Consider two adjacent flat patches P and Q with corners in parameter space (v_0, v_1, v_2, v_3) and (v_1, v_4, v_5, v_3) as shown in Fig. 5 and Fig. 6. Let the corresponding upper case characters, V_i denote the points in 3D space. Assume it is not known if the patch R is flat.

Diagonal lines suffice to generate the triangulation, but they may or may not share a common point (Figs. 6 and 5, respectively). If the triangle defined by the points $\{v_2, v_1, v_5\}$, satisfies tolerances then the two adjacent triangles t^+ and t^- — created when the diagonals share a common point — could be merged, and the midpoint v_3 could be discarded depending on what happens in R. This would result in two triangles less from the total (our models are assumed to be solids therefore all edges are common to exactly two triangles).

The *local merging* procedure will merge triangles (i) if R is not split at all, or (ii) if R is split once in the same direction and the triangles w^+ and w^- can also be merged (see Fig. 6).

The chances of merging depend on how well the original patch is split in the subdivision algorithm and the shape of the patch. The flatness test used here is similar to the one described in [8] except that only the control vertices covered by the triangle are used. This, evidently, does not guarantee tolerances but it is a sensible solution.

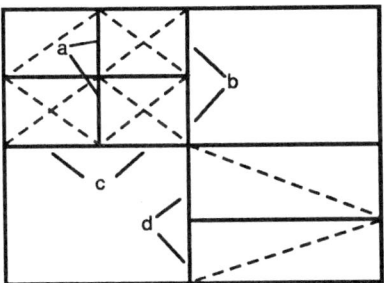

Figure 7. The indicated triangles allow merging

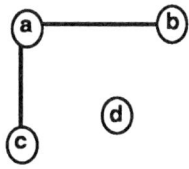

Figure 8. The interference graph

Further optimizations are possible if the triangulation is performed after the subdivisions are done. Given a patch P, the subdivision algorithm might split P as shown in Fig. 7. Let us assume that the line segment pairs a, b, c and d are the ones which allow triangle merging.

Observe that choosing a precludes b and we say that a *interferes* with b. An interference graph can be built (see Fig. 8) and the problem is then to find the *maximum independent set* (MIS) of the given graph. In practice, we have noticed that such interferences are rare and that a brute force method would be acceptable in finding the optimal solution. The method described in [16] Section 3.4 is easy to implement and is expected to exhibit running times which are independent of the size of the graph [2], although finding the MIS of a graph is NP-hard.

It is prohibitive to store all the control mesh information in order to perform a flatness test as described earlier. The best solution we have found so far is to associate to each triangle a normal and the error in the approximation. The additional error incurred by the merging is based on the angle between the normals. If tolerances are still respected then the two triangles are eligible for merging.

We do not attempt to merge more than two triangles because we believe the gains would be negligible.

6   Topology and Trimmed Curves

6.1   Discretizing the trimmed curves

Our interest lies on methods which discretize trimmed curves without using their explicit form, since otherwise one must handle polynomials of very high degree. We use a method [15] (Section 9.1.3) that consists of approximating the curve locally by the osculating circle and then using a truncated Taylor series expansion to obtain an approximate value for the parameter.

Let ε be the tolerance and ρ_i be the radius of curvature of the trimmed curve $\vec{c}$ at t_i. Then the identity $L_i = 2\sqrt{\varepsilon(2\rho_i - \varepsilon)}$ holds true, where $L_i = \vec{c}(t_i + \delta t_i) - \vec{c}(t_i)$, which, in turn, can be approximated by

$$\vec{c}(t_i + \delta t_i) - \vec{c}(t_i) \approx \sum_{j=0}^{k} \frac{(\delta t_i)^j}{j!} \frac{d\vec{c}^{(j)}(t_i)}{dt} + \kappa$$

with $\kappa = \max \|d\vec{c}^{(k+1)}/dt\|(\delta t_i)^{k+1}/(k+1)!$ which is hopefully small enough and, therefore, it may be discarded. The value $k = 2$ is the obvious choice since we need second order derivatives in any case to compute the radius of curvature. Provisions must be made if ρ_i is very large, so L_i must be bounded by some value, e.g. supplied by the user or by L_{i-1}. If the curvature increases abruptly the algorithm backtracks to $t_i + \delta t_i/2$ provided δt_i is not already too small. Points are tagged as knots whenever the behavior of the curvature changes. Finally, if μ is the machining tolerance and ν is the tolerance used when computing the intersection then $2\varepsilon \leq \mu - \nu$, with $\mu > \nu$. (In practice, ν is usually unknown. A rule-of-thumb is to use, by default, $\varepsilon = \mu/2$.)

6.2 Averaging the boundary trimmed curves

In order that no gaps exist between intersecting surfaces, nor that triangles intersect improperly, the trimmed curves associated to a common boundary must be modified, or rather coerced to match. The method we have adopted is to evaluate an average of the set of trimmed curves associated to a common boundary. In what follows, we use $\vec{P}_d$ and $\vec{Q}_d$ to refer to the discretized approximation of the original trimmed curve $\vec{P}$ and $\vec{Q}$, respectively.

We take a reference curve, say $\vec{P}_d$, preferably the one with the most number of line segments in the hope that it is more accurate. Then the following procedure is applied to each point $p_i \in \vec{P}_d$, where p_i are the Euclidian coordinates, and all other curves $\vec{Q}_d$ associated to the given boundary (in that order):

- Find the nearest point $q_i \in \vec{Q}_d$ to p_i. Usually, q_i will not be a line segment endpoint and so we must find the corresponding parameter values (u_{q_i}, v_{q_i}).

- The above inversion problem can be avoided by doing the following. Let $s = (a, b)$ be the line segment where q_i lies with corresponding parameter values $t = (t_a, t_b)$. Take t_q such that t is split in the same proportion as q_i splits s and use $\hat{q}_i = \vec{Q}(t_q)$ instead of q_i.

- All other points of $\vec{Q}_d$ are discarded.

Once the above 3D points and corresponding parameter values have been evaluated the average of all 3D points is taken retaining the corresponding parameter values. Evidently, the values of the curves at the parameter values is not of interest but only the computed average.

For practical purposes the above procedure should give satisfactory results and can be made fast using appropriate data structures. Since $\hat{q}_i$ is *not* the closest point in Q to p_i one cannot guarantee tolerances. In order to do so, one must refine t_q by splitting the line segment s and applying the method again until no substantial improvements are possible. However, we are of the opinion that such refinements are, in practice, unnecessary.

Finally, a data reduction procedure is used to eliminate redundant points generated by the discretizing procedure. It is similar to the one described in [19] §3.2.4 (with scan-along knot selection) except that some knots are already selected by the discretizing procedure (Section 6.1).

6.3   Gathering topological information

There are two problems we must handle with respect to topology: (i) it must be detected which trimmed curves are related to a common intersection, and (ii) whether adjacent surfaces have no gaps. The second problem is briefly discussed in the next section.

In order to bring together different trimmed curves 'around' a common boundary it is necessary (i) either to use a tolerance ϵ stating how near trimmed curves must be in order for them to be considered as describing the same boundary, or (ii) to ignore the trimmed curves from the input data and recompute the intersection(s). Clearly, we must try and avoid the latter alternative.

Let us assume that the machining tolerance μ is greater than the intersection tolerance ν, and that the trimmed curves are first discretized as described in Section 6.1. Then we want to compute the set $N(s, \epsilon)$ for each line segment s of the trimmed curve $c \in C$, where $t \in N(s, \epsilon)$ iff $\|s, t\|_2 \leq \epsilon \wedge t \notin c$. Again, since ν is usually unknown, a suitable value for ϵ is μ. When $\epsilon = \infty$ then $N(s, \infty)$ computes the nearest neighbor of s. The topology of the boundaries is derived from the sets $N(s, \epsilon)$.

There are two problems with this simple solution: (i) it must be implemented efficiently, and (ii) in practice, it will eventually happen that $\mu < \nu$ and it will fail to find the common boundaries.

Fortunately, the first problem can be solved by using data structures which provide fast access to geometrical data (e.g.the EXCELL system [1]). The second problem requires an 'ad hoc' solution if we do not want help from the user. If $\|s, s'\|_2 > \mu$, $s' = N(s, \infty)$ but the surfaces are not further apart than μ then both segments are assigned to the same boundary. Testing if the surfaces are close enough is very expensive.

6.4   Filling the holes

As mentioned earlier, a part may be composed of several surfaces some of which may be 'joined' manually, meaning that the distance between both surfaces is relatively small by still in the order of ≈ 0.5mm. In contrast, trimmed curves may be as close as ≈ 0.005mm. Therefore, it is not acceptable to treat this case as we did with trimmed curves because re-arranging the boundaries may upset tolerances significantly over potentially large areas. The solution is then to *somehow* add triangles.

Since this topic is peripheral to the subject of this article we will only outline one possible solution. Evidently, the first problem is to find the holes, but it is more interesting to fill them.

The curves along the holes are either trimmed curves or patch boundaries. In either case, they have been discretized and are, therefore, piecewise linear. One solution is to enumerate all possible triangulations and take the one in which (i) triangles do not intersect and (ii) the area is minimized. The search space is clearly too large and one must resort to heuristic methods. The reader is directed to [3] and references therein for further details.

7 Conclusions

In this paper we presented the design and implementation of a triangulation algorithm for trimmed parametric surfaces applied to manufacturing. The main contributions of this paper are:

- the handling of surface intersections,

- the optimization of the number of triangles produced,

- a different recursive subdivision scheme,

- and the discussion of implementation issues.

We believe that the problem we were given as stated in the Introduction will be satisfactorily solved using the approach presented here, but final judgement must wait until the implementation is completed.

Acknowledgements

We wish to thank Charles Woodward for his encouragement and useful discussions and suggestions. This work has been partially financed by the Technology Development Center of Finland (TEKES).

References

1. Tamminen, M. and Sulonen, R. (1982). The EXCELL method for efficient geometric access to data. *ACM/IEEE 19th Design Automation Conference*, 345–351.

2. Bron, C. and Kerbosch, J. (1973). Algorithm 457: finding all cliques of an undirected graph. *Comm. ACM*, *16*, 575–577.

3. Ganapathy, S. and Dennehy, T.G. (1982). A new general triangulation method for planar contours. *Computer Graphics*, *16*, 69–74.

4. Hoschek, J. and Schneider, F.-J. (1989). Spline conversion for trimmed rational Bézier- and B-spline surfaces. Technische Hochschule Darmstadt, Fachbereich Mathematik.

5. Weiss, L.E., Gursoz, E.L., Prinz, F.B. Fussel, P.S., Mahalingam, S. and Patrick, E.P. (1990). A rapid tool manufacturing system based on stereolithography and thermal spraying. *Manufacturing Review, 3,* 40–48.

6. Deitz, D. (1990). Stereolithography automates prototyping. *Mechanical Engineering*, 34–39.

7. Garey, M.R., Johnson, D.S., Preparata, F.P. and Tarjan, R.E. (1978). Triangulating a simple polygon. *Information Processing Letters, 7,* 175–179.

8. Lane, J.M. and Riesenfeld, R.F. (1980). A theoretical development for the computer generation and display of piecewise polynomial surfaces. *IEEE Trans. on Pattern Analysis and Machine Intelligence, PAMI-2,* 35–46.

9. Watson, M.A. and Worsey, A.J. (1988). Degree reduction of Bézier curves. *Computer-Aided Design, 20,* 398–405.

10. Duffin, R.J. and Schaeffer, A.C. (1941). A refinement of an inequality of the Brothers Markoff. *Trans. Amer. Math. Soc., 50,* 517–528.

11. Cody, W. J. (1988). Algorithm 665, MACHAR: a subroutine to dynamically determine machine parameters. *ACM Transactions on Mathematical Software, 14,* 303–311.

12. Filip, D., Magedson, R. and Markot, R. (1986). Surface algorithms using bounds on derivatives. *Computer Aided Geometric Design, 3,* 295–311.

13. Lane, J.M. and Carpenter, L. (1979). A generalized scan line algorithm for the computer display of parametrically defined surfaces. *Computer Graphics and Image Processing, 11,* 290–297.

14. Rockwood, A. Heaton, K. and Davis, T. (1989). Real-time rendering of trimmed surfaces. *Computer Graphics, 23,* 107–116.

15. Faux, I.D. and Pratt, M.J. (1979). *Computational Geometry for Design and Manufacture*, Ellis Horwood Ltd.

16. Christofides, N. (1975). *Graph Theory: An Algorithmic Approach*, Academic Press.

17. Mehlhorn, K. (1984). *Data Structures and Algorithms 3: Multidimensional Searching and Computational Geometry*, Springer-Verlag.

18. Rivlin, T.J. (1974). *The Chebyshev Polynomials*, John Wiley & Sons, Inc.

19. Reeves, W.T. (1980). Quantitative representations of complex dynamic shape for motion analysis, PhD Thesis, University of Toronto, Dept. of Computer Science.

A1 Computing β_j^-

The objective is to expand $p_n^-(u-1) = \sum_{i=0}^n \beta_i^- T_i(u-1)$ in terms of the power series so we can equate the coefficients with those of $p_n^+(u) = \sum_{i=0}^n \beta_i^+ T_i(u)$.

$$\sum_{i=0}^n \beta_i^- \left(\sum_{j=0}^i t_{i,j}(u-1)^j \right) = \sum_{i=0}^n \beta_i^- \left(\underbrace{\sum_{j=0}^i t_{i,j} \sum_{k=0}^j u^k (-1)^{j-k} \binom{j}{k}}_{(i)} \right) =$$

$$\sum_{i=0}^n \beta_i^- \sum_{k=0}^i u^k \underbrace{\sum_{j=k}^i t_{i,j}(-1)^{j-k} \binom{j}{k}}_{f_{i,k}} = \sum_{i=0}^n u^i \left(\sum_{j=i}^n \beta_j^- f_{j,i} \right)$$

$$\underbrace{\qquad\qquad\qquad\qquad\qquad}_{(i)}$$

Equating the coefficients of the power series of both polynomials we obtain the desired relation between β_i^- and β_i^+:

$$\beta_i^- = \beta_i^+ + 2^{-e} \left(\sum_{j=i+1}^n \beta_j^+ t_{j,i} - \sum_{j=i+1}^n \beta_j^- f_{j,i} \right)$$

where $2^e = f_{i,i} = t_{i,i}$ for some suitable e, and $\beta_n^- = \beta_n^+$. The terms $f_{j,i}$ must be precomputed and stored for later use.

A Symmetric Domain for a 6-sided Patch

M.A. Sabin

FEGS Limited, Cambridge

1 Symmetric Formulation of n-sided Patch Equations

It should be possible to describe any n-sided patch in terms which are symmetric with respect to the n sides. One way of doing this is to use n parameters, each of which is in some sense a 'distance' from one of the sides. The basis functions from which a surface is built are then simple functions of these distances.

The simplest example is the use of homogeneous coordinates within a triangle.

a, b and c are 'distances' from the three sides relative to the height of the triangle.

$$P = Aa + Bb + Cc$$

expresses in a symmetric way the equation for a generic point in terms of the variable parameters and the control points A, B and C of a linear triangle (see Fig.1).

Because a surface is bivariate, these distances are not independent; there are $n - 2$ normalisation equations which link them. In the case of the triangle $n = 3$, and so we need $3 - 2 = 1$ normalisation equation.

$$a + b + c = 1.$$

The normalisation equations themselves have to exhibit an n-fold rotational symmetry. In the case of the triangle we can deduce symmetrically rotated versions

$$\begin{aligned} b + c + a &= 1 & (1.1) \\ c + a + b &= 1 & (1.2) \end{aligned}$$

from the original.

We can also deduce the mirror symmetry versions

$$c + b + a = 1 \qquad (1.3)$$

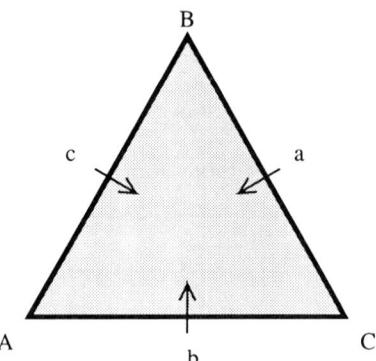

Figure 1.

$$b + a + c \;=\; 1 \tag{1.4}$$
$$a + c + b \;=\; 1. \tag{1.5}$$

The symmetry group required is that which maps edges into edges and adjacent edges into adjacent edges over a closed loop of edges.

In the four-sided case, the normalisation equations for standard bi-Bezier patches are

$$a + c \;=\; 1 \tag{1.6}$$
$$b + d \;=\; 1. \tag{1.7}$$

from which can be derived the symmetric forms

$$c + a \;=\; 1 \tag{1.8}$$
$$d + b \;=\; 1. \tag{1.9}$$

The relationship to normal notation is just

$$a \;=\; u \tag{1.10}$$
$$b \;=\; v \tag{1.11}$$
$$c \;=\; 1 - u \tag{1.12}$$
$$d \;=\; 1 - v. \tag{1.13}$$

The familiar basis functions just become terms like, for example, see Fig. 2

$$3a^2cd^3.$$

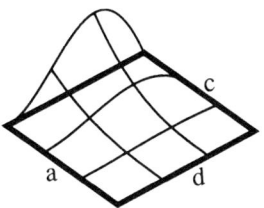

Figure 2.

Note that in both the above cases, each distance is zero on one edge, varies linearly along the two adjacent edges, and takes the value one on the remaining 'opposite' edges (see Fig. 3). This boundary condition is a property which helps to make it easy to find simple basis functions for which it is easy to derive continuity conditions across the patch edges. This is the major difference from the approach taken by Gregory and others [2],[4].

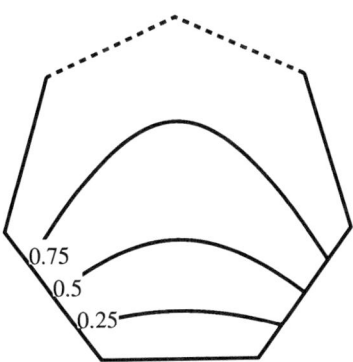

Figure 3.

Another property we require is that it should be possible to use the normalisation equations to compute all of the parameters from any adjacent pair by a well-conditioned, unambiguous, cheap calculation.

The ideal is a simple rational expression.

2   What About Patches of More Than Four Sides ?

We need to identify a set of normalisation equations satisfying both the symmetry conditions and the boundary conditions. Once that is found, a set of basis functions is then needed to give a patch formulation.

The set of normalisation equations is the difficult thing to find. I term such a set, taken together with the bounds $0 \leq a, b, \ldots \leq 1$, a 'symmetric domain'. The domain itself is a bounded region of a bivariate manifold in n-space.

Deriving basis functions is a more mechanistic process. Two heuristics for doing this are

- to try raising a form of the normalisation equations symmetric in all the variables to an appropriate power, and then separate the terms. This is the way the triangle gives basis functions.

- to find a first order basis, raise the sum of the basis functions $(= 1)$ to an appropriate power, and then separate the terms. This works for the square.

In both cases some subsequent fiddling is required to get a basis with the required boundary properties, but this is not particularly difficult.

3   Five Sides

In [1] and [3] I described normalisation equations for a five-sided region

$$
\begin{array}{rcll}
a + cd & = & 1 & \text{(3.1)} \\
b + de & = & 1 & \text{(3.2)} \\
c + ea & = & 1. & \text{(3.3)}
\end{array}
$$

These have all the required boundary properties, and it is interesting to see how simply the symmetry properties may also be shown.

$$
\begin{aligned}
1 &= a + cd & \text{by 3.1} \\
&= a + (1 - ea)d & \text{by 3.3} \\
&= a + d - ead \\
&= a + d - eda \\
&= a + d - (1 - b)a & \text{by 3.2} \\
&= a + d - a + ab \\
&= d + ab = 1.
\end{aligned}
$$

It is left as a really easy exercise for the reader to derive $e + bc = 1$.

4 Six Sides

The result now being reported is that the normalisation equations

$$
c^2(1 - ab)(1 - 2ab) + c(2a - 3a^2b + ab^2) + a^2 = 1 \tag{4.1}
$$
$$
d^2(1 - bc)(1 - 2bc) + d(2b - 3b^2c + bc^2) + b^2 = 1 \tag{4.2}
$$
$$
e^2(1 - cd)(1 - 2cd) + e(2c - 3c^2d + cd^2) + c^2 = 1 \tag{4.3}
$$
$$
f^2(1 - de)(1 - 2de) + f(2d - 3d^2e + de^2) + d^2 = 1 \tag{4.4}
$$

form a set with all the required symmetries and boundary properties.

Confirming the boundary properties is a simple exercise left to the reader.

Confirming the symmetry properties is far from trivial. The annotated transcript of a Reduce session is included in the next section.

This symmetric domain does not have the ideal rational computation property. However, it was shown in [5] that such a property is not possible for $n > 5$. The computation of all of the remaining four parameters from any adjacent pair requires only one square root in addition to rational calculations. The square root required is always the positive one. This can therefore reasonably be claimed to be one of the simplest possible solutions.

5 Demonstration of Symmetry. Annotated Reduce Session

REDUCE Development Version, 5-Sep-90 ...

```
1: procedure φ(a, b, c);
1: c*c*(1−a*b)*(1−2*a*b)+c*(2*a−3*a*a*b+a*b*b)+a*a−1;
2: resultant(φ(a, b, c), φ(b, c, d), c);
```

$4*A^4*B^8 - 2*A^4*B^7*D - 2*A^4*B^6*D^2 - 8*A^4*B^6 + 7*A^4*B^5*D + A^4*B^4*D^2 + 4*A^4*B^4 - 5*A^4*B^3*D + A^4*B^2*D^2 + 6*A^3*B^8*D + 6*A^3*B^7*D^2 - 12*A^3*B^7 - 21*A^3*B^6*D + 3*A^3*B^5*D^2 + 24*A^3*B^5 + 9*A^3*B^4*D - 9*A^3*B^3*D^2 - 12*A^3*B^3 + 6*A^3*B^2*D + 2*A^2*B^8*D^2 +$

$4*A^2*B^7*D^3-8*A^2*B^7*D+2*A^2*B^6*D^4-29*A^2*B^6*D^2+13*A^2*$
$B^6-12*A^2*B^5*D^3+41*A^2*B^5*D-A^2*B^4*D^4+28*A^2*B^4*D^2-26*$
$A^2*B^4+6*A^2*B^3*D^3-31*A^2*B^3*D-A^2*B^2*D^4-A^2*B^2*D^2+13*$
$A^2*B^2+2*A^2*B*D^3-2*A^2*B*D-6*A*B^6*D^3+3*A*B^6*D-6*A*$
$B^5*D^4+21*A*B^5*D^2-6*A*B^5+24*A*B^4*D^3-27*A*B^4*D+6*$
$A*B^3*D^4-27*A*B^3*D^2+12*A*B^3-18*A*B^2*D^3+24*A*B^2*D+$
$6*A*B*D^2-6*A*B+4*B^4*D^4-5*B^4*D^2+B^4-6*B^3*D^3+6*B^3*$
$D-5*B^2*D^4+7*B^2*D^2-2*B^2+6*B*D^3-6*B*D+D^4-2*D^2+1$

3: factorize(ws);
{$B-1,$
$B+1,$

$4*A^2*B^4+2*A^2*B^3*D-4*A^2*B^2+A^2*B*D+2*A*B^4*D+$
$2*A*B^3*D^2-4*A*B^3-11*A*B^2*D+A*B*D^2+4*A*B-$
$4*B^2*D^2+B^2+6*B*D+D^2-1,$
$A^2*B^2-A^2*B*D+A*B^2*D+A*B*D^2-2*A*B-D^2+1/$

At this point the first magic occurs in that the fourth factor of this big expression satisfies the boundary conditions. The other factors are therefore regarded as being spurious and are thrown away.

4: part(ws,4);
$A^2*B^2-A^2*B*D+A*B^2*D+A*B*D^2-2*A*B-D^2+1$
5: procedure $\psi(a,b,d)$;
5: $a*a*b*b-a*a*b*d+a*b*b*d+a*b*d*d-2*a*b-d*d+1;$

This is just a copy of the factor picked out above, and reinput because I was not familiar enough with Reduce to be any cleverer.

6: resultant($\phi(a,b,c),\psi(b,c,e),c$);

6: $2*A^4*B^6*E^2+2*A^4*B^5*E^3-A^4*B^5*E-3*A^4*B^4*E^2-2*A^4*B^3*$
$E^3+A^4*B^3*E+A^2*B^2*E^2-2*A^3*B^6*E^3+3*A^3*B^6*E+2*A^3*B^5*$
$E^4-3*A^3*B^5*E^2-2*A^3*B^4*E^3-A^3*B^4*E-2*A^3*B^3*E^4+3*A^3*$
$B^3*E^2+4*A^3*B^2*E^3-2*A^3*B^2*E-3*A^2*B^6*E^2+A^2*B^6+6*A^2*$
$B^5*E^3-7*A^2*B^5*E-5*A^2*B^4*E^4+10*A^2*B^4*E^2-2*A^2*B^4-4*A^2*$
$B^3*E^3+5*A^2*B^3*E+5*A^2*B^2*E^4-7*A^2*B^2*E^2+A^2*B^2-2*A^2*$
$B*E^3+2*A^2*B*E-A*B^6*E+3*A*B^5*E^2-2*A*B^5-6*A*B^4*E^3+$
$9*A*B^4*E+4*A*B^3*E^4-9*A*B^3*E^2+4*A*B^3+6*A*B^2*E^3-8*$
$A*B^2*E-4*A*B*E^4+6*A*B*E^2-2*A*B-B^4*E^2+B^4+2*B^3*E^3-$
$2*B^3*E-B^2*E^4+3*B^2*E^2-2*B^2-2*B*E^3+2*B*E+E^4-2*E^2+1$

7: factorize(ws(6));
{$B-1,$
$B+1,$

$A^2 * B^2 + A^2 * B * E - A * B^2 * E + A * B * E^2 - 2 * A * B - E^2 + 1,$
$2 * A^2 * B^2 * E^2 - A^2 * B * E + 3 * A * B^2 * E - 3 * A * B * E^2 + B^2 - 2 * B * E + E^2 - 1\}$

Again, the third factor satisfies the boundary conditions on its own, and the rest of the resultant is discarded.

8: part(ws(7),3);
8: $A^2 * B^2 + A^2 * B * E - A * B^2 * E + A * B * E^2 - 2 * A * B - E^2 + 1$
9: ws(8)-$\psi(b, a, e)$;
9: 0

Here is the gist of the demonstration. e is shown to have the same relationship to b and a, as d had to a and b.

One approach to completing the symmetry demonstration would be to continue in like fashion right around.

However, these results are a sufficient indicator to take a short cut which also gives a lot of other results useful in generating basis functions.

We have, in $\phi(a, b, c)$ and $\psi(a, b, d)$ quadratic equations for c and d in terms of a and b, and the above results indicate that we can confidently use $\phi(b, a, f)$ and $\psi(b, a, e)$ as equations for e and f.

Schoolboy solution of these quadratics gives

$$D(a,b) = a^2 b^2 (b-a)^2 - 4(ab-1)^3 \tag{5.1}$$
$$T = \sqrt{D} \tag{5.2}$$
$$c = 2(1-a^2)/(T + (2a - 3a^2 b + ab^2)) \tag{5.3}$$
$$f = 2(1-b^2)/(T + (2b - 3ab^2 + a^2 b)). \tag{5.4}$$

If $a \geq b$

$$d = 2(1-ab)^2/(T + (a-b)ab) \tag{5.5}$$
$$e = (T + (a-b)ab)/2(1-ab). \tag{5.6}$$

If $a \leq b$

$$d = (T + (b-a)ab)/2(1-ab) \tag{5.7}$$
$$e = 2(1-ab)^2/(T + (b-a)ab). \tag{5.8}$$

From these comes directly the quadrivariate relationship

$$de = 1 - ab$$

in which the necessary symmetry is very obvious.

It is then possible to work further to derive much more internally symmetric forms

$$ab + bc + cd + de + ef + fa = 3.$$

$$2(abc + bcd + cde + def + efa + fab) = a + b + c + d + e + f$$

$$abcd + bcde + cdef + defa + efa + fabc = ad + be + cf = 1 + 4abcdef.$$

6 First Order Basis Functions

From these in turn can be derived a first order basis function set containing functions of the form

$$p_{ef}(a, b, c, d, e, f) = abcd + 2abcdef/3.$$

Each such function is zero at all vertices but one, which has the value unity. Along the edges abutting the unit value the basis function varies linearly, and on the other edges it is identically zero, (also a linear form). The $abcdef$ term, which does not influence the values on the boundary is added so that the six functions sum to unity.

It is not proven that this is the only 6-sided symmetric domain. There are certainly alternative triangles, since the form

$$a + b + c + \lambda abc = 1$$

satisfies all the symmetry and boundary conditions, and, indeed, $\lambda = 2$ gives advantage in setting up a 'suitcase corner' patch.

It is not proven that this is in any sense the simplest 6-fold symmetric domain, nor is it claimed that this first order basis is unique or optimal.

Acknowledgements

I am very grateful to James Davenport for introducing me to Reduce, and indeed driving it for me the first time I derived these results and for giving me access to his equipment to reproduce them for this paper.

References

1. Sabin, M.A. (1968). Parametric surface equations for non-rectangular regions. BAC Weybridge report VT0/MS/147.

2. Gregory, J.A. and Charrot, P. (1980). A C1 triangular interpolation patch for computer-aided geometric design. *CGIP*, *13*, 80–87.

3. Sabin, M.M. (1983). Non-rectangular surface patches suitable for inclusion in a B-spline surface. In *Proc. Eurographics 83*, T. Hagen, Editor, Elsevier, pp. 57–69.

4. Charrot, P. and Gregory, J.A. (1984). A pentagonal patch for CAGD. *CAGD*, *1*, 87–94.

5. Sabin, M.A. (1986). Some negative results in N-sided patches. *CAD*, *18*, 38–44.

Computing Gröbner Bases over Rational Function Fields

Stefan Arnborg[1] and Joachim Hollman[2]

[1]*Department of Numerical Analysis and Computing Science,
The Royal Institute of Technology, Stockholm, Sweden and*
[2]*Department of Mathematics, Stockholm University,
Stockholm, Sweden*

1 Introduction

We have studied the computation of Gröbner bases in $\mathbf{Q}(\bar{p})[\bar{x}]$, *i.e.*, with coefficients in rational function fields. Such bases have potential application in various practical geometry problems, such as inverse kinematics in robotics [3] and recovery of 3D geometry from 2D images [1], [8], [15], [18].

It does not take a lot of experimentation to find out that the problem is quite hard. The coefficients, if regarded as expanded polynomials, are enormous in size. The available Gröbner base routines grind to a halt computing g.c.d.s of large intermediate polynomials (which hardly ever gives a result that is not expected). For coefficients in $\mathbf{Z}$, natural problems have a tendency to yield non-trivially factoring H-polynomials which permits use of the efficient decomposition method developed by Davenport [6] and Melenk, Möller and Neun [19]. Natural problems over function fields with many parameters have no such simplification possibilities except when there are symmetries in the problem, and typically these are not enough to make the solution possible in practice.

The solution we tried, and in fact the only solution that appears feasible, is the idea of straight-line program (*slp*) or black box representation of intermediate polynomials [16], [17]. With these techniques it turns out to be feasible to obtain fast evaluation of coefficients for quite large problems (*slp* approach) or to obtain reasonably fast evaluation of coefficients for any problem for which it is feasible to execute Buchberger's algorithm over a finite coefficient field (black box approach).

2   Straight-Line Program Coefficient Representation

Buchberger's algorithm requires very fast access to monomials so we used the *slp* representation only for coefficients, which with the normalization used is polynomials in $\mathbf{Z}[\bar{p}]$.

Although the *slp* representation supports the g.c.d. operation that is normally used to extract the contents of reduced polynomials in Buchberger's algorithm, we have found that this is not advantageous. The reason is that the g.c.d. and cofactors can have rather long *slp* representations in terms of the original polynomials. It turns out that the *slp* representation can be used to find most non-trivial g.c.d.s between polynomials by keeping a factored stub of them. Thus, we assume that f and g have no non-trivial g.c.d. except as indicated by the factorization patterns of f and g (or rather, these are the only common factors we look for). As an example: if we have two coefficient polynomials $c' = t_{17}^5 t_{18}^2$ and $c'' = -t_{17}^5 t_{19}^2$, then $c' + c''$ will be represented by $t_{17}^5 t_{20} t_{21}$, where $t_{20} = t_{18} - t_{19}$ and $t_{21} = t_{18} + t_{19}$ are new variables.

We have checked experimentally that this assumption does not cause a significant blow-up on medium-sized problems (it is not possible to check it on large problems). With this technique we have a method analogous to the GECD [11]: we do not find all g.c.d.s of coefficient polynomials, but enough of them to avoid the explosion of coefficient sizes consequent on ignoring the problem altogether. With this technique the only polynomial operation we must be able to perform decently on *slp* polynomials is to test them for zero.

The implementation of the above ideas in Maple [4] is straightforward. The zero test can be done probabilistically [13], [14] and is not difficult to implement. A few changes were made to Maple's Gröbner package [5] and Gonnet's zero test routine was modified [13] to handle *slp* coefficients. We also had to implement a new version of Maple's expand routine. The new expand routine replaces the expression to be expanded with a polynomial collected in the x_i and the coefficients (in $\mathbf{Z}(p)$) are handled so that each irreducible component of the expression is replaced by a new variable. The zero test is accomplished by successively replacing the signatures of new temporary variables by the signatures of the expressions assigned to them.

As we began to experiment with truly large problems the number of intermediate variables and the size of signature tables became large and the need for a garbage collector became apparent. We therefore added a simple *slp* garbage collector.

These changes to the original Maple code reduced both the run-time (often two orders of magnitude) and memory requirements considerably.

3   Black Box Representation of Coefficients

Kaltofen and Trager's concept of a black box representation [17] is an algorithm which, given values for indeterminates, allows one to compute the value of a function. For our purpose, Buchberger's algorithm with any basis over $\mathbf{Q}(\bar{p})[\bar{x}]$ can be taken as such a representation. The only snag with this choice is that a specialization of the parameters $\bar{p}$ to rational values $\bar{v}$ typically leads to a basis with very large coefficients that are very uncertain if the values are noisy measurements. On the other hand, it is not possible to execute Buchberger's algorithm numerically because the termination criterion is an exact test for zero (of a number of reduced S-polynomials).

It is possible to overcome this problem using Traverso's [20] concept of generic computation traces (used for relating a computation over $\mathbf{Q}$ to a computation over $\mathbf{Z}_q$). That is, with controllably high probability a suitable specialization of $\bar{p}$ to $\bar{v}$ and a prime number q lead to the same sequence of reduced S-polynomials over $\mathbf{Z}_q$ as the general computation over $\mathbf{Q}(\bar{p})$. It is not difficult to obtain the generic trace and to follow it on a set of polynomials with floating-point coefficients. The main purpose of the trace is to identify zero polynomials, without this one would soon start reducing H-polynomials using polynomials that consist only of round-off errors. Similar ideas have been used in numerical analysis, *e.g.* the deflation method for finding roots of univariate polynomials (when dividing out a linear factor, the remainder is known to be zero and can be discarded) and methods for computing Jordan normal forms [12].

Certain values $\bar{v}$ will of course lead to ill-conditioned polynomials whose roots are very uncertain — in the application this means that the chosen variable ordering is unsuitable or that the selected points are unfortunate and cannot by any method be used to compute the answer.

The implementation of a prototype for this method is also straightforward: the signature function is used to get coefficients in the finite field, a few calls to the **mod** function and trace collecting instructions were inserted. The stability of the numerical coefficients can be estimated by executing with different precisions. It is of course desirable also to have hard analyses of error probabilities (unluckiness in the modular computation) and round-off errors. These problems are being studied.

4   Examples and Experiments

In the examples, we try to reconstruct the translation and rotation of a rigid object relative to the camera position from corresponding point pairs measured in two images. That is, from pairs of 3-vectors $(\bar{x}_i, \bar{y}_i)$ we want to construct a translation vector $\bar{t}$, and a rotation matrix R such that $u_i \bar{y}_i = R(v_i \bar{x}_i + \bar{t})$ for all i. Here the u_i and v_i are unknown scale factors, caused by the inability of a 2D image to gauge the distance to a 3D point. This also means that $\bar{t}$ cannot be fully reconstructed, it is only possible to compute its direction.

The question of the multiplicity of the solutions of the motion problem has recently been solved. This result together with a method of computing the solutions is presented in Faugeras and Maybank [8]. The objective here is not to solve the problem in a better or more elegant way, but to use it as a realistic test problem.

Let $M^{(n)}$ for $n = 1, 2, 3$ be matrices obtained by multiplication of n matrices describing rotations around different axes with different angles. $M^{(n)}$ will thus contain sines and cosines of n angles. We want to solve for the n angles in the system

$$\{\bar{x}_i^T M^{(n)} \bar{y}_i = 0\}_{i=1}^n,$$

where $\bar{x}_i$ and $\bar{y}_i$ are n-column vectors representing measurements from the image. We consider problems 2 and 3 to be this system with $n = 2$ and $n = 3$, adjoined with the trigonometric identity $\sin^2 \phi + \cos^2 \phi - 1 = 0$ for the n angles. Problems 2s and 3s are obtained by expressing sine and cosine simultaneously with a rational function of one parameter, $\sin \phi = 2t/(1 + t^2)$ and $\cos \phi = (1 - t^2)/(1 + t^2)$. Computing times for obtaining a total degree, then inverse lexicographic ordered basis, its *slp* representation and the generic trace for the black box representation are given in the table. With our prototype, which also uses the Maple code, obtaining the numerical coefficients with the black box (bb) method (using arbitrary precision floating-point arithmetic) takes time comparable to obtaining the trace itself, but with compiled allocation and vectorization this can be improved considerably.

Table 1.

Problem	2s	2	3s	3
Macsyma time	> 2h	-	-	-
Macsyma size	175 KB	-	-	-
slp time	1m	70m	-	-
slp size	135v	1700v	-	-
bb time	5s	1m	10m	40m
bb size	15nf	38nf	74nf	135nf

Before the changes were made to Maple's Gröbner package, Macsyma was the only computer algebra system (we tried Macsyma, Maple and Reduce) that could solve problem 2s in less than 10 hours and we therefore include Macsyma's result in Table 1 together with those obtained with Maple.

The timing is for a SUN3/75 with 16 MB memory. The units used are: s – seconds, m – minutes, h – hours, KB – kilo bytes, nf – normal form reductions, v – intermediate variables.

To solve a zero-dimensional algebraic system it is known that it is usually best to first compute a Gröbner basis with the order total degree, then inverse lexicographic, and then use this basis to construct a new basis with the desired term ordering using linear algebra [9]. For the problems above it is easy to convert the basis to lexicographical order, since linear algebra is easily performed both with *slp* and floating-point coefficients — the probability of hitting a nearly singular point seems to be low when parameter values are generated randomly. However, our experiments showed that rounding errors are important to consider and that the corresponding ideals are not always prime. For this reason it is necessary in practice to include the structure of the lexicographically ordered base in the generic trace information. Presently, we are investigating alternatives in the collection and use of trace information with the objective of finding methods that allow computation with single precision arithmetic.

Acknowledgements

We gratefully acknowledge valuable advice given by James Davenport and Germund Dahlquist. The problems were communicated to us by Magnus Andersson. This work was financed by the Swedish Board for Technical Development.

References

1. Aggarwal, J. and Mitiche, A. (1985). Structure and motion from images: fact and fiction. *Third IEEE Workshop on Vision*, pp. 127–128.

2. Buchberger, B. (1985). Gröbner bases: an algorithmic method in polynomial ideal theory. In N. K. Bose, editor, *Recent Trends in Multidimensional Systems Theory*, chapter 6, pp. 184–232, D. Reidel Publishing Co., New York.

3. Buchberger, B. (1987). Applications of Gröbner bases in non-linear computational geometry. In *Trends in Computer Algebra*, Volume 296 of Lecture Notes in Computer Science, Springer-Verlag, Heidelberg, pp. 52–80.

4. Char, B.W., Geddes, K.O., Gonnet, G.H. and Watt, S.M. (1985). *Maple User's Guide*, WATCOM Publications, Waterloo.

5. Czapor, S.R. and Geddes, K.O. (1986). On implementing Buchberger's algorithm for Grobner bases. In *Proc. SYMSAC '86*, ACM, pp. 233–238.

6. Davenport, J.H. (1986). Looking at a set of equations. Bath CS Tech. Report 87–06

7. Davenport, J.H., Siret, Y. and Tournier, E. (1988). *Computer Algebra*, Academic Press.

8. Faugeras, O.D. and Maybank, S. (1989). Motion from point matches: multiplicity of solutions. *IEEE Workshop on Visual Motion*, pp. 248–255.

9. Faugère, J.C., Gianni, P., Lazard, D. and Mora, T. (1989). Efficient computation of zero-dimensional Gröbner bases by change of ordering. Preprint.

10. Freeman, T.S., Imirzian, M., Kaltofen, E. and Lakshman, Y. (1988). Dagwood: a system for manipulating polynomials given as straight-line programs. *ACM Trans. Math. Software, 14*, pp. 218–240.

11. Galligo, A., Pottier, L. and Traverso, C. (1988). *Greater Easy Common Divisor and Standard Basis Completion Algorithm*, Volume 358 of Lecture Notes in Computer Science, Springer-Verlag, Heidelberg, pp. 162–176.

12. Golub, G.H. and Van Loan , C.F. (1989). *Matrix Computations*. The Johns Hopkins University Press Ltd., 2nd edn.

13. Gonnet, G.H. (1984). Determining equivalence of expressions in random polynomial time. *ACM STOC 1984*, pp. 334–341.

14. Ibarra, O.H. and Moran, S. (1983). Probabilistic algorithms for deciding equivalence of straight-line programs. *Journal of the ACM, 30* pp. 217–228.

15. Jerian, C. and Jain, R. (1988). Polynomial methods for structure from motion. *ICCV*, IEEE, pp. 197–206.

16. Kaltofen, E. (1988). Greatest common divisors of polynomials given by straight-line programs. *Journal of the ACM, 35*, pp. 231–264.

17. Kaltofen, E. and Trager, B. (1988). Computing with polynomials given by black boxes for their evaluations: greatest common divisors, factorization, separation of numerators and denominators. *IEEE STOC*, pp. 296–305.

18. Longuet-Higgins, H.C. (1981). A computer algorithm for reconstructing a scene from two projections. *Nature, 293*.

19. Melenk, H., Möller, H.M. and Neun, W. (1988). Symbolic solution of large stationary chemical kinetics problems. Preprint SC 88–7, Konrad-Zuse-Zentrum für Informationstechnik Berlin.

20. Traverso, C. (1988). *Gröbner Trace Algorithms*. Volume 358 of Lecture Notes in Computer Science, Springer-Verlag, Heidelberg, pp. 125–138.

The Cylindrical Algebraic Decomposition Algorithm and Multiple Algebraic Extensions

Lars Langemyr

Universität Tübingen, Germany

1 Introduction

The algorithm for *Cylindrical Algebraic Decomposition* (CAD) was developed by Collins [5] as part of a quantifier elimination procedure for the first order theory of real closed fields. Given a set of polynomials with integer coefficients in n variables, the CAD algorithm decomposes the real n-dimensional Euclidean space into disjoint cells in which all the polynomials have constant signs. The cells obtained are semi-algebraic varieties, which means that they can be described by a finite set of polynomial equalities and inequalities, combined by using logical connectives. The algorithm constructs a sample point in each such cell, and can optionally construct formulas which describe each cell. A bibliography on CAD can be found in [3].

Arnon [2] made the first complete implementation of the original Collins algorithm, contributing his own improvements. Usage reports [2], [15] show that much of the computation time is often spent in the final extension phase in which a sample point in each cell is constructed. Arnon improved the computing time for the extension phase by using a method called clustering. We thereby avoid constructing certain sample points which may be costly to compute.

We approach the problem of improving the computing time of the extension phase in another way. For cells of dimension lower than n we may need algebraic numbers for the representation of the sample point coordinates. In Collins' original algorithm the coordinates are represented as elements of a single algebraic extension. This extension is formed by iteratively computing primitive elements. In our new implementation we directly use the multiple extension which is naturally formed in the extension phase of the algorithm. This has been previously suggested [7], [17], but as far as we are aware of no implementation comprising this idea has been made. In this paper we do not give a computing time bound for our implementation, but instead we apply the algorithm to some previously attempted problems. It turns out that the practical computing times obtained are often improved by several orders of magnitudes. In the last section we give some empirical computing times to substantiate the claim.

We have also made some other important improvements to the algorithm while working on our implementation. Arnon used rational arithmetic for algebraic number computations. We suggest that we should instead use integer arithmetic throughout. This means that algebraic integers have to be used for representing the algebraic extension fields which are constructed.

2   Original CAD Algorithm

We briefly describe the original CAD algorithm [5]. We are given a set A of polynomials in $\mathbf{Z}[x_1, \ldots, x_n]$. In the first phase (the *projection phase*) we eliminate the variables of the polynomials A by successively computing resultants, discriminants and certain other projection polynomials. We obtain a sequence of set polynomials P_i in $\mathbf{Z}[x_1, \ldots, x_i]$, for $i = n, n - 1$, $\ldots, 1$, where $P_n = A$.

We discuss the second phase (the *extension phase*) in more detail. In this phase the coordinates of the sample points are computed by iteratively constructing the x_1 coordinates of all sample points, then the x_2 coordinates, and so on up to x_n.

We first compute a square-free basis $\tilde{P}_1$ of the set P_1 of univariate polynomials over the integers. We then perform real root isolation on $\tilde{P}_1$. For each real root $\tilde{\alpha}_1$ we obtain a polynomial $\tilde{r}_1 \in \tilde{P}_1$ and an isolating interval $\tilde{I}_1$. The isolating interval has rational endpoints and uniquely separates $\tilde{\alpha}_1$ from all other real roots in $\tilde{P}_1$. The set of real roots of the set $\tilde{P}_1$ constitutes a part of the x_1 coordinates of the sample points. The complete set of x_1 coordinates of the sample points is obtained by choosing rational points between each real root, and a further two on each side of all real roots. We will not describe how we handle the rational points since in principle they can be represented as the root of a linear polynomial. In the implementation we take care in treating rational points in an efficient way (cf. Section 7).

We then go on constructing the x_2 coordinates of the sample points. We substitute $\tilde{\alpha}_1$ for x_1 in P_2, and compute a square-free basis $\tilde{P}_2$ of P_2 viewed as a set of polynomials in x_2 over $\mathbf{Q}(\tilde{\alpha}_1)$. We then perform real root isolation on $\tilde{P}_2$, also over $\mathbf{Q}(\tilde{\alpha}_1)$, for the particular real root of $\tilde{r}_1$ characterized by the interval $\tilde{I}_1$. For each real root $\tilde{\alpha}_2$ we obtain a square-free polynomial $\tilde{r}_2(x_2) \in \tilde{P}_2$, and an isolating interval $\tilde{I}_2$. For each $\tilde{\alpha}_2$ we then compute a primitive element $\hat{\alpha}_2$, root of the integral square-free polynomial $\hat{r}_2$, generating the field $\mathbf{Q}(\tilde{\alpha}_1, \tilde{\alpha}_2)$. We select an isolating interval $\hat{I}_2$ uniquely separating $\hat{\alpha}_2$ from other roots of $\hat{r}_2(x_2)$. We also compute the representation of $\tilde{\alpha}_1$ and $\tilde{\alpha}_2$ in $\mathbf{Q}(\hat{\alpha}_2)$.

At the ith extension stage we do the following: We have expressed $\tilde{\alpha}_1$, $\ldots$, $\tilde{\alpha}_{i-1}$ in terms of the primitive element $\hat{\alpha}_{i-1}$. We substitute these values for x_1, $\ldots$, x_{i-1} in P_i, and we then compute the square-free basis $\tilde{P}_i$ of P_i viewed as a set of polynomials in x_i over $\mathbf{Q}(\hat{\alpha}_{i-1})$. We then perform real root isolation on $\tilde{P}_i$. We obtain a square-free polynomial $\tilde{r}_i$ and an isolating interval $\tilde{I}_i$ for each real root $\tilde{\alpha}_i$. For each $\tilde{\alpha}_i$ we select a primitive element $\hat{\alpha}_i$, root of a square-free univariate integral polynomial $\hat{r}_i(x_i)$, with isolating interval $\hat{I}_i$, for which we have $\mathbf{Q}(\hat{\alpha}_i) = \mathbf{Q}(\hat{\alpha}_{i-1}, \tilde{\alpha}_i)$.

Having computed $\hat{\alpha}_n$, and the representation of $\tilde{\alpha}_1$, $\ldots$, $\tilde{\alpha}_n$ in $\mathbf{Q}(\hat{\alpha}_n)$, we can determine the sign of A on the coordinates $\tilde{\alpha}_1$, $\ldots$, $\tilde{\alpha}_n$.

References to the algorithms used to perform algebraic number arithmetic can be found in Collins' original paper [5]. Collins did not require that the polynomials $\hat{r}_i$ were irreducible since integer polynomial factoring was not then shown to be in polynomial time. Instead the algorithms which Collins used (e.g., Rubald [16]), were carefully designed to cope with the possibility that $\hat{r}_i$ may be reducible. In the implementation [2] minimal polynomials were used instead. This simplifies the implementation of the algebraic number arithmetic.

Thus in the original CAD algorithm Collins represented the sample point coordinates as elements of an algebraic number field generated by a primitive element, i.e., he determined a single univariate rational polynomial which generated all solutions to the above system. Such representation can incur a large coefficient blowup (Arnon [2] pages 171–173).

3 Multiple Algebraic Extensions

Instead of primitive element construction we will use multiple algebraic extensions. We therefore outline the basics about such algebraic extensions. The usage of ring extensions of $\mathbf{Z}$ instead of field extensions of $\mathbf{Q}$ helps us avoid rational arithmetic.

We consider a multiple algebraic extension of the rational numbers. We have $K_i = K_{i-1}(\alpha_i)$, for $i = 1, \ldots, n$, where $K_0 = \mathbf{Q}$, and where α_i is a root of the monic polynomial $r_i(x_i) \in \mathbf{Z}[\alpha_1, \ldots, \alpha_{i-1}][x_i]$ of degree m_i in x_i irreducible over K_{i-1}. Thus we insist that the algebraic numbers used to build the algebraic extension are algebraic integers. We thus have a corresponding tower of ring extensions $A_i = A_{i-1}[\alpha_i]$, for $i = 1, \ldots, n$, where $A_0 = \mathbf{Z}$. These rings are integral domains.

We now discuss how an element of this multiple algebraic extension is represented in the computer. First we note that any element of K_n can be represented by a pair consisting of a rational number and a representative of an element in the ring A_n. We therefore restrict our attention to how to represent an element of A_n. We can view the polynomials $r_n(x_n), \ldots, r_1(x_1)$

as an ideal $J_n = (r_n(x_n), \ldots, r_1(x_1))$ in the polynomial ring $\mathbf{Q}[x_1, \ldots, x_n]$, and then apply the theory of Gröbner bases [4]. Since J_n is a Gröbner basis in the lexicographical variable ordering $x_n > x_{n-1} > \cdots > x_2 > x_1$ we see that an element of $\mathbf{Q}[x_1, \ldots, x_n]$ has a canonical form modulo the ideal J_n, which can be obtained by reducing modulo the minimal polynomials until no further reductions are possible. We thus obtain a canonical representation of an element in A_i as an element of the quotient ring $R_i = \mathbf{Z}[x_1, \ldots, x_i]/J_i$. We denote $\mathbf{Z}$ by R_0.

4   Our Usage of Multiple Extensions

We want to explain how multiple extensions can be used instead of primitive element construction. In the projection phase a projection set P_i is computed for $i = n, n-1, \ldots, 1$, having $A = P_n$.

We now describe our new extension phase. The treatment of isolating intervals is very similar to the original algorithm, and we therefore omit them from the description. We first compute a square-free basis $\bar{P}_1$ of P_1, and then perform real root isolation of the polynomials $\bar{P}_1$. Polynomials with real roots in $\bar{P}_1$ of degree greater than one are factorized, and the factors corresponding to the roots are extracted. Consider $\bar{\alpha}_1$, root of a positive, primitive, irreducible polynomial $\bar{r}_1(x_1)$. In order to obtain an algebraic number field generated by an algebraic integer we choose $\alpha_1 = d_1\bar{\alpha}_1$, where d_1 is the leading coefficient of $\bar{r}_1(x_1)$, to be the generator of an algebraic number field in which $\bar{\alpha}_1$ lies. Then α_1 is a root of the monic irreducible polynomial $r_1(x_1) = d_1^{m_1-1}\bar{r}_1(x_1/d_1)$, where m_1 is the degree of $\bar{r}_1$. For the next step in the extension phase, where $\bar{\alpha}_1$ is substituted into the polynomials in P_2, we substitute $\bar{\alpha}_1 = \alpha_1/d_1$ for x_1. We arrange the substitution so that an associate of the result with integer coefficients is returned. Using the notation of the previous section we have thus constructed the ring extensions $R_0 = \mathbf{Z}$, and $R_1 = \mathbf{Z}[\alpha_1]$.

We can continue by applying real root isolation to the square-free basis $\bar{P}_2$ of P_2 over $\mathbf{Z}[\alpha_1]$. Let $\bar{\alpha}_2$ be a real root of a polynomial in $\bar{P}_2$. Let $\bar{r}_2$ be the irreducible factor corresponding to the root, with positive minimal integer leading coefficient. We let $\alpha_2 = d_2\bar{\alpha}_2$, where d_2 is the leading coefficient of $\bar{r}_2(x_2)$, be the generator of an algebraic number field in which $\bar{\alpha}_2$ lies. Then α_2 is a root of the monic irreducible polynomial $r_2(x_2) = d_2^{m_2-1}\bar{r}_2(x_2/d_2)$, where m_2 is the degree of $\bar{r}_2$. We have constructed the extension $R_2 = \mathbf{Z}[\alpha_1, \alpha_2]$.

We now have a look at the general case: We assume an algebraic integer extension $R_{i-1} = \mathbf{Z}[\alpha_1, \ldots, \alpha_{i-1}]$, given by the roots $\alpha_1, \ldots, \alpha_{i-1}$ of the monic irreducible polynomials $r_1(x_1) \in R_0[x_1]$, $r_2(x_2) \in R_1[x_2]$, $\ldots$, $r_{i-1}(x_{i-1}) \in R_{i-2}[x_{i-1}]$. We also have a translation array $d_1, \ldots, d_{i-1}$. Here we want to substitute the roots $\bar{a}_1, \ldots, \bar{a}_{i-1}$ into each polynomial in P_i. Since $r_1(x_1)$, ..., $r_{i-1}(x_{i-1})$ are minimal polynomials of $d_1\bar{\alpha}_1$, ..., $d_{i-1}\bar{\alpha}_{i-1}$, respectively, we substitute α_1/d_1, ..., α_{i-1}/d_{i-1} for x_1, ..., x_{i-1} in P_i. We then compute a square-free basis $\bar{P}_i$ of P_i. Each polynomial in the basis with real roots is then factorized over $\mathbf{Q}(\alpha_1, \ldots, \alpha_{i-1})$, obtaining factors in $R_{i-1}[x_i]$, with integral leading coefficients using the factorization algorithm from Trager [18]. Denote one such factor with a real root by $\bar{r}_i$. We let $\alpha_i = d_i\bar{\alpha}_i$, where d_i is the integral leading coefficient of $\bar{r}_i$ and $r_i(x_i) = d_i^{m_i - 1}\bar{r}_i(x_i/d_i)$, where m_i is the degree of $\bar{r}_i$. We obtain the ring extension $R_i = \mathbf{Z}[\alpha_1, \ldots, \alpha_i]$.

5   Other New Ideas

In order to avoid that the same computation is performed several times Arnon used a complicated scheme for partitioning the cells into conjugacy equivalence classes. We use another scheme based on an idea originally found in Maple, i.e., we use a hash table for looking up previously calculated results. This simplifies the implementation. The hash table approach has the further advantage that it potentially catches all previously computed results. This would be difficult to obtain with the conjugacy equivalence class approach. This idea was independently developed by Collins and Hong [6].

Since the modular algorithms for computing the inverse of an element in a multiple extension and the GCD of two polynomials over a multiple extension both require that all the minimal polynomials are irreducible over all underlying extensions, we have to try to factorize the polynomials over the multiple extension. For this purpose we first implemented the Kronecker factorization algorithm [18].

We also consider the approach of extending the modular algorithms for inverse and GCD not to require that the polynomials defining algebraic extensions are irreducible. The problem is that we may find a factor in the modular image of a defining polynomial because of two reasons. Firstly, because the factor only exists in the modular image. Secondly, because a true non-modular factor has been found. To determine what case we are in we use the following probabilistic method. Try a small number of modular images; if we obtain the same factor in all of them we assume that we are in the second case, and we apply a non-modular method. Otherwise we go on by using a modular image in which no factor occurred. This idea has not yet been incorporated into the system.

We claim that the improvements to CAD from Arnon [2], McCallum [15] and this paper independently improve the computing time and thus that their respective improvements multiply. We believe that together these improvements make the CAD algorithm so feasible that it could be included in standard computer algebra systems for practical calculations, just like the Gröbner base package now present in most systems. In the future we hope to make a unified implementation where the recent improvement for quantifier elimination problems by Collins and Hong [6] is also included.

6   Implementation of Algorithms for a Multiple Algebraic Extension

The basis for the new implementation is a package for arithmetic in a multiple algebraic extension, which has recently been implemented in SAC-2. The algorithms work only in the ring R_n described in Section 3. We thereby avoid rational arithmetic.

There is an algorithm for computing the discriminant of the basis α_1, ..., α_n. We refer to Abbott [1] for a definition of the discriminant for a multiple algebraic extension. There are algorithms for inverse, negation, norm, exponent, product, reduction modulo minimal polynomials, and sum for elements in R_n. For polynomials over R_n there are algorithms for different types of evaluation, element product, product, reduction modulo minimal polynomials, and sum. For univariate polynomials over R_n there are algorithms for transformations needed for the Uspensky root isolation algorithm, factorization, and GCD computations, and an algorithm which returns an associate of a polynomial with integer leading coefficient.

6.1   Asymptotically fast algorithms

We say that an algorithm is almost optimal if it has a computing time bound which is $O(d^{1+\delta})$, for all $\delta > 0$, where d is the best known a priori bound on the size of the output. We have given almost optimal algorithms for product, inverse and univariate polynomial product for a single algebraic extension of the rational numbers [12]. These results were in turn used to obtain an almost optimal probabilistic algorithm for polynomial greatest common divisor (GCD) over a single algebraic extension of the rational numbers [12], a refinement of an algorithm [14] which did not fully take advantage of dynamic evaluation [9]. Recently, we have generalized these results to multiple algebraic extensions. In [10] we obtained almost optimal algorithms for computing product and inverse in a multiple algebraic extension. In [11], we extend the results also to univariate GCD computations over multiple algebraic extensions. For our multiple extension inverse and the GCD algorithm the application of dynamic evaluation seems essential.

In Langemyr and McCallum [14] it was observed that factors of the minimal polynomial (only a single extension was treated) were almost never discovered if the prime numbers are of the size of a normal computer word (32 bits). It turns out that this is also true in our case. We can therefore in

practice skip implementing dynamic evaluation, and just ignore the particular prime if a factor is discovered. The principle of dynamic evaluation is
theoretically important, because it seems difficult to obtain almost optimal
algorithms if it is not utilized [12].

7 Data Structures

The implementation constructs a cylindrical algebraic decomposition of n-
space. We represent this decomposition by using a tree data structure.
It is recursive and the first level contains the x_1 coordinate of all sample
points, and some further information. The next level adds the different x_2
coordinate values for each x_1 value. Each such x_2 value adds a branch from
the node representing the x_1 coordinate. This goes on up to x_n. The data
structure has the following syntax:

$$
\begin{aligned}
C_j &\rightarrow (L_j, \ldots, L_j) \\
C_{n+1} &\rightarrow () \\
L_j &\rightarrow (n_j, C_{j+1}, S_j, D_j, N_j) \\
S_j &\rightarrow (0,0) \\
S_j &\rightarrow (0, d_j, c_j), &&\text{where} && d_1 \neq 0 \\
S_j &\rightarrow (m_j, d_j, b_j), &&\text{where} && m_j < 0 \\
S_j &\rightarrow (m_j, d_j, r_j, I_j), &&\text{where} && m_j > 0,
\end{aligned}
$$

for $j = 1, \ldots, n$, where C_j is the decomposition of the real Euclidean $(n-
j+1)$-space. L_i contains information about the cell index n_j, decomposition
above the sample point C_{j+1}, a description of the sample point S_j, D_j
optionally contains defining formulas, and N_j optionally contains the signs
of P_j at the sample points. There are four different types of sample points.
$(0,0)$ is the sample point 0. $(0, d_j, c_j)$, where $d_j, c_j \in \mathbf{Z}$ is a rational sample
point c_j / d_j. When $m_j < 0$ (m_j, d_j, b_j) is a sample point with $b_j \in R_{-m_j}$. d_j
is the translation. Similarly for m_j (m_j, d_j, r_j, I_j) is the most general case
of sample point. We have a translation d_j, a monic irreducible polynomial
$r_j \in R_{m_{j-1}}[x_{m_j}]$ and an isolating interval I_j.

8 Implementation

We briefly describe the program structure. A main algorithm CAD constructs the projection polynomials and tree data structure according to
Section 7. It uses an algorithm for projection (IPPROJ) and algorithms for
construction of the cylindrical algebraic structure (CASREC and CASEXT).
The algorithm CASREC uses recursion to evaluate B_i at the coordinates
α_j / d_j for $j = 1, \ldots, i-1$, and then finally calls CASEXT to extend the tree
one level further. For a three dimensional CAD the call structure would
be the following:

```
CAD     IPPROJ
        CAD     IPPROJ
                CAD     CASEXT
                CASREC  CASEXT
        CASREC  CASREC  CASEXT
```

9 Empirical Computing Times

Several practical test examples have been run, and the result is that our implementation performs better than the previous implementation by Arnon. The example in [8] now takes 65 seconds compared to 1480 seconds previously. Both tests were made on the same hardware (Sun-3/280) using the lisp implementation of SAC-2 [13] and Sun Common Lisp. We suspect that this particular improvement is due to more efficient handling of rational arithmetic and lookups of previous results since algebraic extensions are little used in this example.

We have also tried the 'tacnode' example which Arnon reported ([2]pages 169–173) took 1508 seconds for doing full CAD on a VAX-780. We have been able to do the computation in 76 seconds in the Lisp system on a Sun-3/280.

Acknowledgements

Thanks to Rüdiger Loos for valuable discussions and to Joachim Hollman for comments. Supported by ESPRIT BRA 3012 CompuLog and Fakultetsnämnden KTH.

References

1. Abbott, J.A. (1989). Factorization of polynomials over algebraic number fields. Ph. D. Thesis, University of Bath.

2. Arnon, D.S. (1981). Algorithms for the geometry of semi-algebraic sets. Ph. D. Thesis, University of Wisconsin, Madison.

3. Arnon, D.S. (1988). A bibliography of quantifier elimination for real closed fields. *J. Symbolic Comp.*, *5*, 267–274.

4. Buchberger, B. (1986). Basic features and development of the critical pair completion procedure. In *Rewriting Techniques and Applications*, J.P. Jouannaud, editor, Springer-Verlag, LNCS *202*, 1–45.

5. Collins, G.E. (1975). Quantifier elimination for real closed fields by cylindrical algebraic decomposition. In *Second GI Conf. Automata Theory and Formal Languages*, Springer-Verlag, LNCS *33*, 134–183.

6. Collins, G.E. and Hong, H. (1989). Partial CAD construction in quantifier elimination. Technical Report OSU-CISRC-10/89 TR 45, Computer Science Dept., Ohio State University.

7. Davenport, J.H. (1985). Computer algebra for cylindrical algebraic decomposition. Technical Report TRITA-NA-8511, NADA, Royal Institute of Technology, Stockholm

8. Davenport, J.H. and Heintz, J. (1988). Real quantifier elimination is doubly exponential. *J. Symbolic Comp.*, *5*, 29–35.

9. Duval, D. (1987). Diverse questions relatives au CALCUL FORMEL AVEC DES NOMBRES ALGÉBRIQUES. Ph. D. Thesis, L'université scientifique, technologique, et médicale de Grenoble.

10. Langemyr, L. (1991). Algorithms for a multiple algebraic extension. *Proc. Effective Methods in Algebraic Geometry (MEGA-90)*, T. Mora and C. Traverso, editors, Birkhäuser, Progress in Mathematics, *94*, 235–248.

11. Langemyr, L. (1991). Algorithms for a multiple algebraic extension II. T. Mora, editor, *Proc. 9th International Conference on Applied Algebra, Algebraic Algorithms and Error Correcting Codes (AAECC-9)*, Lecture Notes in Computer Science *539*, 224–233, Springer Verlag, Berlin-Heidelberg-New York.

12. Langemyr, L. (1991). An asymptotically fast probabilistic algorithm for computing polynomial GCDs over an algebraic number field. S. Sakata, editor, *Proc. 8th International Conference on Applied Algebra, Algebraic Algorithms and Error Correcting Codes (AAECC-8)*, Lecture Notes in Computer Science *508*, 222–233, Springer Verlag, Berlin-Heidelberg-New York.

13. Langemyr, L. (1989). Converting SAC-2 code to lisp. In *Proc. EUROCAL '87*, J.H. Davenport, editor, Springer-Verlag, LNCS *378*, 50–51.

14. Langemyr, L. and McCallum, S. (1989). The computation of polynomial greatest common divisors over an algebraic number field. *J. Symbolic Comp.*, *8*, 429–448.

15. McCallum, S. (1985). An improved projection operation for cylindrical algebraic decomposition. Ph. D. Thesis, University of Wisconsin, Madison.

16. Rubald, C.M. (1974). Algorithms for polynomials over a real algebraic number field. Ph. D. Thesis, University of Wisconsin, Madison.

17. Schwartz, J.T. and Sharir, M. (1983). On the "piano movers" problem. II. General techniques for computing topological properties of real algebraic manifolds. *Advances in Applied Maths.*, *4*, 298–351.

18. Trager, B. (1976). Algebraic factoring and rational function integration. In *Proc. SYMSAC '76*, ACM, New York, 219–226.

Simple Singularities in Surface-Surface Intersections

Gábor Lukács

Computer and Automation Institute, Hungarian Academy of Sciences, Budapest

Abstract Surface-surface intersections are degenerated if the two surfaces are tangent in a common point. Most of the traditional methods in CAD for intersection fail to analyse the behaviour of the intersection set in the environment of these points. This paper tries to resolve the problem in case of first order singularities. Finally a test is shown for singular intersection in case of second degree implicit surfaces.

1 Singularities—When Do They Occur?

The determination of the intersection curve of two free-form surfaces is one of the central mathematical problems in computational geometry (in the Faux-Pratt's [4] sense). Most "classical" solutions to this problem exclude cases when the two surfaces are tangent. These situations are usually considered rare and of no importance from the "practical" point of view. However, in pratice these problems arise quite often as a consequence of deliberate design, especially if "traditional" surfaces like cylinders, spheres, cones or torii are considered as well. If we try to establish the "frequency" of such cases on a heuristic basis, we can state that in order to obtain singular intersection it is sufficient to move one surface in the appropriate position in the 3 dimensional space, i.e. there need not be any special correspondence between the inner geometry of the two surfaces. As a matter of fact, this may occur relatively frequently and considering that near-to-singular cases must be usually handled by similar methods being used for singular intersections, we can say that a reliable CAD/CAM system must not leave singular cases out of consideration. It is especially true for solid modelling systems dealing with curved surfaces, for during Boolean operations indispensably a great number of intersections are calculated which possibly do not appear in the final object. Because of this it is much more difficult to "avoid" singularities.

There have been several attempts for solving the problem of singular surface-intersections in case of Newton-Raphson like methods [12], [9] or in some other way [2]. Most papers do not qualify the behaviour of the intersection set in the vicinity of singular points in details. This problem is going to be analysed in this paper.

In this paper surfaces are considered differentiable continuously up to the 3rd derivatives (smooth). They are supposed to be set up correctly, that is in case of parametric surfaces the parametric derivatives are never parallel; in case of implicit surfaces the gradient vector of the surface is never zero. Real numbers (and functions denoting the left hand side of implicit surfaces) are denoted by lower case italic letters or by lower case Greek letters (y, u, f or σ). Vectors (and parametric surfaces) are denoted by bold face lower case letters ($\mathbf{n}$ or $\mathbf{p}$). Derivation (of any degree) is denoted by upper indices ($\mathbf{p}^u$ or g^{xz}).

2   First Order Singularities Between Two Parametric Surfaces

In this section we are going to analyse the singularities between parametric surfaces. The results for the implicit-implicit and for the implicit-parametric intersections will be similar, but the basic technique can be explained somewhat easier in this case.

Let us determine the intersection curve of two parametric surfaces $\mathbf{p}(u, v)$ and $\mathbf{q}(s, t)$. The condition of intersection can be formulated as:

$$\mathbf{p}(u, v) - \mathbf{q}(s, t) = 0. \tag{2.1}$$

Suppose that we have an initial solution of (2.1) (u_0, v_0, s_0, t_0). Assume that the two surfaces are tangent in the point defined by the parameters (or (s_0, t_0)). This can be expressed as:

$$(\mathbf{p}^u \times \mathbf{p}^v) \parallel (\mathbf{q}^s \times \mathbf{q}^t).$$

We shall suppose further on that the two surfaces are set up correctly, i.e.

$$(\mathbf{p}^u \times \mathbf{p}^v) \neq 0 \neq (\mathbf{q}^s \times \mathbf{q}^t). \tag{2.2}$$

Let

$$\mathbf{n} = \frac{(\mathbf{p}^u \times \mathbf{p}^v)}{|(\mathbf{p}^u \times \mathbf{p}^v)|}.$$

Let one branch of the intersection set starting from $\mathbf{p}(u_0, v_0)$ be parameterised by its arclength parameter σ. Let us denote it by $\mathbf{r}(\sigma)$.

$$\mathbf{r}(\sigma) = \mathbf{p}(u(\sigma), v(\sigma)) = \mathbf{q}(s(\sigma), t(\sigma)).$$

Differentiating by σ, we get

$$\mathbf{e} := \mathbf{p}^u \cdot u^\sigma + \mathbf{p}^v \cdot v^\sigma = \mathbf{q}^s \cdot s^\sigma + \mathbf{q}^t \cdot t^\sigma. \tag{2.3}$$

However, $\mathbf{e}$ is the tangent vector of this branch of the intersection curve. In consequence of (2.2) (u^σ, v^σ) can be expressed by (s^σ, t^σ) and vice versa. For example by multiplying vectorially by $\mathbf{p}^u$ we get:

$$(\mathbf{p}^u \times \mathbf{p}^v) \cdot v^\sigma = (\mathbf{p}^u \times \mathbf{q}^s) \cdot s^\sigma + (\mathbf{p}^u \times \mathbf{q}^t) \cdot t^\sigma.$$

In order to get v^σ here, we can divide by the coordinate of the largest magnitude of $\mathbf{p}^u \times \mathbf{p}^v$. Thus

$$\begin{aligned} s^\sigma &= \alpha \cdot u^\sigma + \beta \cdot v^\sigma \\ t^\sigma &= \gamma \cdot u^\sigma + \delta \cdot v^\sigma \end{aligned} \quad \text{where} \quad \alpha\delta - \beta\gamma \neq 0. \tag{2.4}$$

By derivating (2.3) once more we get

$$\ddot{\mathbf{r}}(\sigma) = \mathbf{r}^{\sigma\sigma} =$$

$$\mathbf{p}^{uu} u^{\sigma\,2} + 2\mathbf{p}^{uv} u^\sigma v^\sigma + \mathbf{p}^{vv} v^{\sigma\,2} + \mathbf{p}^u u^{\sigma\sigma} + \mathbf{p}^v v^{\sigma\sigma} = \tag{2.5}$$

$$\mathbf{q}^{ss} s^{\sigma\,2} + 2\mathbf{q}^{st} s^\sigma t^\sigma + \mathbf{q}^{tt} t^{\sigma\,2} + \mathbf{q}^s s^{\sigma\sigma} + \mathbf{q}^t t^{\sigma\sigma}.$$

Let us dot-multiply this by $\mathbf{n}$ on both sides. In this way the terms with $^{\sigma\sigma}$ fall out and we get the normal curvature looking in the given direction [11] on the corresponding surfaces, i.e.

$$\langle \mathbf{p}^{uu}, \mathbf{n} \rangle u^{\sigma\,2} + 2\langle \mathbf{p}^{uv}, \mathbf{n} \rangle u^\sigma v^\sigma + \langle \mathbf{p}^{vv}, \mathbf{n} \rangle v^{\sigma\,2} =$$

$$\langle \mathbf{q}^{ss}, \mathbf{n} \rangle s^{\sigma\,2} + 2\langle \mathbf{q}^{st}, \mathbf{n} \rangle s^\sigma t^\sigma + \langle \mathbf{q}^{tt}, \mathbf{n} \rangle t^{\sigma\,2}. \tag{2.6}$$

With more conventional notations:

$$L_\mathbf{p} \cdot du^2 + 2 \cdot M_\mathbf{p} \cdot du \cdot dv + N_\mathbf{p} \cdot dv^2 = L_\mathbf{q} \cdot ds^2 + 2 \cdot M_\mathbf{q} \cdot ds \cdot dt + N_\mathbf{q} \cdot dt^2.$$

Note that the condition of the equality of the normal curvatures can be clearly realized without any special argumentation as well. Let us substitute (2.4) into (2.6). In this way we get the following relation:

$$\lambda u^{\sigma\,2} + 2\mu u^\sigma v^\sigma + \nu v^{\sigma\,2} = 0. \tag{2.7}$$

Let us perform the principal axis transformation to this quadratic form (see [5] or [3]).

$$\lambda_+ \xi^2 + \lambda_- \eta^2 = 0 \quad \text{where} \quad \lambda_+ \geq \lambda_-. \tag{2.8}$$

Here ξ and η are linear orthogonal combinations of u^σ and v^σ. One can clearly see that we can have non-trivial solutions only if $\lambda_+ \geq 0$ and $\lambda_- \leq 0$.

This means that the behaviour of the solution in the environment of a tangential point depends basically on the definiteness of the the the quadratic form defined by (2.7). If the quadratic form is indefinite (or semi-definite) then there are two (or one resp.) directions where the intersection curve can proceed. Note that λ_+ and λ_- are the eigenvalues of the matrix

$$\Lambda = \begin{pmatrix} \lambda & \mu \\ \mu & \nu \end{pmatrix}$$

that is

$$\lambda_+ = \frac{\lambda+\nu}{2} + \sqrt{\left(\frac{\lambda-\nu}{2}\right)^2 + \mu^2}$$
$$\lambda_- = \frac{\lambda+\nu}{2} - \sqrt{\left(\frac{\lambda-\nu}{2}\right)^2 + \mu^2}.$$

If $\mathbf{v}_+$ and $\mathbf{v}_-$ are the corresponding eigenvectors (in the u–v parametric space) of the matrix Λ, then the critical directions (where the intersection curve goes on) are:

$$\sqrt{-\lambda_-}\,\mathbf{v}_+ + \sqrt{\lambda_+}\,\mathbf{v}_-$$

and   (2.9)

$$\sqrt{-\lambda_-}\,\mathbf{v}_+ - \sqrt{\lambda_+}\,\mathbf{v}_-.$$

Note that e in (2.3) must be a unitary vector. From this fact the magnitude of the vectors in (2.9) can be computed. Thus the required $(u^\sigma, v^\sigma, s^\sigma, t^\sigma)$ quantities can be easily calculated. In this way the following theorem can be formulated.

Theorem 1

If two smooth parametric surfaces $\mathbf{p}$ *and* $\mathbf{q}$ *are tangent in a point, then in the direction of a branch of the intersection curve the normal curvatures in the two surfaces must coincide. The behaviour of the two surfaces in the environment of the singular point can be described by the quadratic form (2.7). (That is by the eigenvalues* λ_+ *and* λ_-.) *The following four cases can be distinguished:*

(i) *The form is definite* $(\lambda_+ \cdot \lambda_- > 0)$. *The common point is isolated (no branch starts from here). (Like a sphere touching a plane.)*

(ii) *The form is indefinite* $(\lambda_+ \cdot \lambda_- < 0)$. *Two branches start from the singular point. The directions are to be computed by (2.9). (Like a hyperbolic paraboloid touching a plane.)*

(iii) *The form is semi-definite* $(\lambda_+ = 0$ *or* $\lambda_- = 0$ *but not both). One branch passes at the point of tangency. The direction of the branch can be calculated by one of the formulae in (2.9). (Like a cylinder touching a plane.)*

(iv) The form is zero. ($\lambda_+ = 0$ and $\lambda_- = 0$). There is a second order singularity at this point. Higher degree derivatives must be computed in order to analyse the behaviour at this critical point. More advanced methods are necessary here—basically those of catastrophe theory [10],[1].

Further on we shall exclude cases (iv). Geometrically it means that we do not deal with situations when all the normal curvatures of the two surfaces coincide in the given point, i.e. the curvatures do not help us where to "continue". These cases are really special—the inner geometry (the second fundamental form) of the two surfaces must be in accordance. The two surfaces are likely to have been designed in a special way in order to obtain this kind of meeting. Here it is adviseable to use special information from the design phase in order to establish the set of common points.

2.1 Curvature at singularities between two parametric surfaces

In this section the computation of the curvature at a singular point of a branch of the intersection curve is derived. The curvature is necessary for establishing an additional condition for the search of the "next" point on the intersection curve [8]. Let us differentiate (2.5) by σ once more.

$$\dddot{\mathbf{r}}(\sigma) = \mathbf{r}^{\sigma\sigma\sigma} =$$

$$\mathbf{p}^{uuu}u^{\sigma 3} + 3\mathbf{p}^{uuv}u^{\sigma 2}v^{\sigma} + 3\mathbf{p}^{uvv}u^{\sigma}v^{\sigma 2} + \mathbf{p}^{vvv}v^{\sigma 3}+$$

$$3(\mathbf{p}^{uu}u^{\sigma}u^{\sigma\sigma} + \mathbf{p}^{uv}(u^{\sigma\sigma}v^{\sigma} + u^{\sigma}v^{\sigma\sigma}) + \mathbf{p}^{vv}v^{\sigma}v^{\sigma\sigma}) + \mathbf{p}^{u}u^{\sigma\sigma\sigma} + \mathbf{p}^{v}v^{\sigma\sigma\sigma} =$$

$$\text{(2.10)}$$

$$\mathbf{q}^{sss}s^{\sigma 3} + 3\mathbf{q}^{sst}s^{\sigma 2}t^{\sigma} + 3\mathbf{q}^{stt}s^{\sigma}t^{\sigma 2} + \mathbf{q}^{ttt}t^{\sigma 3}+$$

$$3(\mathbf{q}^{ss}s^{\sigma}s^{\sigma\sigma} + \mathbf{q}^{st}(s^{\sigma\sigma}t^{\sigma} + s^{\sigma}t^{\sigma\sigma}) + \mathbf{q}^{tt}t^{\sigma}t^{\sigma\sigma}) + \mathbf{q}^{s}s^{\sigma\sigma\sigma} + \mathbf{q}^{t}t^{\sigma\sigma\sigma}.$$

Again, let us dot-multiply this by $\mathbf{n}$ on both sides. In this way the last terms with $^{\sigma\sigma\sigma}$ will fall out.

$$\langle\mathbf{n}, \mathbf{p}^{uuu}\rangle u^{\sigma 3} + 3\langle\mathbf{n}, \mathbf{p}^{uuv}\rangle u^{\sigma 2}v^{\sigma} + 3\langle\mathbf{n}, \mathbf{p}^{uvv}\rangle u^{\sigma}v^{\sigma 2} + \langle\mathbf{n}, \mathbf{p}^{vvv}\rangle v^{\sigma 3}+$$

$$3(\langle\mathbf{n}, \mathbf{p}^{uu}\rangle u^{\sigma}u^{\sigma\sigma} + \langle\mathbf{n}, \mathbf{p}^{uv}\rangle(u^{\sigma\sigma}v^{\sigma} + u^{\sigma}v^{\sigma\sigma}) + \langle\mathbf{n}, \mathbf{p}^{vv}\rangle v^{\sigma}v^{\sigma\sigma}) = \quad \text{(2.11)}$$

$$\langle\mathbf{n}, \mathbf{q}^{sss}\rangle s^{\sigma 3} + 3\langle\mathbf{n}q^{sst}\rangle s^{\sigma 2}t^{\sigma} + 3\langle\mathbf{n}q^{stt}\rangle s^{\sigma}t^{\sigma 2} + \langle\mathbf{n}, \mathbf{q}^{ttt}\rangle t^{\sigma 3}+$$

$$3(\langle\mathbf{n}, \mathbf{q}^{ss}\rangle s^{\sigma}s^{\sigma\sigma} + \langle\mathbf{n}, \mathbf{q}^{st}\rangle(s^{\sigma\sigma}t^{\sigma} + s^{\sigma}t^{\sigma\sigma}) + \langle\mathbf{n}, \mathbf{q}^{tt}\rangle t^{\sigma}t^{\sigma\sigma}).$$

What we need are the quantities $(u^{\sigma\sigma}, v^{\sigma\sigma}, s^{\sigma\sigma}, t^{\sigma\sigma})$. Here (2.5) covers two linear equations. (Notice that $\mathbf{p}^u, \mathbf{p}^v, \mathbf{q}^s, \mathbf{q}^t$ are all perpendicular to $\mathbf{n}$). (2.11) means one more linear equation. The fourth equation can be obtained if one takes into consideration the arclength parameterisation, i.e. that $\mathbf{r}^{\sigma\sigma}$ is perpendicular to $\mathbf{e}$ (to $\mathbf{r}^{\sigma}$). Thus

$$\langle \mathbf{e}, \mathbf{r}^{\sigma\sigma} \rangle = 0.$$

If we write this into the left side of (2.5) we get:

$$\langle \mathbf{e}, \mathbf{p}^{uu} \rangle u^{\sigma\,2} + 2\langle \mathbf{e}, \mathbf{p}^{uv} \rangle u^{\sigma} v^{\sigma} + \langle \mathbf{e}, \mathbf{p}^{vv} \rangle v^{\sigma\,2} + \langle \mathbf{e}, \mathbf{p}^{u} \rangle u^{\sigma\sigma} + \langle \mathbf{e}, \mathbf{p}^{v} \rangle v^{\sigma\sigma} = 0.$$
$$(2.12)$$

In this way we have four linear equations for four unknowns. By solving it we can substitute $u^{\sigma\sigma}$ and $v^{\sigma\sigma}$ into (2.5) and get $\mathbf{r}^{\sigma\sigma}$. The curvature is the magnitude of this vector.

3 First Order Singularities Between Two Implicit Surfaces

Let us consider the intersection of two implicit surfaces

$$\begin{aligned} f(x,y,z) &= 0 \\ g(x,y,z) &= 0. \end{aligned} \qquad (3.1)$$

Let us denote as usual by ∇f the gradient vector of the function f.

$$\nabla f = \left(\frac{\partial f}{\partial x}, \frac{\partial f}{\partial y}, \frac{\partial f}{\partial z} \right) = (f^x, f^y, f^z).$$

Suppose that we have an initial solution of (3.1) (x_0, y_0, z_0). Assume that the two surfaces are tangent in this point. This can be expressed as:

$$(\nabla f) \parallel (\nabla g)$$

that is, with some ϑ

$$\nabla f = \vartheta \cdot \nabla g. \qquad (3.2)$$

We shall suppose further on that the two surfaces are themselves not degenerated, i.e.

$$\nabla f \neq 0 \neq \nabla g.$$

Let one branch of the intersection set starting from (x_0, y_0, z_0) be parameterised by its arclength parameter σ. Let us denote it by $\mathbf{r}(\sigma)$.

$$\begin{aligned} f(\mathbf{r}(\sigma)) &= f(x(\sigma), y(\sigma), z(\sigma)) = 0 \\ g(\mathbf{r}(\sigma)) &= g(x(\sigma), y(\sigma), z(\sigma)) = 0. \end{aligned}$$

Differentiating by σ, we get

$$f^{\sigma} = \langle \nabla f, \mathbf{r}^{\sigma} \rangle = f^x x^{\sigma} + f^y y^{\sigma} + f^z z^{\sigma} = 0 \qquad (3.3)$$

$$g^\sigma = \langle \nabla g, \mathbf{r}^\sigma \rangle = g^x x^\sigma + g^y y^\sigma + g^z z^\sigma = 0. \tag{3.4}$$

Let us define h in the following way

$$h = f - \vartheta g \qquad \text{where } \vartheta \text{ is from (3.2)}.$$

We can observe that $\nabla h = 0$ in the critical point. Let us differentiate h twice by σ! (Note that terms with $^{\sigma\sigma}$ fall out in (x_0, y_0, z_0).) Now using (3.3) and (3.4) we get the following equation:

$$h^{xx} x^{\sigma 2} + h^{yy} y^{\sigma 2} + h^{zz} z^{\sigma 2} + 2h^{xy} x^\sigma y^\sigma + 2h^{xy} x^\sigma y^\sigma + 2h^{xy} x^\sigma y^\sigma = 0. \tag{3.5}$$

Suppose that the coordinate of the largest magnitude of ∇f (and ∇g) is f^z. Let us express z^σ from (3.3) and substitute in (3.5). We get:

$$\tilde{h}_{xx} x^{\sigma 2} + 2\tilde{h}_{xy} x^\sigma y^\sigma + \tilde{h}_{yy} y^{\sigma 2} = 0. \tag{3.6}$$

This equation plays the same role as (2.7) in the previous chapter. By performing the principal axis transformation to this quadratic form we have:

$$\psi_+ \xi^2 + \psi_- \eta^2 = 0 \qquad \text{where} \qquad \psi_+ \geq \psi_-. \tag{3.7}$$

As previously ξ and η are linear orthogonal combinations of x^σ and y^σ. We can have non-trivial solutions only if $\psi_+ \geq 0$ and $\psi_- \leq 0$. Here ψ_+ and ψ_- are the eigenvalues of the matrix

$$\Psi = \begin{pmatrix} \tilde{h}_{xx} & \tilde{h}_{xy} \\ \tilde{h}_{xy} & \tilde{h}_{yy} \end{pmatrix}$$

that is

$$\psi_+ = \frac{\tilde{h}_{xx} + \tilde{h}_{yy}}{2} + \sqrt{\left(\frac{\tilde{h}_{xx} - \tilde{h}_{yy}}{2}\right)^2 + \tilde{h}_{xy}^2}$$

$$\psi_- = \frac{\tilde{h}_{xx} + \tilde{h}_{yy}}{2} - \sqrt{\left(\frac{\tilde{h}_{xx} - \tilde{h}_{yy}}{2}\right)^2 + \tilde{h}_{xy}^2}.$$

As in the previous chapter if s_+ and s_- are the corresponding eigenvectors (in the x-y plane) of the matrix Ψ, then the critical directions (where the intersection curve goes on) are:

$$\sqrt{-\psi_-} s_+ + \sqrt{\psi_+} s_-$$

$$\text{and} \tag{3.8}$$

$$\sqrt{-\psi_-} s_+ - \sqrt{\psi_+} s_-.$$

Note that this is still in the x-y "parametric" plane. We can get 3 dimensional vectors by calculating z^σ using the relation (3.3). This vector must be scaled in order to get a unitary vector (arclength parameterisation!) for

$$x^{\sigma 2} + y^{\sigma 2} + z^{\sigma 2} = 0.$$

From this fact the magnitude of the vectors in (3.8) can be computed. Thus the required $(x^\sigma, y^\sigma, z^\sigma)$ quantities can be calculated easily. In this way the following theorem can be formulated.

Theorem 2 *If two smooth implicit surfaces $f = 0$ and $g = 0$ are tangent in a point, then in the direction of a branch of the intersection curve the normal curvatures in the two surfaces must coincide. Suppose that the surfaces are not vertical in the singular point ($f^z \neq 0$) then the behaviour of the two surfaces in the environment of the singular point can be described by the quadratic form (3.6). (That is by the eigenvalues ψ_+ and ψ_-.) The following four cases can be distinguished:*

(i)   *The form is definite ($\psi_+ \cdot \psi_- > 0$). The common point is isolated (no branch starts from here). (Like a sphere touching a plane.)*

(ii)   *The form is indefinite ($\psi_+ \cdot \psi_- < 0$). Two branches start from the singular point. The directions are to be computed by (3.8). (Like a hyperbolic paraboloid touching a plane.)*

(iii)   *The form is semi-definite ($\psi_+ = 0$ or $\psi_- = 0$ but not both). One branch passes at the point of tangency. The direction of the branch can be calculated by one of the formulae in (3.8). (Like a cylinder touching a plane.)*

(iv)   *The form is zero. ($\psi_+ = 0$ and $\psi_- = 0$). There is a second order singularity at this point. Higher degree derivatives must be computed in order to analyse the behaviour at this critical point. More advanced methods are necessary here—basically those of catastrophe theory [10],[1].*

As in the parametric case we shall exclude cases (iv). The same considerations are valid as in the previous chapter.

3.1   Curvature at singularities between two implicit surfaces

This chapter is completely analogous to 2.1. For establishing the curvature we need the 3 order derivatives of the surfaces. Let us differentiate the equation (3.3) by σ. We get something similar to (3.5) for f but with $^{\sigma\sigma}$-s.

$$f^{xx}x^{\sigma 2} + f^{yy}y^{\sigma 2} + f^{zz}z^{\sigma 2} + 2f^{xy}x^\sigma y^\sigma + 2f^{yz}y^\sigma z^\sigma + 2f^{zx}z^\sigma x^\sigma +$$

$$f^x x^{\sigma\sigma} + f^y y^{\sigma\sigma} + f^z z^{\sigma\sigma} = 0. \tag{3.9}$$

Moreover, let us differentiate $h = f - \vartheta g$ by σ once more. Note that $\nabla h(x_0, y_0, z_0) = 0$.

$$h^{\sigma\sigma\sigma}(x_0, y_0, z_0) =$$

$$h^{xxx}x^{\sigma 3} + h^{yyy}y^{\sigma 3} + h^{zzz}z^{\sigma 3}+$$

$$3[h^{xxy}x^{\sigma 2}y^{\sigma} + h^{xyy}x^{\sigma}y^{\sigma 2} + h^{yyz}y^{\sigma 2}z^{\sigma}+$$

$$h^{yzz}y^{\sigma}z^{\sigma 2} + h^{xxz}x^{\sigma 2}z^{\sigma} + h^{xzz}x^{\sigma}z^{\sigma 2}] + \qquad (3.10)$$

$$3[h^{xx}x^{\sigma}x^{\sigma\sigma} + h^{yy}y^{\sigma}y^{\sigma\sigma} + h^{zz}z^{\sigma}z^{\sigma\sigma}+$$

$$h^{xy}(x^{\sigma\sigma}y^{\sigma} + y^{\sigma\sigma}x^{\sigma}) + h^{yz}(y^{\sigma\sigma}z^{\sigma} + z^{\sigma\sigma}y^{\sigma}) + h^{yz}(z^{\sigma\sigma}x^{\sigma}$$

$$+x^{\sigma\sigma}z^{\sigma})] = 0.$$

We need the quantities $(x^{\sigma\sigma}, y^{\sigma\sigma}, z^{\sigma\sigma})$. Here (3.9) and (3.10) are two linear equations. The third equation can be obtained if one takes into consideration the arclength parameterisation. That is that $\mathbf{r}^{\sigma\sigma}$ is perpendicular to $\mathbf{r}^{\sigma}$. Thus

$$x^{\sigma}x^{\sigma\sigma} + y^{\sigma}y^{\sigma\sigma} + z^{\sigma}z^{\sigma\sigma} = 0.$$

In this way we have three linear equations for three unknowns. By solving it we can get $\mathbf{r}^{\sigma\sigma}$. The curvature is the magnitude of this vector.

4 First Order Singularities Between Parametric and Implicit Surfaces

Let us consider the intersection of a parametric and an implicit surface.

$$\mathbf{p}(u, v) \qquad \text{and} \qquad g(x, y, z) = 0$$

we can luckily substitute $\mathbf{p}$ into g, that is $\qquad (4.1)$

$$g(\mathbf{p}(u, v)) = 0.$$

This is a fairly delightful situation because the whole intersection problem can be reduced to the tracking of a 2 dimensional function. Let us denote by k the composition of g and $\mathbf{p}$ above, that is:

$$k := g \circ \mathbf{p} \qquad k(u, v) = g(\mathbf{p}(u, v)).$$

We can easily see that this reduction is justified even in the case of singular points. More exactly the following lemma holds.

Lemma 1. Singularity Substitution *Let* $\mathbf{p}(u, v)$ *be a continuously differentiable parametric and* $g(x, y, z) = 0$ *a continuously differentiable implicit surface. Suppose that* $\mathbf{r}_0 = \mathbf{p}(u_0, v_0)$ *is a common singular point, that is:*

(i) $g(\mathbf{p}(u_0, v_0)) = 0$

(ii) $\mathbf{p}^u(u_0, v_0) \times \mathbf{p}^v(u_0, v_0) \parallel \nabla g(\mathbf{r}_0)$.

Under these assumptions the (u_0, v_0) point is singular at $k(u, v) = g(\mathbf{p}(u, v))$, i.e.

$$k(u_0, v_0) = 0 \qquad k^u(u_0, v_0) = 0 \qquad k^v(u_0, v_0) = 0. \qquad (4.2)$$

Moreover the other way round: assuming that the u–v parameter lines are not parallel in the given point, i.e.

$$\mathbf{p}^u(u_0, v_0) \times \mathbf{p}^v(u_0, v_0) \neq 0.$$

If as above in (4.2) (u_0, v_0) is singular at k, then (i)-(ii) hold.

Proof Let us consider the derivatives of k by u and v by the chain rule:

$$
\begin{aligned}
k^u(u_0, v_0) &= \langle \nabla g(\mathbf{r}_0), \mathbf{p}^u(u_0, v_0) \rangle = 0 \\
k^v(u_0, v_0) &= \langle \nabla g(\mathbf{r}_0), \mathbf{p}^v(u_0, v_0) \rangle = 0.
\end{aligned}
$$

From here the statement of the lemma is clear. Notice in general that the singular points of k can be points where the parameter lines of $\mathbf{p}$ are parallel or points where the gradient of g vanishes.

Suppose that we have an initial solution of (4.1) (u_0, v_0). Assume that the two surfaces are tangent in this point, that is (4.2) holds by Lemma 1. As in the previous chapters let one branch of the intersection set starting from $\mathbf{p}(u_0, v_0)$ be parameterised by its arclength parameter σ. Let us denote the branch in the 3 dimensional space as usual by $\mathbf{r}(\sigma)$:

$$k(u(\sigma), v(\sigma)) = 0.$$

Differentiating by σ, we get

$$k^\sigma = k^u u^\sigma + k^v v^\sigma = 0. \qquad (4.3)$$

Let us compute the second derivative of k by σ:

$$k^{uu} u^{\sigma 2} + 2k^{uv} u^\sigma v^\sigma + k^{vv} v^{\sigma 2} + k^u u^{\sigma\sigma} + k^v v^{\sigma\sigma} = 0. \qquad (4.4)$$

If one takes into consideration (4.2), then the last two terms fall out in (u_0, v_0). Thus in this point we get a quadratic form similar to (2.7) and (3.6).

$$k^{uu} u^{\sigma 2} + 2k^{uv} u^\sigma v^\sigma + k^{vv} v^{\sigma 2} = 0. \qquad (4.5)$$

Let us perform the principal axis transformation to this quadratic form.

$$\varphi_+ \xi^2 + \varphi_- \eta^2 = 0 \qquad \text{where} \qquad \varphi_+ \geq \varphi_-. \qquad (4.6)$$

As previously ξ and η are linear orthogonal combinations of u^σ and v^σ. We can have non-trivial solutions only if $\varphi_+ \geq 0$ and $\varphi_- \leq 0$. Here φ_+ and φ_- are the eigenvalues of the matrix

$$\Phi = \begin{pmatrix} k^{uu} & k^{uv} \\ k^{uv} & k^{vv} \end{pmatrix}$$

that is

$$\begin{aligned} \varphi_+ &= \frac{k^{uu}+k^{vv}}{2} + \sqrt{\left(\frac{k^{uu}-k^{vv}}{2}\right)^2 + k^{uv\,2}} \\ \varphi_- &= \frac{k^{uu}+k^{vv}}{2} - \sqrt{\left(\frac{k^{uu}-k^{vv}}{2}\right)^2 + k^{uv\,2}}. \end{aligned}$$

As in the previous chapters if $\mathbf{w}_+$ and $\mathbf{w}_-$ are the corresponding eigenvectors (in the u–v plane) of the matrix Φ, then the critical directions (where the intersection curve goes on) are:

$$\sqrt{-\varphi_-}\,\mathbf{w}_+ + \sqrt{\varphi_+}\,\mathbf{w}_-$$

$$\text{and} \tag{4.7}$$

$$\sqrt{-\varphi_-}\,\mathbf{w}_+ - \sqrt{\varphi_+}\,\mathbf{w}_-.$$

Now this is still in the u–v parametric plane, meaning only directions where to continue. In order to get proper values for u^σ and v^σ one has to take into consideration that for arclength parameterisation the tangent vector

$$\mathbf{e} := \mathbf{r}^\sigma = \mathbf{p}^u u^\sigma + \mathbf{p}^v v^\sigma \tag{4.8}$$

is unitary. Analogously to the previous chapters the following theorem can be formulated.

Theorem 3 *If a smooth parametric surface $\mathbf{p}$ and a smooth implicit surface $g = 0$ are tangent in a point, then in the direction of a branch of the intersection curve the normal curvatures in the two surfaces must coincide. The behaviour of the two surfaces in the environment of the singular point can be described by the quadratic form (4.5). (That is by the eigenvalues φ_+ and φ_-.) The following four cases can be distinguished:*

(i) *The form is definite ($\varphi_+ \cdot \varphi_- > 0$). The common point is isolated (no branch starts from here). (Like a sphere touching a plane.)*

(ii) *The form is indefinite ($\varphi_+ \cdot \varphi_- < 0$). Two branches start from the singular point. The directions are to be computed by (4.7). (Like a hyperbolic paraboloid touching a plane.)*

(iii) *The form is semi-definite ($\varphi_+ = 0$ or $\varphi_- = 0$ but not both). One branch passes at the point of tangency. The direction of the branch can be calculated by one of the formulae in (4.7). (Like a cylinder touching a plane.)*

(iv) *The form is zero.* $(\varphi_+ = 0$ *and* $\varphi_- = 0)$. *There is a second order singularity at this point. Higher degree derivatives must be computed in order to analyse the behaviour at this critical point. More advanced methods are necessary here—basically those of catastrophe theory. (See [10], [1].)*

As in the parametric and implicit cases we shall exclude cases (iv). The same considerations are valid as in the previous chapters.

4.1   Curvature at singularities between a parametric and an implicit surface

This chapter is again completely analogous to 2.1 and 3.1. We need the third order derivatives of the surfaces. Let us differentiate (4.4) by σ once more.

$$k^{uuu}u^{\sigma 3} + 3(k^{uuv}u^{\sigma 2}v^\sigma + k^{uvv}u^\sigma v^{\sigma 2}) + k^{vvv}v^{\sigma 3} +$$

$$3(k^{uu}u^\sigma u^{\sigma\sigma} + k^{uv}(u^{\sigma\sigma}v^\sigma + v^{\sigma\sigma}u^\sigma) + k^{vv}v^\sigma v^{\sigma\sigma}) = 0. \qquad (4.9)$$

We need the quantities $(u^{\sigma\sigma}, v^{\sigma\sigma})$. As in 2.1 in addition to (4.9) another equation is necessary in order to establish these values. This can be obtained from the arclength parameterisation. The second derivative of the intersection curve branch as in (2.5):

$$\ddot{\mathbf{r}}(\sigma) = \mathbf{r}^{\sigma\sigma} =$$

$$\mathbf{p}^{uu}u^{\sigma 2} + 2\mathbf{p}^{uv}u^\sigma v^\sigma + \mathbf{p}^{vv}v^{\sigma 2} + \mathbf{p}^u u^{\sigma\sigma} + \mathbf{p}^v v^{\sigma\sigma}. \qquad (4.10)$$

Because of the arclength parameterisation $\mathbf{r}^{\sigma\sigma}$ is perpendicular to e in (4.8). (to $\mathbf{r}^\sigma$). Thus

$$\langle e, \mathbf{p}^{uu} \rangle u^{\sigma 2} + 2\langle e, \mathbf{p}^{uv} \rangle u^\sigma v^\sigma + \langle e, \mathbf{p}^{vv} \rangle v^{\sigma 2} + \langle e, \mathbf{p}^u \rangle u^{\sigma\sigma} + \langle e, \mathbf{p}^v \rangle v^{\sigma\sigma} = 0.$$

In this way we have two linear equations for two unknowns. By solving it we can substitute $u^{\sigma\sigma}$ and $v^{\sigma\sigma}$ into (4.10) and get $\mathbf{r}^{\sigma\sigma}$. The curvature is the magnitude of this vector.

5   Application: Non-planar Quadric-quadric Intersections

A quadric surface in the 3 dimensional Euclidean space is an implicit surface defined in the following way:

$$a_{11}x^2 + 2a_{12}xy + 2a_{13}xz + 2a_{14}x +$$
$$a_{22}y^2 + 2a_{23}yz + 2a_{24}y +$$
$$a_{33}z^2 + 2a_{34}z +$$
$$a_{44} = 0.$$

Or it can be written in a matrix form as:

$$(x \quad y \quad z \quad 1) \cdot \mathbf{A} \cdot (x \quad y \quad z \quad 1)^T = 0$$

where $\mathbf{A}$ is a symmetric matrix:

$$\mathbf{A} = \begin{pmatrix} a_{11} & a_{12} & a_{13} & a_{14} \\ a_{12} & a_{22} & a_{23} & a_{24} \\ a_{13} & a_{23} & a_{33} & a_{34} \\ a_{14} & a_{24} & a_{34} & a_{44} \end{pmatrix}.$$

Well known second order surfaces as cylinders, cones, ellipsoids (spheres), paraboloids, hyperboloids are all quadrics. Planes or pairs of planes can be described as quadrics as well. The following theorem serves as the basis of the intersection calculation of two quadric surfaces. It can be found in [6], [7] or [13].

Theorem *If the intersection of two quadric surfaces*

$$(x \quad y \quad z \quad 1) \cdot \mathbf{P} \cdot (x \quad y \quad z \quad 1)^T = 0 \tag{5.1}$$

and

$$(x \quad y \quad z \quad 1) \cdot \mathbf{Q} \cdot (x \quad y \quad z \quad 1)^T = 0 \tag{5.2}$$

is non-planar, then the intersection set lies in a hyperbolic paraboloid or in a cylinder (elliptic, parabolic or hyperbolic).

In consequence of this theorem we can assume that $\mathbf{Q}$ in (5.2) defines a parabolic hyperboloid or a cylinder. Let us examine these cases in detail.

If $\mathbf{Q}$ defines a parabolic hyperboloid then in an appropriate coordinate system $\mathbf{Q}$ has the following form:

$$\mathbf{Q} = \begin{pmatrix} A & 0 & 0 & 0 \\ 0 & -B & 0 & 0 \\ 0 & 0 & 0 & J \\ 0 & 0 & J & 0 \end{pmatrix}.$$

That is if we substitute

$$x = \alpha(t+s) \qquad y = \beta(t-s) \qquad z = \gamma ts \tag{5.3}$$

where

$$\alpha = \frac{1}{\sqrt{A}} \qquad \beta = \frac{1}{\sqrt{B}} \qquad \gamma = -\frac{2}{J}$$

we can readily parameterise the implicit surface defined by $\mathbf{Q}$. If we write (5.3) into (5.1) as has been done in (4.1), the resulted k function will be a quadratic polynomial in both of its variables t and s.

$$k(t, s) = f_2(t)s^2 + f_1(t)s + f_0(t) = 0 \tag{5.4}$$

where

$$f_i(t) = f_{i2}t^2 + f_{i1}t + f_{i0} \qquad i = 0, 1, 2.$$

Similarly if $\mathbf{Q}$ defines a cylinder then in an appropriate coordinate system $\mathbf{Q}$ has the following form:

$$\mathbf{Q} \begin{pmatrix} A & 0 & 0 & 0 \\ 0 & B & 0 & 0 \\ 0 & 0 & 0 & 0 \\ 0 & 0 & 0 & K \end{pmatrix} \qquad \text{Elliptic or hyperbolic cylinder}$$

or

$$\mathbf{Q} \begin{pmatrix} A & 0 & 0 & 0 \\ 0 & 0 & 0 & H \\ 0 & 0 & 0 & 0 \\ 0 & H & 0 & 0 \end{pmatrix} \qquad \text{Parabolic cylinder.}$$

Let us consider the first (elliptic or hyperbolic cylinder) case. We can suppose that $A > 0$ and $K < 0$ (if A, B and K have the same sign then the surface is imaginary). Denote by

$$a = -\frac{A}{K} \qquad b = -\frac{B}{K}. \tag{5.5}$$

The cylinder is elliptic if $b > 0$. The following t–z parameterisation will do:

$$x = \frac{2t}{(1 + t^2)\sqrt{a}} \qquad y = \frac{\pm(1 - t^2)}{(1 + t^2)\sqrt{b}} \qquad z = z.$$

Here $\pm$ denotes the upper and lower branch of the ellipse.

If in (5.5) $b < 0$ then the cylinder is hyperbolic. Now similarly to the elliptic case the following t–z parameterisation is satisfactory:

$$x = \frac{1 + t^2}{(1 - t^2)\sqrt{a}} \qquad y = \frac{2t}{(1 - t^2)\sqrt{-b}} \qquad z = z. \tag{5.6}$$

If we write (5.6) into (5.1) as has been done in (5.4), the resulted κ function will be a rational quartic polynomial in t and a quadratic polynomial in z.

$$\kappa(t, z) = \varphi_2(t)z^2 + \varphi_1(t)z + \varphi_0(t) = 0 \tag{5.7}$$

where

$$\varphi_i(t) = \frac{\sum_{k=0}^{2j} \varphi_{ik}t^k}{(1 \pm t^2)^j} \qquad i = 0, 1, 2; \qquad j = 2 - i.$$

For analysing the elliptic and hyperbolic case let us consider the following lemma:

Lemma 2 *Let f and e be continuously differentiable functions in a $\mathcal{G}$ domain of the 2 dimensional plane. Let $(u_0, v_0) \in \mathcal{G}$ be a point where $e(u_0, v_0) \neq 0$. In this case (u_0, v_0) is a singular point of f if and only if it is a singular point of $f \cdot e$.*

Proof Simple. (Let us compute $\frac{\partial (f \cdot e)}{\partial u}$ and $\frac{\partial (f \cdot e)}{\partial v}$.)

Applying this statement to $\kappa(t, z)$ as f and $\left(1 \pm t^2\right)^2$ as e, we are allowed to multiply the equation (5.7) by $\left(1 \pm t^2\right)^2$ on both sides and get:

$$k(t, z) = f_2(t)z^2 + f_1(t)z + f_0(t) = 0 \tag{5.8}$$

where

$$f_i(t) = f_{i2}t^4 + f_{i2}t^3 + f_{i2}t^2 + f_{i1}t + f_{i0} \qquad i = 0, 1, 2.$$

Finally if the cylinder is parabolic a simple t–z parameterisation is striking:

$$x = t \qquad y = -\frac{At^2}{2H} \qquad z = z.$$

Analogously to (5.4) we can write this into (5.1), the resulted k function will be a quadratic polynomial in both of its variables t and z.

$$k(t, z) = f_2(t)z^2 + f_1(t)z + f_0(t) = 0 \tag{5.9}$$

where

$$f_i(t) = f_{i2}t^2 + f_{i1}t + f_{i0} \qquad i = 0, 1, 2.$$

Summarizing (5.4), (5.8) and (5.9) we can state the following theorem:

Theorem 4 *The non-planar intersection of two quadric surfaces can be described by the set of roots of a suitable 2 variables polynomial*

$$k(\tau, \zeta) = f_2(\tau)\zeta^2 + f_1(\tau)\zeta + f_0(\tau) = 0 \tag{5.10}$$

where

$$f_i(\tau) = \sum_{j=0}^{n} f_{ij}\tau^j \qquad n = 2 \ or \ 4 \qquad i = 0, 1, 2$$

where the points in the space can be obtained by simple rational quadratic parameterisation:

$$\mathbf{q} = \mathbf{q}(\tau, \zeta).$$

Let us consider (5.10) as a quadratic polynomial in ζ for fixed τ. Obviously it has a zero if and only if the discriminant of this polynomial is non-negative. The discriminant is a four or eight degree polynomial of τ:

$$w(\tau) := f_1(\tau)^2 - 4f_2(\tau)f_0(\tau) \geq 0. \tag{5.11}$$

The solution for τ must be looked for in the intervals between the roots of $w(\tau)$ where this function is positive. (Actually in elliptic and hyperbolic cases $w(\tau)$ is always divisible by $\left(1 \pm t^2\right)^2$—so in fact at most quartic polynomials are to be handled.) Now we can state the main result of this chapter:

Theorem 5 *Let us suppose that for a given τ_0 f_2 does not vanish ($f_2(\tau_0) \neq 0$). This value corresponds to a singular point on a non-planar intersection curve of two quadric surfaces if and only if τ_0 is a multiple root of $w(\tau)$ of (5.11), that is*

$$w(\tau_0) = 0 \qquad and \qquad w'(\tau_0) = 0 \tag{5.12}$$

Proof Let us use lemma 1, from chapter 4, and consider

$$\begin{align}
k(\tau, \zeta) &= f_2(\tau)\zeta^2 + f_1(\tau)\zeta + f_0(\tau) = 0 \tag{5.13}\\
k^\tau(\tau, \zeta) &= f_2'(\tau)\zeta^2 + f_1'(\tau)\zeta + f_0'(\tau) = 0 \tag{5.14}\\
k^\zeta(\tau, \zeta) &= 2f_2(\tau)\zeta + f_1(\tau) = 0. \tag{5.15}
\end{align}$$

Denote by $\zeta_0 = -\frac{f_1(\tau_0)}{2f_2(\tau_0)}$. Assume that τ_0 satisfies (5.12). Now (τ_0, ζ_0) of course satisfies (5.15) and owing to $w(\tau_0) = 0$ this meets (5.13) as well. If we write it into (5.14) we get

$$\frac{f_2'(\tau_0)w(\tau_0) - f_2(\tau_0)w'(\tau_0)}{f_2(\tau_0)^2} = 0.$$

On the other way round if τ_0 satisfies (5.13), (5.14) and (5.15), then by substituting ζ_0 into (5.14) we get the left side of the equation above; by substituting it into (5.13) we get $\frac{-w(\tau_0)}{4f_2(\tau_0)} = 0$.

If $f_2(\tau_0) = 0$ then there exists a tedious but not too difficult case analysis whether it can be a singular point or not. It will not be scrutinized here; for it depends on the degree of f_i polynomials. In general it can be recorded that singular points τ-s are either roots of w and w' or of f_2. This criterion is enough for establishing the required points.

Finally let us calculate the matrix of quadratic form (4.5) where the directions of the intersection curve branches can be computed from.

Theorem 6 *The tangent vectors of the intersection curve branches of a non-planar intersection curve of two quadric surfaces can be computed by (4.7) theorem 3 from the eigenvectors of the matrix (4.5) of the form:*

$$\Phi(\tau, \zeta) = \begin{pmatrix} f_2''(\tau)\zeta^2 + f_1''(\tau)\zeta + f_0''(\tau) & 2f_2'(\tau)\zeta + f_1'(\tau) \\ 2f_2'(\tau)\zeta + f_1'(\tau) & 2f_2(\tau) \end{pmatrix}$$

where $f_2(\tau)$, $f_1(\tau)$ and $f_0(\tau)$ are the four or eight degree polynomials defined by (5.10).

Proof Obvious.

Notice that Φ above does not depend on the linear and constant terms of k (on f_{00}, f_{01} and f_{10}).

6 Conclusion

In this paper an analysis of the behaviour of the intersection set in the environment of a singular point has been introduced. Under certain not too strict assumptions the directions and the curvatures of the intersection curve branches have been derived for parametric-parametric, implicit-implicit and implicit-parametric surface-surface intersections. Using this information the iteration to the "next" intersection point on a given branch can be done by generalized inverse methods, as introduced for example in [8]. Note that classical Newton-Raphson type methods fail to converge here owing to the rank deficiency of the Jacobian in singular points.

Acknowledgements

This work has been mostly supported by a tender of the grant for basic research of the Computer and Automation Institute, Budapest in 1989. I am indebted to the board of the grant for their help. Special thanks to my colleagues in the Department for CAD in Mechanical Engineering of CAI for their initiative and assistance.

References

1. Arnold, V.I., Gusein-Zade, S.M. and Varchenko, A.N. (1985). *Singularities of Differentiable Maps. Volume 1. The Classification of Critical Points, Caustics and Wave Fronts*, Monographs in Mathematics Vol. 82. Birkhäuser, Boston-Basel-Stuttgart.

2. Cheng, K.P. (1989). Using plane vector fields to obtain all the intersection curves of two general surfaces. In: *Theory and Practice of Geometric Modeling*, (Eds. W. Straßer and H.P. Seidel), Springer-Verlag Berlin, Heidelberg, New York, 187–204.

3. Gel'fand, I.M. (1971). *Lekcii po lineĭnoĭ algebre (Lectures on linear algebra. In Russian)*, Nauka Publishers, Moscow.

4. Faux, I.D. and Pratt, M.J. (1979). *Computational Geometry for Design and Manufacture*, Ellis Horwood Limited, ch. 9.1. 257–267; ch. A4 295–300.

5. Lancaster, P.(1969). *Theory of Matrices*, Academic Press, New York and London.

6. Levin, J. (1976). A parametric algorithm for drawing pictures of solid objects composed of quadric surfaces. *Communications of the ACM*, *19*, No. 10., 555–563.

7. Levin, J. (1979). Mathematical models for determining the intersections of quadric surfaces. *Computer Graphics and Image Processing*, *11*.

8. Lukács, G. (1989). The generalized inverse matrix and the surface-surface intersection problem. In: *Theory and Practice of Geometric Modeling*, (Eds. W. Straßer and H.P. Seidel), Springer-Verlag Berlin, Heidelberg, New York, 167–186.

9. Markot, R.P. and Magedson, R.L. (1989). Solutions of tangential surface and curve intersections. *CAD*, *21*, No. 7, 412–429.

10. Morse, M. and Cairns, S.S. (1969). *Critical Point Theory in Global Analysis and Differential Topology Series: Pure and Applied Mathematics*, Academic Press, New York and London.

11. Nutbourne, A.N. and Martin, R.R. (1988). *Differential Geometry Applied to Curve and Surface Design Volume 1: Foundations*, Ellis Horwood Limited, Chichester.

12. Owen, J.C. and Rockwood, A.P. (1987). Intersection of general implicit surfaces. In: *Geometric Modeling: Algorithms and New Trends*, (Ed. G.E. Farin), Siam, Tempe, Arizona, 335–345.

13. Solomon, B.J. (1985). Surface Intersections for Solid Modelling. PhD Thesis, Clare College, Cambridge.

A System for Parametric Surface Intersection

U. Cugini[1], S. Radi[2] and C. Rizzi[2]

[1]*Università degli Studi di Parma, Italy and* [2]*IMU-CNR, Milano, Italy*

1 Introduction

In many applications of CAD/CAM systems a solid object is defined by joining and/or intersecting primitive parametric surfaces: this procedure allows the representation of objects having extremely complex shapes. In order to define these surfaces many mathematical models have been proposed; each having its own peculiarities both from a theoretical and from an application point of view. This paper describes a system for calculating intersecton curves of parametric surfaces, based on a combination of the subdivision and the curve tracing techniques [1] [2] [3]. It allows the handling of all parametric surfaces $S(u, v)$, belonging to class C^2, defined through a rectangular parametrization. It is easily adaptable to surfaces C^1 although it involves the linear approximation of the calculated intersection curve.

The independence of the surface description model, in the form described in [1], has been realized through the implementation of a module which evaluates parametric surfaces, providing the values of points and derivatives of any surface using an identifier for the surface and the pair of parametric values u and v.

This means that a surface is seen as a single object which contains both the geometrical data and the routines for the calculation of its points and its derivatives. These values, together with the parametric coordinates of a point belonging to the intersection curve, are the only data necessary to solve a problem of surface intersection through a marching method.

The intersection of two surfaces is represented by [1]:

a) an empty set
b) a point collection
c) a curve collection
d) a surface collection
e) some combination of the above.

The implemented algorithm solves intersection problems, the solution of which is of the type a), b), c) and e).

The intersection curve is found point by point, and is represented in R^3 through a linear interpolation.

This type of representation provides sufficient information for utilizing the curve in numerical control applications.

2   System Architecture

Fig. 1 shows the architecture of the implemented system.

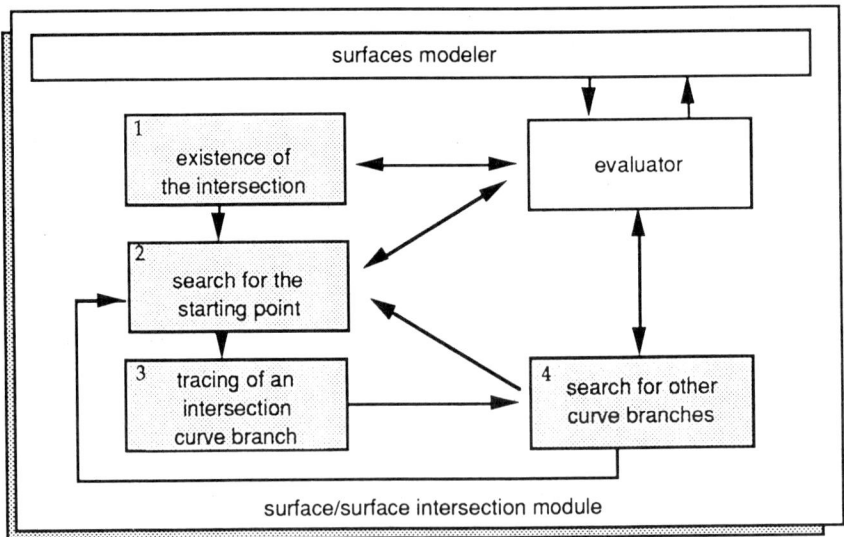

Figure 1.

The search for the intersection curve takes place in four phases denoted in Fig. 1 by shadowed boxes:

1) existence of the intersection: it consists of an initial test to check if the two surfaces are disjointed or not.

2) search for a starting point: having adopted a marching technique, it is necessary to have a starting point for the loop generating the sequence of points belonging to the intersection curve.

3) tracing of an intersection curve branch: a sequence of points belonging or "very near" to the sought curve is generated, starting from a point generated in the previous phase, and moving in a direction defined in function of the local geometry of the involved surfaces.

4) search of other curve branches: in order to have a robust algorithm it is essential that all the curve branches be found.

In the following each phase will be described in more detail.

3 Existence of Intersection

The test has been realized calculating the bounding box of the two surfaces and verifying if the boundary boxes are disjointed.

In literature several methods [4] can be found to calculate a bounding box. The method based on the calculation of parallelepipeds oriented with axes has been adopted.

In order to calculate the bounding box a table (Table 1) is created with the points of the surface $P(u,v)$, obtained by sampling:

Table 1.

$P(0,0)$	$P(1/n,0)$	$P(2/n,0)$	...	$P(1,0)$
$P(0,1/n)$	$P(1/n,1/n)$	$P(2/n,1/n)$	...	$P(1,1/n)$
$P(0,2/n)$	$P(1/n,2/n)$	$P(2/n,2/n)$	...	$P(1,2/n)$
.	.	.		.
.	.	.		.
.	.	.		.
$P(0,1)$	$P(1/n,1)$	$P(2/n,1)$	...	$P(1,1)$

Then the maximum and minimum values of the x, y and z coordinates are selected.

With these values a box (oriented with the axes) is defined to contain the surface. The dimensions of the box along the directions x, y and z are respectively:

$$x_{max} - x_{min}, y_{max} - y_{min}, z_{max} - z_{min}.$$

In this way the two bounding boxes containing the surfaces $P(u,v)$ and $Q(s,t)$ are defined.

If the two bounding boxes do not intersect, the algorithm stops here. However it is not determined in these cases if the surfaces present cusps.

4   Search of the Starting Point

The aim of this phase is the search for a starting point for the iteration generating the chain of points belonging to the intersection curve.

Given two points $P_0 = P(u_0, v_0)$ and $Q_0 = Q(s_0, t_0)$ belonging respectively to the first $P(u, v)$ and second $Q(s, t)$ surfaces, a vector $w_0 = [u_0, v_0, s_0, t_0]$ represents a starting point if

$$P_0 - Q_0 \leq \varepsilon$$

ε being the tolerance used to indicate when a pair of points represents a true intersection point.

w_0 is calculated by minimizing the distance between two points, $P(u, v)$ and $Q(s, t)$, in the parameter spaces of the two surfaces.

The algorithm used to minimize the distance is that described in [2]. It requires a starting point w_0^* to be determined in order to find w_0. w_0^* is calculated as follows.

First of all the bounding box V containing both surfaces is calculated and decomposed into voxels.

V is determined starting from the bounding boxes of the two surfaces obtained in the previous phase, through determination of the end points $(x_{max}, y_{max}, z_{max})$ and $(x_{min}, y_{min}, z_{min})$ of a V diagonal.

Let (Px_i, Py_i, Pz_i) and (Qx_i, Qy_i, Qz_i) be generic vertices of the bounding boxes containing the surface $P(u, v)$ and $Q(s, t)$ respectively, then we have that:

$$x_{max} = MAX(Px_{max}, Qx_{max})$$

$$y_{max} = MAX(Py_{max}, Qy_{max})$$

$$z_{max} = MAX(Pz_{max}, Qz_{max})$$

and

$$x_{min} = MIN(Px_{min}, Qx_{min})$$

$$y_{min} = MIN(Py_{min}, Qy_{min})$$

$$z_{min} = MIN(Pz_{min}, Qz_{min})$$

where Px_{max} is the maximum component chosen among the vertices of the bounding box built on the surface P, and so on for Qx_{max}, Py_{min}, etc.

The result of this operation is represented in Fig. 2.

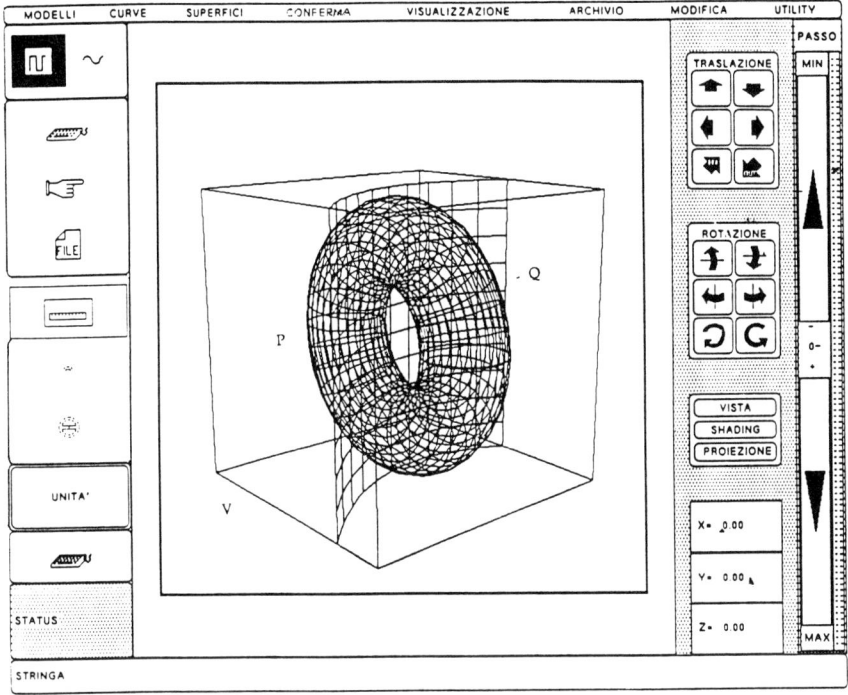

Figure 2.

Notice that the lengths of the edges of V are given by:

$$L_x = x_{max} - x_{min}$$

$$L_y = y_{max} - y_{min}$$

$$L_z = z_{max} - z_{min}$$

Now V must be decomposed into voxels, the number and dimension of which are controlled through three integers $NVOX_x$, $NVOX_y$, $NVOX_z$, respectively the number of voxels along the three axes.

The decomposition of V into voxels is represented by a $NVOX_x * NVOX_y * NVOX_z$ array.

The length of each voxel edge is given by:

$$\ell_x = \frac{L_x}{NVOX_x} = \frac{(x_{max} - x_{min})}{NVOX_x};$$

$$\ell_y = \frac{L_y}{NVOX_y} = \frac{(y_{max} - y_{min})}{NVOX_y};$$

$$\ell_z = \frac{L_z}{NVOX_z} = \frac{(z_{max} - z_{min})}{NVOX_z}.$$

A function $f : R^3 -- > I^3$ corresponds to this decomposition.

The function f associates each generic point with the voxel containing it.

Now we define two grids, one for each surface. Let $P_{ij} = P(u_i, v_j)$, the nodes of the P grid and $Q_{hk} = Q(s_h, t_k)$, the nodes of the Q grid.

Applying f to each P_{ij} we obtain a set $V_p = \{f(P_{ij})\}$ whose elements are those voxels of V containing at least a point P_{ij}.

In the same way we obtain a set $V_Q = \{f(Q_{hk})\}$.

Calculating $V_{int} = V_P \bigcap V_Q$ those voxels which contain at least a P_{ij} and a Q_{hk} are selected (Fig. 3).

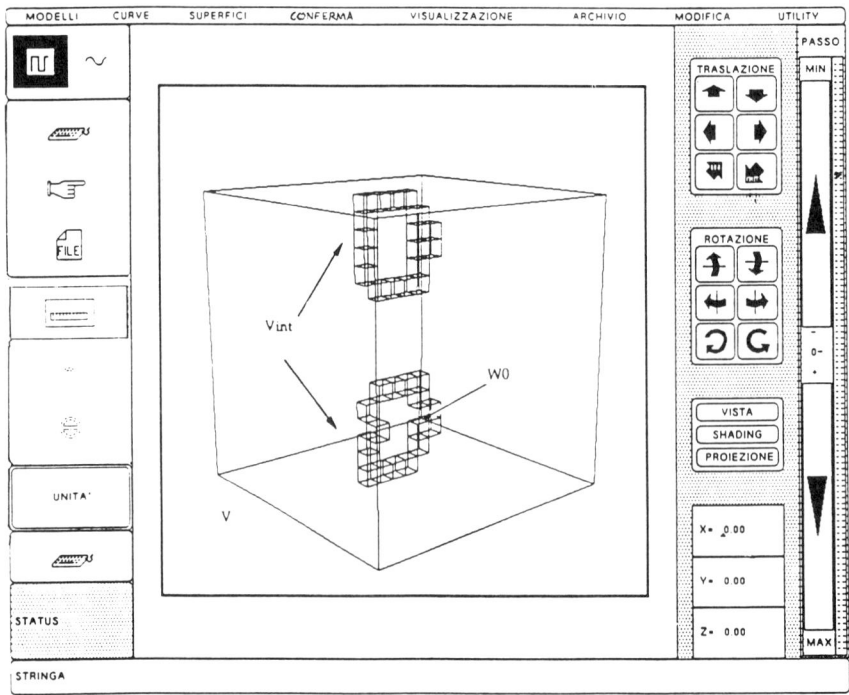

Figure 3.

Let $v \varepsilon V_{int}$ then v corresponds to a pair of points P_{ij} and Q_{hk}.

Let $w_0^* = [u_i, v_j, s_h, t_k]$ and using an algorithm like that described in [2] we obtain $w_0 = [u_0, v_0, s_0, t_0]$ such that $P(u_0, v_0) = Q(s_0, t_0)$, (Fig. 3).

The algorithm effectiveness is connected to the choice of $NVOX_x$, $NVOX_y$ and $NVOX_z$, and to the choice of the dimensions of the parametric region sampling steps, dimension controlling the approximation degree of the surfaces.

The main advantages of our algorithm are:

• the calculation of a starting point in a fully automatic way;

• the possibility of obtaining a set of points allows the problem of determining all the curve branches, composing the intersection of two surfaces, to be solved.

5 Tracing of the Intersection Curve

Once the starting point w_0 is calculated, the subsequent point must be determined. A marching method has been chosen.

The intersection condition can be formulated as follows [5]:

$$P(u, v) - Q(s, t) = 0 \qquad (5.1)$$

The intersection curve is composed of all points corresponding to the solution of the system (5.1).

Given w_0, the starting point calculated in the previous phase, and considering (5.1) we have:

$$P(u_0, v_0) = Q(s_0, t_0)$$

The intersection condition (5.1) leads to a non-determined system (three equations and four unknown values), therefore it is necessary to introduce a fourth equation, the so called step constraint equation.

Geometrically, it is equivalent to the following requirements: the new generated point must not only belong to the intersection curve (intersection condition $P - Q = 0$), but it must also lie in a plane, whose distance from the current point is c (step length), and whose normal vector is given by the tangent vector of the intersection curve evaluated in the current point [6] (Fig. 4).

The choice of the step length depends on the application and on the type of interpolation that must be consequently utilized.

In this work the step length depends on the curve curvature and a linear interpolation has been chosen since this technique is utilized by most numerical control machines.

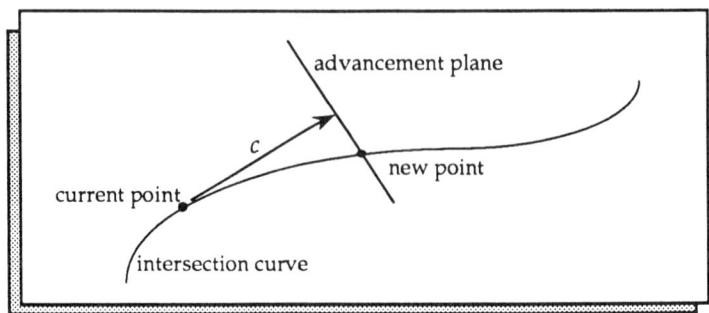

Figure 4.

The problem then becomes the search for the solution of the following determined system:

$$\begin{cases} P(u,v) - Q(s,t) &= 0 \\ G_1(u,v) &= 0 \end{cases} \qquad (5.2)$$

or also:

$$\begin{cases} P(u,v) - Q(s,t) &= 0 \\ G_2(s,t) &= 0 \end{cases} \qquad (5.3)$$

where G_1, G_2 are the step constraints.

The aim of the fourth equation is the determination of the step length along the intersection curve.

So that, let:

$$
\begin{array}{lcl}
P_0 = P(u_0, v_0) &=& \text{current point of the intersection curve} \\
T &=& \text{unit vector of the intersection curve in } P_0; \text{ nor-} \\
  & & \text{mal to the advancement plane} \\
c &=& \text{advancement plane distance from the current} \\
  & & \text{point } P_0 \\
P = P_0 + cT &=& \text{point belonging to the advancement plane} \\
P(u,v) &=& \text{generic surface point}
\end{array}
$$

G_1 is as follows:

$$G(u,v) = G_1(u,v) = [P(u,v) - P_0] \cdot T - c = 0.$$

The coefficients of G_1 are either known (as P_0) or can be fixed, depending on the required accuracy degree (as for c coefficient).

The unit tangent T has not been determined yet.

If the two surfaces are not tangent, then [8]

$$T = \frac{(P^u \times P^v) \times (Q^s \times Q^t)}{|(P^u \times P^v) \times (Q^s \times Q^t)|}$$

The system of equations is iteratively solved through the Newton-Raphson method, considering $P = P_0 + cT$ as initial point [2].

The system solution, w_{new}, is the new point belonging to the intersection curve.

This point is now the current point P_0.

Now, set $w_0 = w_{new}$, the process is iterated until the curve branch is completely determined (Fig. 5).

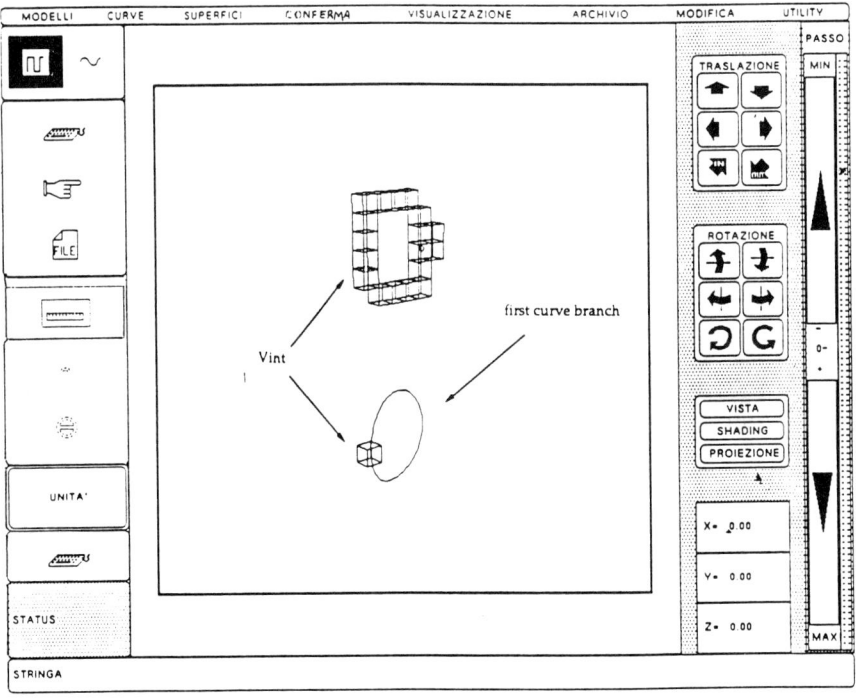

Figure 5.

Anomalous situations, for example when the surfaces $P(u, v)$ and $Q(s, t)$ are tangent, are solved through the method proposed in [2], based on a more sophisticated technique for the numerical solution of the system 5.2 with the modified Newton system described by Deuflhard [7].

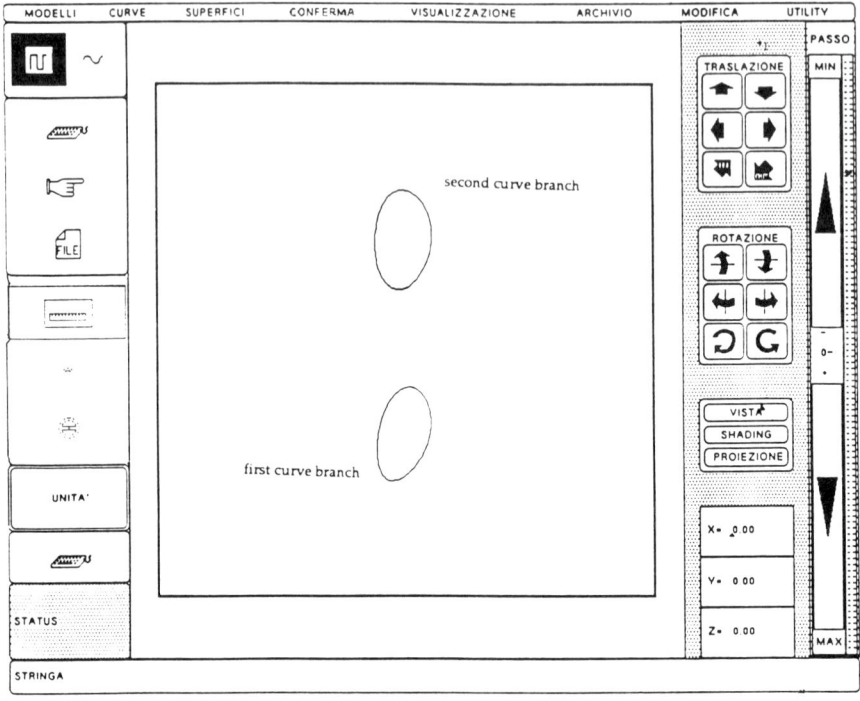

Figure 6.

6   Search for Other Curve Branches

In this case the solution consists of a set of disjointed curve branches, each produced in a different marching phase.

No curve branch of the intersection must be lost. This is equivalent to needing to know at least one starting point for each curve branch [1].

Another critical situation to be avoided is the generation of the same curve branch starting from two different starting points.

These problems are connected to the quantity and quality of the considered starting points.

Assume that the intersection curve is composed by n different curve branches.

Then:

- quantitative aspect: the number of starting points must be at least equal to n

- qualitative aspect: each of the n starting points must belong to a different curve branch

The number of starting points, i.e. V_{int} cardinality, is no greater than $NVOX_x * NVOX_y * NVOX_z$. Moreover $|V_{int}|$ depends on the number of nodes of the grids defined on $P(u,v)$ and $Q(s,t)$.

Generally $|V_{int}|$ is greater than the number of curve branches of the intersection curve.

The search for other curve branches takes place as follows.

Function f is applied to each point P_i generated in the tracing phase. If $f(P_i) \in V_{int}$ then $V_{int} = V_{int} - \{f(P_i)\}$.

Moreover all voxels v crossed by the segment joining two following intersection points are deleted from V_{int}.

At the end of the tracing phase, all voxels $v \in V_{int}$ crossed by the found curve branch have been deleted, so that all starting points belonging to the curve branch are indirectly deleted (Fig. 5).

The algorithm stops when $V_{int} = \phi$, otherwise a new starting point is generated and the tracing phase is repeated (Fig. 6).

7 Conclusion

The independence of the surface description model, in the form described in [1], has been realized through the implementation of a module valuing parametric surfaces, providing the values of points and derivatives of any surface starting from the surface indentifier and from the pair of parametric values u and v.

In addition to the realization of the independence of the mathematical model, the implemented algorithm has the following characteristics:

- The search for a starting point phase allows the determination of all branches composing the intersection curve. In fact, the subdivision phase for the search of the starting points generates at least one intersection point for each curve branch.

- The algorithm allows a check of the accuracy degree of the approximation of the sought intersection curve, starting from estimates of local geometrical properties of the curve itself, such as the radius of curvature and the tangent vector. It is possible to thicken the points in the regions where the curve has a small radius of curvature, while in the regions with almost linear behaviour it is possible, and also convenient, to reduce the density of the points.

The system has been tested by intersecting surfaces generated by means of Bezier and B-spline mathematical representations, as well as parametric analytic surfaces such as sphere, cylinder, torus and so on.

In order to prove the actual independence of the modeler, the implemented module has been also utilized through connecting it to two different surface modelers (MISS a surface modeler, developed at I.M.U-C.N.R., and SERIES7000, a CAD system released by Italcad and Autotrol).

The system has been implemented on Apollo Workstation, with AEGIS (Unix like) as operative system, by utilizing the *C* programming language and the graphic package GMR-3D, Apollo implementation of the standard PHIGS.

References

1. Barnhill, R.E., Farin, G., Jordan, M. and Piper, B.R. (1987). Surface/surface intersection. *Computer Aided Geometric Design*, *4*, 3–16.

2. Lukacs, G. (1989). The generalized inverse matrix and the surface-surface intersection problem. *Theory and Practice of Geometric Modeling*, W. Strasser and H.-P. Seidel (Eds.).

3. Pratt, M.J. and Geisow, A.D. (1986). Surface/surface intersection problems. *The Mathematics of Surfaces*, Oxford University Press.

4. Factor, J.D. and Sabharval, C.L. (1988). Cross intersection between any two C^0 parametric surfaces.

5. Chen, J.J. and Ozsoy, T.M. (1986). An intersection algorithm for C^2 parametric surface. *Proceedings of CAD86*.

6. Faux, I.D. and Pratt, M.J. (1979). *Computational Geometry for Design and Manufacture*, Ellis Horwood Limited.

7. Choi, B.K. and Ju, S.Y. (1989). Constant radius blending in surface modelling. *Computer Aided Design*, *21*, 213–220.

8. Deuflhard, P. (1974). A modified Newton method for the solution of ill-conditioned systems of nonlinear equations with applications to multiple shooting. *Numerische Matematik*, *22*, 289–315.

The Volume Function

Philip Milne

School of Mathematical Sciences, University of Bath

Abstract Given a number of real polynomials in the same number of variables, the algorithm presented here computes a sequence of polynomials which, in a way analogous to the sequences of Sturm in one dimension, may be used to count the number of zero-dimensional distinct real common solutions of the equations in any co-ordinate aligned box. Through recursive subdivision of a bounding box this technique may be used either to isolate each of the solutions or to approximate them to a given accuracy. In particular we are able to distinguish a multiple root from two close roots, by a generalisation of the concept of "square-free".

Precise algorithms for testing the sign of a single multivariate polynomial throughout a box also follow from this as do sequences for locating complex roots.

1 Introduction

What follows is a description of a technique for locating the zero-dimensional or "point-like" solutions of systems of simultaneous polynomial equations, assuming that there are no higher-dimensional solutions. Each solution is counted just once regardless of its multiplicity, allowing the notion of root isolation to be generalised to higher dimensions.

Most of the proofs are given for the two dimensional case since it is normally easier to generalise a two dimensional theorem than it is to understand an N-dimensional one. Unlike the author's previous work, no part of this account is believed to be limited to two dimensions.[1]

2 Real Roots of Systems of Equations

Cauchy's theorem relating:

$$\int \frac{f'}{f} dz$$

---

[1] Formal N-dimensional proofs of these ideas have since been published in Milne [5].

to the number of roots of f is, of course, quite remarkable. This extraordinary result, that the values of f'/f on a contour in $\mathbf{C}$ determine the values of f inside the contour, is a consequence of the condition that f is analytic inside the appropriate domain. The associated Sturm sequence, which begins with the two terms f and f', has the same wonderful dimension-cheating property.

The aim of this paper is to demonstrate that, by encapsulating the positions of the roots of the system in a single characteristic polynomial, both of these notions may easily be generalised to higher dimensions. In two real dimensions we could view the sequence as something derived from a single polynomial mapping F from $\mathbf{R}^2$ to $\mathbf{R}^2$. The sequence may then be used to count two dimensional "roots", $\alpha = (\alpha_x, \alpha_y)$, which are points which map to the origin.

Many, including Hermite [3] have studied the following integral over a surface in $\mathbf{C}^2$ as a two dimensional analogue of Cauchy's root counting integral for functions in one variable:

$$\int \int \frac{J(P,Q)}{PQ} dx dy$$

where $F(x,y) = (P(x,y), Q(x,y))$ and

$$F' = \left(\begin{array}{cc} P_x & P_y \\ Q_x & Q_y \end{array} \right)$$

so that the Jacobian satisfies

$$J(P,Q) = \det(F').$$

Cauchy's $\frac{f'}{f}$ has the following property[2],

$$\lim_{x \to \alpha} \left(\frac{f'(x)}{f(x)} (x - \alpha) \right) = \left\{ \begin{array}{ll} 1 & \text{if } f(\alpha) = 0 \\ 0 & \text{otherwise} \end{array} \right. .$$

The two dimensional function $\frac{J(P,Q)}{PQ}$ is nearly analogous to this:

$$\lim_{(x,y) \to \alpha} \left(\frac{J(P,Q)(x,y)}{P(x,y)Q(x,y)} (x - \alpha_x)(y - \alpha_y) \right) =$$
$$\left\{ \begin{array}{ll} \omega & \text{if } P(\alpha_x, \alpha_y) = Q(\alpha_x, \alpha_y) = 0 \\ 0 & \text{otherwise} \end{array} \right. ,$$

---

[2] We assume that the function is square-free so that each root is of multiplicity 1. Later we see that the notion of square-freeness, in the sense of ensuring that all roots are of multiplicity 1, *does* extend to higher dimensions through the definition of the volume function.

but there is this funny constant ω which varies with the orientation of the axes.

In some sense, we do not want the Jacobian here, but instead,

$$\frac{(P_x P_y)\left(\begin{array}{c}dx\\dy\end{array}\right)(Q_x Q_y)\left(\begin{array}{c}dx\\dy\end{array}\right)}{PQ}.$$

Here, $\omega = 1$ in the above, but this is not a simple volume integral of the sort above. One fix to this is to "orthogonalise" the system so that instead of P and Q we have $\hat{P}, \hat{Q}$ such that $\hat{P}_y = \hat{Q}_x = 0$, then the systems above are equivalent. Perhaps this is motivation enough for the definition of a new function which, like the u-resultant [7], uniquely characterises the point-like roots of a system of N equations in N variables.

The naïve analogue of the univariate factorisation:

$$f(x) \propto \prod_\alpha (x - \alpha)$$

might be

$$f(x, y) \propto \prod_\alpha ((x - \alpha_x)(y - \alpha_y))$$

where $\propto$ is used to mean "has the same zero set as". But this, of course, suffers from the problem that the roots (α_x, α_y), (β_x, β_y) are not distinguished from the spurious intersections of the factors of f, so that the points (α_x, β_y) appear to be roots as well. One solution to this is to add another variable, u, to this product, so as to bind, in an algebraic way, the components of the factors into their associated pairs.

Definition. In $\mathbf{R}^2$ we define[3] the *volume function* as follows:

$$V(u, x, y) \propto \prod_\alpha (u + (x - \alpha_x)(y - \alpha_y)).$$

The extension to $\mathbf{R}^n$ is done in the obvious way:

$$V(u, x) \propto \prod_\alpha \left(u + \prod_i (x_i - \alpha_i)\right)$$

where, $x = (x_1, x_2, ..., x_n)$ and $\alpha = (\alpha_1, \alpha_2, ..., \alpha_n)$.

Returning to the two dimensional case we note that the volume function has two useful properties. Firstly that when $u = 0$ it is the "orthogonalised" root system we had above and secondly that its derivative w.r.t. u satisfies,

$$\frac{V_u(0, x, y)}{V(0, x, y)} = \sum_\alpha \frac{1}{(x - \alpha_x)(y - \alpha_y)},$$

---

[3] Up to a constant, for now.

the numerator of which is zero at the spurious intersection points mentioned above.

We can now set $\hat{P}\hat{Q} = V(0, x, y)$ and $\hat{J} = V_u(0, x, y)$ so that:

$$\lim_{(x,y)\to\alpha} \left(\frac{\hat{J}(x, y)}{\hat{P}(x, y)\hat{Q}(x, y)}(x - \alpha_x)(y - \alpha_y) \right) =$$

$$\begin{cases} 1 & \text{if } P(\alpha_x, \alpha_y) = Q(\alpha_x, \alpha_y) = 0 \\ 0 & \text{otherwise} \end{cases}.$$

3   Computation of the Volume Function

In essence the volume function is computed by eliminating a and b from the following three expressions: $P(a, b), Q(a, b)$ and $u + (x - a)(y - b)$. It is defined to be zero when there are infinitely many roots.

Define an *associate* of the volume function to be any polynomial with the same zero set. For the algorithms which follow it suffices to be able to compute the square-free part of the volume function and this may be computed from any of its associates. The subresultant algorithm may be used to compute an associate of the volume function as follows.

$$V(u, x, y) \propto \frac{Res_b(Res_a(P, R), Res_a(Q, R))}{u^{deg_a(P)deg_b(Q)}},$$

where $R = u + (x - a)(y - b)$ and P and Q are polynomials in a and b.

Although it is currently less efficient, Buchberger's algorithm may also be used to perform the elimination, and in this form the extension to larger numbers of variables is more straightforward. Denote by G the reduced Gröbner basis of the ideal generated by $P(a, b)$, $Q(a, b)$ and $u+(x-a)(y-b)$ with respect to the purely lexicographic ordering $a > b > u$,

$$G(\{P(a, b), Q(a, b), u + (x - a)(y - b)\}, \{a, b, u\}).$$

When there are finitely many roots, an associate of the volume function is the unique member of the Gröbner basis which is of degree zero in both a and b.

$$V(u, x, y) \propto \mathbf{R}[u, x, y] \cap G(\{P(a, b), Q(a, b), u + (x - a)(y - b)\}, \{a, b, u\}).$$

At least in the absence of special cases, the obvious generalisation to larger numbers of variables is the correct one. For three variables we compute:

$$V(u, x, y, z) \propto \mathbf{R}[u, x, y, z] \cap G(\{P, Q, R, u+(x-a)(y-b)(z-c)\}, \{a, b, c, u\}),$$

where P, Q and R are all functions of a, b and c.

4   Cauchy's Theorem and the Volume Function

We will now see how the volume function may be used in a trivial extension of Cauchy's theorem to polynomial maps in N-dimensions.

Theorem (Cauchy *et al*). Let C be a rectifiable Jordan curve in $\mathbf{C}$ with inner region S. If f is analytic both on C and in S then the number, n, of roots α which satisfy $f(\alpha) = 0$ and $\alpha \in S$ is related to the integral of $\frac{f_z}{f}$ around C by the following formula:

$$2\pi i n = \oint_C \frac{f_z(z)}{f(z)} dz.$$

One possible analogue of this is the following. Call the cross product of two rectifiable Jordan curves a *rectifiable Jordan surface*.[4]

Lemma 1. Let $\underline{C} = (C_1 \times C_2)$ be a rectifiable Jordan surface in $\mathbf{C}^2$ with inner region $\underline{S} = (S_1 \times S_2)$. If $\underline{f}$ is polynomial both on $\underline{C}$ and in $\underline{S}$ then number, n, of roots $\underline{\alpha}$ which satisfy $\underline{f}(\underline{\alpha}) = \underline{0}$ and $\underline{\alpha} \in \underline{S}$ is related to the integral over $\underline{C}$ of the volume function V of $\underline{f}$ by the following formula:

$$-4\pi^2 n = \int_C \int \frac{V_u(0, z_1, z_2)}{V(0, z_1, z_2)} dz_1 dz_2.$$

Proof.

$$\int_C \int \frac{V_u(0, z_1, z_2)}{V(0, z_1, z_2)} dz_1 dz_2 = \oint_{C_1} \oint_{C_2} \frac{V_u(0, z_1, z_2)}{V(0, z_1, z_2)} dz_1 dz_2.$$

$$= \oint_{C_1} \oint_{C_2} \sum_\alpha \frac{1}{(z_1 - \alpha_1)(z_2 - \alpha_2)} dz_1 dz_2$$

$$= \sum_\alpha \oint_{C_1} \oint_{C_2} \frac{1}{(z_1 - \alpha_1)(z_2 - \alpha_2)} dz_1 dz_2$$

$$= \sum_\alpha \oint_{C_1} \frac{1}{(z_1 - \alpha_1)} dz_1 \oint_{C_2} \frac{1}{(z_2 - \alpha_2)} dz_2$$

$$= \sum_\alpha (2\pi i)\delta_1(\alpha)(2\pi i)\delta_2(\alpha)$$

where

---

[4]Note that the surface is not the *envelope* of a set of points in $\mathbf{C}^2 \equiv \mathbf{R}^4$ as it is of real co-dimension two.

$$\delta_1(\alpha) = \left\{ \begin{array}{ll} 1 & \text{if } \alpha_1 \in S_1 \\ 0 & \text{otherwise} \end{array} \right. \quad \delta_2(\alpha) = \left\{ \begin{array}{ll} 1 & \text{if } \alpha_2 \in S_2 \\ 0 & \text{otherwise} \end{array} \right.$$

$$= \sum_\alpha (2\pi i)^2 \delta(\underline{\alpha})$$

where

$$\delta(\underline{\alpha}) = \left\{ \begin{array}{ll} 1 & \text{if } \underline{\alpha} \in S \\ 0 & \text{otherwise} \end{array} \right.$$

$$= -4\pi^2 n \qquad \square$$

Just as one might expect, it turns out that the sequence of multivariate polynomials which may be used to count the distinct roots of the P, Q system in $\mathbf{R}^2$ begins with the two functions $V(0, x, y)$ and $V_u(0, x, y)$.

5   Univariate Sturm Sequences

Throughout this text we make use of two new functions var and sim[5] which respectively count the number of sign changes and the number of similarly signed adjacent pairs amongst the elements of a sequence of reals. As long as the convention is consistent a zero may be taken as either positive or negative since it will be straddled by terms of opposite sign[6]. We note that:

$$sim(S(x)) + var(S(x)) = l - 1, \tag{5.1}$$

where l is the length of the sequence. The sim primitive therefore makes no important contribution to this construction; var may be used instead with the sole effect that a minus sign is introduced throughout. The sim operator is, in some sense more natural however, and makes evaluations of these sequences analogous to those of definite integration.

The following theorem is in essence Sturm's theorem except that we use sim instead of var and try to find exactly what $sim(S(x))$ is, rather than just how it changes. We will restrict ourselves to *normal* sequences: those in which the degree drop between consecutive terms is unity.

Lemma 2. If $S(u)$ is a *normal* polynomial remainder sequence obtained using the negated subresultant algorithm on $\{V(u), V'(u)\}$ then, provided the last term in the sequence $S_n(a) \neq 0$, $sim(S(a))$ is the number of distinct real roots of V which are less than a plus the number of distinct pairs of complex roots of V.

---

[5] In subsequent work the conventional *per* operator is used instead of sim.

[6] Except at the endpoints of the sequence, where the choice of sign for zero defines closure or openness of the interval for that root.

Proof. The number of distinct real roots of V in the interval $(a, b]$ is given by Sturm's theorem as $var(S(a)) - var(S(b))$. So that the number of real roots in the interval $[a, b)$ is given by $sim(S(b)) - sim(S(a))$. Consequently,

$$sim(S(\infty)) - sim(S(-\infty)) = r \qquad (5.2)$$

where r is the total number of distinct real roots.

Now, $sim(S(-\infty))$ is the number of similarly signed adjacent pairs of leading coefficients after alternate terms have been negated; this is just the number of sign variations in $S(\infty)$, so:

$$sim(S(-\infty)) = var(S(\infty)).$$

Summing equation 5.1 evaluated at $x = \infty$ and the above we have,

$$sim(S(\infty)) + sim(S(-\infty)) = l - 1,$$

where l is the length of the sequence.

If the last term is of degree 0 then the length of the sequence is $deg(V) + 1$, where $deg(V)$ is, from the fundamental theorem of algebra, equal to the total number of roots. If G is the product of all the factors common to V and V' ie. $G = GCD(V, V')$ then G divides each element of the sequence. The sequence is therefore of the same length as a sequence pertinent to the square-free part of V and therefore has a number of elements equal to one plus the number of distinct roots. Thus,

$$sim(S(\infty)) + sim(S(-\infty)) = n$$

where n is the number of distinct roots. Subtracting equation 5.2 from the above gives,

$$2sim(S(-\infty)) = n - r = c$$

where c is the number of distinct complex roots. Since for a polynomial with real coefficients each complex root has a conjugate, $sim(S(-\infty))$ counts the number of such conjugate pairs.

So, using this modified version of Sturm's theorem, we have that: $sim(S(a)) - sim(S(-\infty))$ is the number of distinct real roots less than a. $sim(S(a))$ is therefore the number of distinct real roots less than a plus the number of distinct pairs of complex roots.

$\square$

Corollary 1. Provided S is *normal* and $S_n(0) \neq 0$, $sim(S(0))$ is equal to the number of distinct negative real roots of a polynomial plus the number of distinct pairs of complex roots.

6   An Evaluation Function for the Sequence

We now define an evaluation function $E(M, I)$ which given a sequence of multivariate polynomials M and a coordinate aligned box I evaluates the sequence at the corners of the region and uses the signs of the polynomials at the corners to return a single integer.

In $\mathbf{R}$,

$$E(M, I) = sim(M(x_u)) - sim(M(x_l)).$$

In $\mathbf{R}^2$, $E(M, I) =$

$$\frac{1}{2}\left(sim(M(x_l, y_l)) + sim(M(x_u, y_u)) - sim(M(x_u, y_l)) - sim(M(x_l, y_u))\right).$$

where the subscripts u and l denote the upper and lower bounds on the value of each variable in I.

In $\mathbf{R}^n$, the evaluation function is always: evaluate the sequence at each vertex of the box which makes a positive volume relative to the centroid, calculate the number of adjacent pairs of elements in the sequence which have similar signs and sum these results, do the same for the vertices which make a negative volume relative to the centroid, subtract the previous two integers and divide by 2^{d-1}, where d is the number of dimensions.

7   The Two Dimensional Case

Given two bivariate polynomials $P(x, y)$ and $Q(x, y)$ we may compute their associated volume function, which will be non-zero if and only if they have finitely many solutions, and in this case,

$$V(u, x, y) = \prod_\alpha (u + (x - \alpha_x)(y - \alpha_y)).$$

Treating u as the main variable of V we can now compute a multivariate sequence $S(u, x, y)$ of $V(u, x, y)$ by using the negated subresultant algorithm on V and $V_u{}^7$,

$$S(u, x, y) = nprs_u(V(u, x, y), V_u(u, x, y)).$$

Define M to be the value of the sequence at $u = 0$,

$$M(x, y) = S(0, x, y).$$

---

[7] Since the leading coefficients of both V and V_u are integers, multiplications in the coefficient domain are all by perfect squares of polynomials in $\mathbf{R}[x][y]$ unless the sequence is *abnormal*. It seems likely that there is an algorithm similar to the subresultant algorithm which can generate precisely those terms of section 9 regardless of anomalies in the execution of the Euclidean algorithm.

Theorem 1. The number of distinct simultaneous real roots of a pair of bivariate polynomials with *normal* sequence $M(x,y)$ in any coordinate aligned rectangle I in $\mathbf{R}^2$ is precisely $E(M,I)$ provided that the last term of the sequence M_n does not vanish at any of the vertices of the rectangle.

Proof. We have from corollary 1 that if S is a normal subresultant sequence then $sim(S(0))$ is the number of distinct negative real roots plus the number of distinct pairs of complex roots. $M(x,y)$ is just such a sequence; for the volume function evaluated at $u = 0$. $sim(M(x,y))$ is therefore the number of distinct roots α of the system for which $-(x - \alpha_x)(y - \alpha_y)$ is distinct, negative and real plus half the number of distinct roots for which this expression is complex.

Firstly we note that any complex root, $\alpha \notin \mathbf{R}^2$, for which $-(x - \alpha_x)(y - \alpha_y) \notin \mathbf{R}$ at any of the corners of the box, makes the same contribution at each corner and, in combination with the evaluation function $E(M,I)$, contributes zero to the final root count.

Secondly we consider $\alpha \notin \mathbf{R}^2$ and $-(x - \alpha_x)(y - \alpha_y) \in \mathbf{R}$. Since the polynomial map has real coefficients, any complex root, $\alpha \notin \mathbf{R}^2$, has a conjugate $\alpha^* = (\alpha_x^*, \alpha_y^*)$ which is also a root of the system. So, for the conjugate $-(x - \alpha_x^*)(y - \alpha_y^*) = (-(x - \alpha_x)(y - \alpha_y))^* = -(x - \alpha_x)(y - \alpha_y) \in \mathbf{R}$. This, therefore, corresponds to a double root of the volume function and S_n will be zero.

For the real roots then, $\alpha \in \mathbf{R}^2$, $sim(M(0))$ counts just those which have a *positive volume* with respect to an origin at (x,y). A root α in $\mathbf{R}^2$ has positive real volume if it lies in the upper right or lower left quadrants.

Imagine now that we have a new sequence B, like M, but which counts just those roots which lie in quadrant one: where both Cartesian components are positive. Now changing x and y simply allows us to shift this origin and calculate the number of real roots in quadrant one with respect to an origin at (x,y) ie. to count the number of roots α for which $\alpha_x > x$ and $\alpha_y > y$.

We can now evaluate sequence $B(x,y)$ at each corner of a coordinate aligned rectangle. A study of the areas involved reveals that the number of simultaneous roots in the rectangle is precisely:

$$sim(M(x_l, y_l)) + sim(M(x_u, y_u)) - sim(M(x_u, y_l)) - sim(M(x_l, y_u))$$

where the subscripts l and u denote the upper and lower bounds on the value of each variable in I. Now imagine that instead of calculating the number of roots in quadrant one, a new sequence B' in fact calculates the number of roots in quadrant three relative to an origin at (x,y), ie. $\alpha_x < x$ and $\alpha_y < y$. Well, by symmetry this sequence in conjunction with the evaluation function above counts just the same thing as B: the number of roots in the rectangle.

Unfortunately we cannot compute either sequence B or B', but the sequence M counts precisely the *sum* of the integers given by the sequences B and B'. Since these are the same, the sequence M counts twice the number of roots of the P, Q system which lie in the rectangle. Thus, the number of roots in the rectangle is:

$$\frac{1}{2}(sim(M(x_l, y_l)) + sim(M(x_u, y_u)) - sim(M(x_u, y_l)) - sim(M(x_l, y_u)))$$

$$= E(M, I). \qquad \square$$

8   An Example

As a simple example, let us take the two polynomials $P(x, y) = x^2 + y^2 - 2$ and $Q(x, y) = x - y$. This we might think of as a circle and a line, intersecting at the points $(1, 1)$ and $(-1, -1)$. To compute the volume function we need to eliminate two variables, a and b say, from the following system

$$\{P(a, b), Q(a, b), u + (x - a)(y - b)\}.$$

This is,

$$\{a^2 + b^2 - 2, a - b, u + (x - a)(y - b)\}$$

so we can use the second polynomial to substitute a for b in the other two and this gives,

$$\{2a^2 - 2, u + (x - a)(y - a)\}$$

or

$$\{a^2 - 1, u + xy - a(x + y) + a^2\}.$$

Now use the first term to eliminate a^2 in the second,

$$\{a^2 - 1, u + xy + 1 - a(x + y)\}.$$

then multiply the first term by $(x + y)$, the second by a and add them,

$$\{u + xy + 1 - a(x + y), a(u + xy + 1) - (x + y)\}.$$

Finally multiply the first term by $(u + xy + 1)$, the second by $(x + y)$ and sum to give

$$\{(u + xy + 1)^2 - (x + y)^2\},$$

the volume function.

Its derivative with respect to u is: $2(u+xy+1)$, and minus the remainder of the previous two terms is $4(x + y)^2$. Evaluating at $u = 0$ and removing numeric contents gives the sequence as:

$$((x^2 - 1)(y^2 - 1), (xy + 1), (x + y)^2).$$

We may not evaluate the sequence anywhere along the line $-x = y$ but anywhere else it will serve to count the number of roots.

For example, the region $[-3, 3] \times [-2, 2]$ has corners at $(-3, -2)$, $(-3, 2)$, $(3, -2)$ and $(3, 2)$, where the sequence evaluates to $(24, 7, 25)$, $(24, -5, 1)$, $(24, -5, 1)$ and $(24, 7, 25)$ respectively. The number of similarly signed adjacent pairs in each of the terms are 2, 0, 0 and 2 respectively so that the number of roots in the region is computed as two.

9 A Sturm Sequence in Terms of the Roots

Sylvester's identity may be applied to the terms of the Sturm sequence, giving each polynomial in terms of the n roots of S_0. We use the convention that $\sum_{\alpha\beta} \theta(\alpha, \beta) \equiv \sum_{i=1}^{n} \sum_{j=i+1}^{n} \theta(\alpha_i, \alpha_j)$ etc.

$$S_0 \propto \prod_{\alpha}(x - \alpha)$$

$$\frac{S_1}{S_0} \propto \sum_{\alpha} \frac{1}{(x - \alpha)}$$

$$\frac{S_2}{S_0} \propto \sum_{\alpha\beta} \frac{(\alpha - \beta)^2}{(x - \alpha)(x - \beta)}$$

$$\frac{S_3}{S_0} \propto \sum_{\alpha\beta\gamma} \frac{(\alpha - \beta)^2(\alpha - \gamma)^2(\beta - \gamma)^2}{(x - \alpha)(x - \beta)(x - \gamma)}$$

$$\vdots$$

$$S_n \propto \prod_{\alpha\beta}(\alpha - \beta)^2$$

10 The Multivariate Sequence in Terms of the Roots

Making substitutions:

$$x \to 0,$$

$$\alpha \to -(x - \alpha_x)(y - \alpha_y),$$

$$\beta \to -(x - \beta_x)(y - \beta_y),$$

$$\vdots$$

into the expressions above yield the bivariate sequence as a function of the Cartesian components of roots.

$$M_0 \propto \prod_\alpha (x - \alpha_x)(y - \alpha_y)$$

$$\frac{M_1}{M_0} \propto \sum_\alpha \frac{1}{(x - \alpha_x)(y - \alpha_y)}$$

$$\frac{M_2}{M_0} \propto \sum_{\alpha\beta} \frac{((x - \alpha_x)(y - \alpha_y) - (x - \beta_x)(y - \beta_y))^2}{(x - \alpha_x)(y - \alpha_y)(x - \beta_x)(y - \beta_y)}$$

$$\vdots$$

$$M_n \propto \prod_{\alpha\beta} ((x - \alpha_x)(y - \alpha_y) - (x - \beta_x)(y - \beta_y))^2$$

Hermite gave the first two terms of this sequence between 1852 and 1853 and derived all of the important properties of the remaining terms for any root counting sequence in $\mathbf{R}^2$ [1,2]. Note that the sequence above is a special case of one of Hermite's constructions but, in one dimension, specialises to the original sequence of Sturm.

11 Closure

The phrase "roots in the rectangle" has been used to describe, rather loosely, how each root in $\mathbf{R}^2$ contributes to the count ignoring the possibility of a root appearing at either the edges or corners of the region of interest. In the following section we describe the precise contribution each root makes to the computed count in terms of the set theoretic structure of the region.

Denote the hypercuboidal subset of $\mathbf{R}^n$ by the vector of intervals $\underline{I}$ ie.

$$\underline{I} = (I_1, I_2, ..., I_n)$$

where each interval $I_i = [l_i, u_i]$ is closed at both ends. Denote the left and right half-open counterparts of $\underline{I}$, respectively, by $\underline{J}$ and $\underline{K}$,

$$\underline{J} = (J_1, J_2, ..., J_n)$$

$$\underline{K} = (K_1, K_2, ..., K_n)$$

such that $J_i = [l_i, u_i)$ and $K_i = (l_i, u_i]$.

In **R** we have,

$$E(M, I) = |\{\alpha \in J \mid f(\alpha) = 0\}|.$$

In **R**^{2}

$$2E(M, I) = \left|\{\underline{\alpha} \in J_1 \times J_2 \mid \underline{f}(\underline{\alpha}) = \underline{0}\}\right| + \left|\{\underline{\alpha} \in K_1 \times K_2 \mid \underline{f}(\underline{\alpha}) = \underline{0}\}\right|$$

Roots in the open rectangle are counted twice, roots on the edges are counted once, roots at the NE and SW corners are counted once and any roots at the NW and SE corners (those with negative volume relative to the centroid) are not counted at all.

This pattern extends to higher dimensions: in **R**^{3}

$$4E(M, I) = \left|\{\underline{\alpha} \in J_1 \times J_2 \times J_3 \mid \underline{f}(\underline{\alpha}) = \underline{0}\}\right| +$$

$$\left|\{\underline{\alpha} \in J_1 \times K_2 \times K_3 \mid \underline{f}(\underline{\alpha}) = \underline{0}\}\right| +$$

$$\left|\{\underline{\alpha} \in K_1 \times J_2 \times K_3 \mid \underline{f}(\underline{\alpha}) = \underline{0}\}\right| +$$

$$\left|\{\underline{\alpha} \in K_1 \times K_2 \times J_3 \mid \underline{f}(\underline{\alpha}) = \underline{0}\}\right|.$$

Roots in the open box are counted four times, roots on the faces are counted twice, roots on the edges are counted once and any roots lying at the vertices which make a negative volume w.r.t. the centroid (the vertices of a tetrahedron) are counted just once.

12 Definiteness of a Single Real Polynomial

The definiteness of a single polynomial P in a box is the application for which this work was undertaken. A single sequence pertinent to the simultaneous real roots of the partial derivatives of P may be modified to count $+1$ or -1 depending on the sign of P at these points. This is done simply by using $u + (x - a)(y - b)P(a, b)$ as the projecting function instead of $u + (x - a)(y - b)$.

The strategy above finds "bubbles" inside the box. Definiteness follows from a study of the definiteness of the function on the surface of the box: a problem in $d - 1$ dimensions. Both this question and the problem of finding efficient evaluation mechanisms have been given only a preliminary study, the details will be the subject of future research.

13   The Algorithm

procedure nprs(p, q);
       if degree(q, u) = 0 **then**
           if q = 0 **then** list p **else** list(p, q)
       else p . nprs(q, $-$ rem(lc(q)^2*p, q) / lc(p)^2);

procedure M2(p, q, v1, v2);
begin local v, s;
       v := last(groebner(set(p, q, u + (xp $-$ v1)*(yp $-$ v1)), list(v1, v2, u)));
       s := nprs(v, differentiate(v, u));
       substitute(list(0, v1, v2), list(u, xp, yp), s);
end;

procedure M(f: set, vars: list);
begin local v, s, p, ans;
       p := **for each** var **in** vars **product** (var' $-$ var);
       v := last(groebner(union(f, u + p), append(vars, list(u))));
       s := nprs(v, differentiate(v, u));
       ans := substitute(0, u, s);
       substitute(vars, vars', ans);
end;

14   Historical Note

I read somewhere that it is a great shame that no hint is ever given of the
various contorted means by which new mathematical devices are stumbled
upon. In the inevitable refining and the tidying up that one goes through
to present a piece of work it is often the case that the original sequence of
arguments is quite lost in the cleaner presentation. Such is the case with
this work where the volume function appears for reasons that only become
clear much later on. In fact, the sequence of steps that led to the creation
of the volume function were a great deal less inspired than it appears and
for this reason, as well as to issue the appropriate credits, the original line
of thought will be outlined.

    Given a black box for counting the number of roots in the first quadrant,
$x > 0$ and $y > 0$ of R^2, the number of roots inside a rectangle can easily
be deduced by translating each of the corners of the rectangle to the origin
and using an evaluation function like E to compute the number of roots.
The problem of finding the number of roots satisfying $\alpha_x > 0$ is essentially
solved by eliminating the y variable using the subresultant algorithm. Sim-

ilarly one can deduce the number of roots satisfying $\alpha_y > 0$ by eliminating the x variable.

The problem now seems to be one of "intersecting" these two results. A solution to this problem came, somewhat surprisingly, not from my work on the two dimensional sequence nor from a subsequent introduction to Hermite's inspired papers, but from a paper by Pinkert [6] which deals with the problem of finding the number of complex roots of a polynomial inside a rectangle in **C**. In his paper there is a stroke of absolute genius which Pinkert attributes to Bobby Caviness in the acknowledgements.

Number the quadrants one to four in an anticlockwise direction starting where both components are positive. We know the number of roots in quadrants one and two put together and the number of roots in quadrants one and four put together[8]. We also know the number of roots in all four quadrants put together for this is equal to the degree of the polynomial. One more independent equation, then, would suffice to find the number of roots in quadrant one. Well, we can square the roots of the polynomial simply by manipulating its coefficients. If we then count the number of roots of this polynomial with positive imaginary parts then this gives a count of the number of roots of the original polynomial which lie in quadrants one and three. This is a fourth independent equation and allows the number of roots in quadrant one to be deduced exactly.

Because of the nature of the evaluation function E it is not actually necessary to perform the first three calculations above as all of them will contribute zero to the final answer. All that is needed are the polynomials with roots which are squared with respect to an origin at each corner of the rectangle. But this is bound to be true because quadrant three in combination with the evaluation function E will count just the same as quadrant one — by symmetry. So we can simply count the number of squared[9] roots with positive imaginary parts and divide it by two.

Now the question is an infinitely easier one: is there an analogy of "root-squaring" in $\mathbf{R}^n$? What we need is a sort of "twist" at the origin. Well, how about the following mapping $(x, y) \rightarrow (x, \frac{y}{x})$? This will negate the y coordinate just when x is negative. This works and may be computed as follows:

$$Elim_x(P(x, \frac{y}{x}), Q(x, \frac{y}{x})),$$

where $Elim$ just eliminates variables - but it doesn't seem very symmetric. The symmetric nature of this strategy becomes apparent when it is written like this:

$$Elim_{xy}(P(x, y), Q(x, y), u - xy)).$$

---

[8]These results come from the sequences of Routh.

[9]w.r.t. each vertex.

This was the first sight of the volume function which here is a polynomial in a single variable whose roots are the products of the Cartesian components of the roots of the original system.

It is no coincidence then, that when this technique is applied to the real and imaginary parts of a problem in a single complex variable, it generates the root squaring strategy from which it was derived.

Acknowledgements

Thanks to my supervisor James Davenport, to Geoff Smith and to Dan Richardson, all of whom have made invaluable contributions to this work.

Special thanks also to Adrian Bowyer for his consistent enthusiasm in this work and help in producing this paper.

The author would also like to thank the SERC's ACME Directorate[10] and the IBM UK Scientific Centre, who jointly funded the Computer Algebra and Solid Modeling project in the Schools of Mathematical Sciences and Mechanical Engineering at Bath. The work reported here forms a part of that project, and of the author's Ph.D. thesis [4].

References

1. Hermite. (1852). Sur L'extension du Théorème de M.Sturm à un Système D'Équations Simultanées. Comptes rendus des séances de l'Académie des Sciences. Tome XXXV.

2. Hermite. (1853). Remarques sur Le Théorème De M.Sturm. Comptes rendus des séances de l'Académie des Sciences. Tome XXXVI, pp. 294–297.

3. Hermite. (1880) Œuvres de Charles Hermite, Tome III. Sur L'extension du Théorème de M.Sturm à un Système D'Équations Simultanées. Mémoire inédit.

4. Milne, P.S. (1990). On the algorithms and implementation of a geometric algebra system. Ph.D. thesis, University of Bath.

5. Milne, P.S. (1992). On the solutions of a set of polynomial equations. *Symbolic and Numerical Computation for Artificial Intelligence*. Academic Press.

6. Pinkert, J.R. (1976). Finding the roots of a complex polynomial. *ACM TOMS 2*, pp. 351–363.

---

[10]Grants GR/E/05209 and GR/F/77425

7. van der Waerden, B.L. (1950). *Modern Algebra*. 3rd edn. F. Ungar Publishing Co., New York.

Ray Casting Set-theoretic Rolling Sphere Blends

Alan E. Middleditch

Brunel University

Abstract This paper is concerned with faces which blend between other faces of arbitrary set theoretic volume models. The blend faces have constant curvature cross sections and join the connected faces with tangent continuity. They are defined as the bounding surfaces of volumes defined via vector sum operators and continuous tone images are generated by a ray casting algorithm. The proposed techniques have wider application than blending and should be very useful in a CAD environment.

1 Introduction and Background

Volume modelling systems represent, manipulate, analyse and display computer models of solid objects. Various techniques are described in [2], [13], [14] and [15]. The two most common techniques use objects represented by their boundaries [1] and objects represented by set theoretic combinations of primitive volumes [12] respectively.

A face of an object is considered to be a bounded connected set of points on its boundary which lie in an unbounded analytic surface with no first derivative discontinuities. This paper is concerned with blend faces which provide tangent continuous transitions between a set of adjacent 'base' faces. Such faces are inevitable in cast, moulded, and machined components, and are useful to realise aesthetic designs. They are rarely functional except when reducing stress concentration along concave edges.

Practical volume modelling systems used for design or analysis must incorporate blend faces. Surveys of blending techniques relevant to volume modelling systems can be found in [10], [24] and [27]. Varady & Pratt [24] concentrates on methods relevant to boundary based volume modelling systems and includes more general free form surface techniques. This paper is concerned with set theoretic techniques.

Some applications require blend faces to have cross sections with constant curvature in the direction 'orthogonal' to the base surface intersection curve. Some applications require blends whose constant curvature cross

261

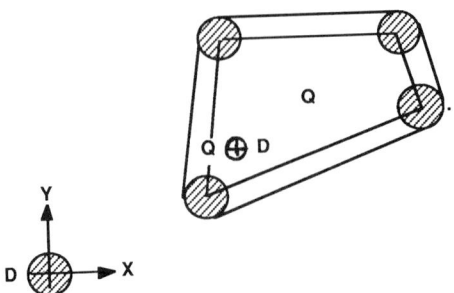

Figure 1. Vector sum

sections all have the same radius. The latter are often called rolling sphere blends; they are parts of the surface of the volume traced by a sphere rolling along an edge while maintaining contact with the adjacent faces. The blends of [10] and [6] have constant curvature in the blending direction only when the base surfaces are planes and the appropriate fullness parameter is used. Analytic solutions for the general case may be found using envelope theory [18] but the resulting high degree surfaces require the solution of high degree polynomial equations which are often unstable [25]. A technique for generating approximate rolling sphere blends of the parametric surfaces found in boundary modellers is described in [3].

This paper introduces an algorithmic method for the creation of constant curvature blend surfaces between arbitrary base surfaces in set theoretic models. The proposed technique uses the vector sum and a related operator to directly define volumes whose boundaries include blend faces. Continuous tone images of these volumes are generated by a ray casting algorithm. The material is extracted from [11] which provides a more detailed analysis and also considers boundary evaluation for line drawing.

2   The Vector Sum Operators

The vector or Minkowski sum operator is defined by

$$Q \oplus D = \{\mathbf{s} \mid \exists \mathbf{q} \in Q \cdot (\mathbf{s} - \mathbf{q}) \in D\}.$$

Fig. 1 illustrates its use to inflate polygon Q into the larger rounded polygon.

Rolling sphere blends are defined using the vector sum operator by the blending formula introduced in [9]. Thus, the convex corners of a 2D region P are rounded by the operation:

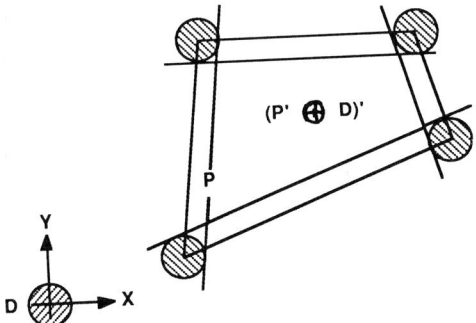

Figure 2. Complementary vector sum

$$\text{Convex blend}(P) = (P' \oplus D(r))' \oplus D(r),$$

where $D(r)$ is the set of points in a disc of radius r, centred at the origin and P' represents the set complement of P. The expression in parantheses deflates the outer polygon of Fig. 2 into the inner polygon. Since this is the polygon Q of Fig. 1, the two diagrams together illustrate the rounding operation for convex polygons.

Concave corners of an arbitrary polygon are rounded by the analogous operation:

$$\text{Concave blend}(R) = ((R \oplus D(r))' \oplus D(r))'.$$

By concantenating the concave and convex blending operations, all corners of an arbitrary polygon are rounded by

$$\text{Blend}(R) = ((R \oplus D(r))' \oplus D(2r))' \oplus D(r).$$

This equation applies equally well to the blending of a 3D volume R by a sphere D of radius r.

Most set theoretic volume modelling systems define volumes by set combinations of bounded primitive volumes using the set operators union, intersection and difference. The use of bounded primitives and set difference in preference to complementation ensures that all volume sub-expressions represent valid bounded volumes. All algorithms can then be restricted to accept and produce only valid volumes. This philosophy can be extended to the vector sum blending of the previous equation by the introduction of a 'complementary' vector sum operator $\oplus'$, defined by

$$P \oplus' D = (P' \oplus D)'.$$

This operator produces a bounded set whenever its operands are bounded. The equation which defines a volume R with all its edges and vertices rounded, can now be written with all it sub-expressions bounded:

$$\text{Blend}(R) = ((R \oplus D(r)) \oplus' D(2r)) \oplus D(r).$$

Some geometric configurations are sensitive to the order in which convex and concave edges are rounded [9]. Thus, the previous equation can give a different result to

$$\text{Alternative Blend}(R) = ((R \oplus' D(r)) \oplus D(2r)) \oplus' D(r).$$

Blend volumes defined by the vector sum operators can be incorporated in set theoretic volume modelling systems either by

1. replacing the vector sum operators with equivalent combinations of the conventional set operators union, intersection and difference,

2. expanding volume defining set expressions into a canonical form where the vector sums operate only on primitive volumes, or

3. using evaluation algorithms which accommodate the vector sum operators.

The first approach is used by Rossignac & Requicha [16], but their technique does not always provide exact tangency. Tangent blends are produced only if the vector sum of a sphere and the intersection curve between base surface offsets has the form of a primitive volume supported by the system. In most systems, this restricts surfaces to cylinders and tori and the base surfaces to planes or a plane and an orthogonal cylinder, cone or sphere.

The second approach requires explicit support for blended primitive volumes. Blends of common primitive volumes (cube, cylinder, cone, sphere and torus) are easily transformed into conventional set combinations of the same primitives. Unfortunately, the appropriate canonical form cannot always be generated. Although the vector sum distributes over set union, it does not in general distribute over set intersection [9]:

$$R \oplus (P \cup Q) = (R \oplus P) \cup (R \oplus Q),$$

but

$$R \oplus (P \cap Q) \neq (R \oplus P) \cap (R \oplus Q).$$

This is illustrated in Fig. 3; $(R \oplus P) \cap (R \oplus Q)$ includes the two small dark triangles, but $R \oplus (P \cap Q)$ does not.

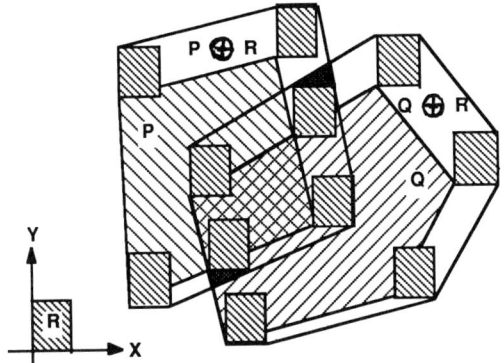

Figure 3. Vector sum of a set intersection

In contrast, the complementary vector sum distributes over set intersection but not union:

$$R \oplus' (P \cap Q) = (R \oplus' P) \cap (R \oplus' Q),$$

but

$$R \oplus' (P \cup Q) \neq (R \oplus' P) \cup (R \oplus' Q).$$

The third approach accommodates the vector sum operators like the conventional set operators in the tree data structures which represent the volume set expressions. Algorithms which operate on such structures are discussed in the remainder of this paper.

3 Ray Casting

Ray casting algorithms are used to generate shaded images [17] and to calculate mass properties such as volume or moment of inertia [7]. Such algorithms use the classification function described by Tilove [21,22] to find those segments of a set of straight lines which are inside a volume. Rays are usually represented by parametric equations and the classifications by a set of parameter intervals corresponding to the line segments inside the volume. Classifications together with the surface normals at the ray penetration points are used for image generation.

A set theoretic volume model consists of a set combination of component volumes. A ray is classified with respect to such volumes by combining its classifications with respect to their similarly represented component volumes. Recursion terminates when a component volume is a primitive volume of the modelling system. This approach is described in [7] and [17] for objects defined entirely using the conventional set operators union, intersection and difference. Ray classifications can be combined under these operators by a 1D merging algorithm [19]. Combinations under the vector sum operator may be achieved by the method developed in the remainder of this section.

If a ray L is defined parametrically in terms of a reference point $\mathbf{q}$ and direction vector $\mathbf{v}$, the set of all points on that ray is given by

$$L = \cup_{k \in \Re} \{\mathbf{q} + \mathbf{v}k\}.$$

Segments of the ray L which lie inside a volume $R \oplus S$ are given [11] by the set intersection

$$(R \oplus S) \cap L = \cup_{\mathbf{a} \in A} \{\{\mathbf{q}\} \oplus \cup_{m \in M(\mathbf{q} - \mathbf{a})} \{\mathbf{v}m\} \oplus \cup_{n \in N(\mathbf{a})} \{\mathbf{v}n\}\},$$

where A is the set of all points $\mathbf{a}$ in a plane not containing the ray $\mathbf{a} + \mathbf{v}n$, $M(\mathbf{q} - \mathbf{a})$ is the set of parameter values m corresponding to the segments of the ray $\mathbf{a} - \mathbf{v}m$ inside R, and $N(\mathbf{a})$ is the set of parameter values n corresponding to the segments of the ray $\mathbf{a} + \mathbf{v}n$ inside S. Using the notation $\mathbf{v}K$ to represent the set of vectors obtained by multiplying the vector $\mathbf{v}$ by each element of the set K of scalar parameter values, this becomes

$$(R \oplus S) \cap L = \{\mathbf{q}\} \oplus \mathbf{v}K,$$

where

$$K = \cup_{\mathbf{a} \in A} \{M(\mathbf{q} - \mathbf{a}) \oplus N(\mathbf{a})\}.$$

This equation embodies a separation of dimensions; the vector sum of 3D point sets R and S is replaced by a 2D vector sum and a 1D vector sum. The 1D sets $M(\mathbf{q} - \mathbf{a})$ and $N(\mathbf{a})$ consist of ray parameter intervals and are the results of single ray classifications against R and S respectively. This is illustrated without loss of generality by the 2D situation of Fig. 4 where the plane A and ray direction $\mathbf{v}$ are represented by the $\mathbf{X}$ axis and $\mathbf{Z}$ direction respectively and a ray parameter value is equal to the z coordinate of the corresponding point. $M(\mathbf{q} - \mathbf{a})$ and $N(\mathbf{a})$ are parameter intervals corresponding to the segments of the rays through $\mathbf{a}$ and $\mathbf{q} - \mathbf{a}$ which lie in the rectangle and triangle respectively. The parameter interval K corresponds to the set of points on the ray through $\mathbf{q}$ which are in the large rectangle. The vector sum of $M(\mathbf{q} - \mathbf{a})$ and $N(\mathbf{a})$ is a subset of K, and K is the union of all such vector sums.

A similar treatment of the complementary vector sum yields ray segments inside the volume $R \oplus' S$ given by

$$(R \oplus' S) \cap L = \{\mathbf{q}\} \oplus \mathbf{v}H,$$

where

$$H = \cap_{\mathbf{a} \in A} \{M(\mathbf{q} - \mathbf{a}) \oplus' N(\mathbf{a})\}.$$

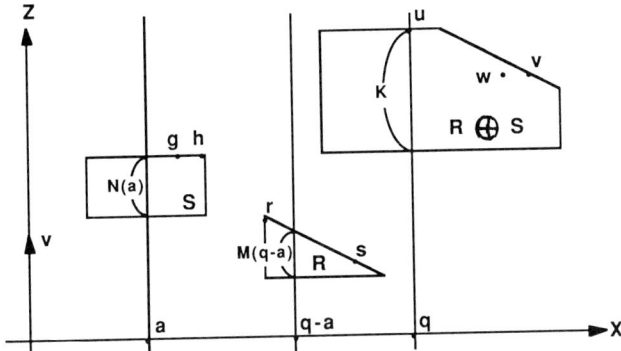

Figure 4. Vector sum of 2D regions

3.1 Vector sum of 1D ray classifications

The vector sum of ray classifications represented as the disjoint union of multiple parameter intervals can be written in terms of vector sums of single intervals. If parameter sets $M(\mathbf{q} - \mathbf{a})$ and $N(\mathbf{a})$ consist of intervals $M_1(\mathbf{q} - \mathbf{a}) \ldots M_I(\mathbf{q} - \mathbf{a})$ and $N_1(\mathbf{a}) \ldots N_J(\mathbf{a})$ respectively, and classification intervals $M_i(\mathbf{q} - \mathbf{a})$ and $N_j(\mathbf{a})$ are represented by M_i and N_j respectively,

$$M(\mathbf{q} - \mathbf{a}) \oplus N(\mathbf{a}) = \cup_{1 \leq i \leq I} \{M_i\} \oplus \cup_{1 \leq j \leq J} \{N_j\} = \cup_{1 \leq i \leq I} \cup_{1 \leq j \leq J} \{M_i \oplus N_j\}.$$

The derivation of a similar expression for the complementary vector sum is a little more complicated:

$$
\begin{aligned}
M(\mathbf{q} - \mathbf{a}) \oplus' N(\mathbf{a}) &= ((\cup_{1 \leq i \leq I} \{M_i\})' \oplus \cup_{1 \leq j \leq J} \{N_j\})', \\
&= \cap_{1 \leq j \leq J} \{((\cap_{1 \leq i \leq I} \{M_i'\}) \oplus N_j)'\},
\end{aligned}
$$

but this equation cannot be expanded because the set intersection does not in general distribute over vector sum. There are additional terms [8]:

$$(A \oplus R) \cap (C \oplus R) = ((A \cap C) \oplus R) \cup (((A - C) \oplus R)) \cap ((C - A) \oplus R))).$$

Intersection does distribute over vector sum for 1D parameter intervals which overlap. If A contains $C, C - A = \{\}, (C - A) \oplus R = \{\}$ and the previous equation reduces to

$$(A \oplus R) \cap (C \oplus R) = (A \cap C) \oplus R.$$

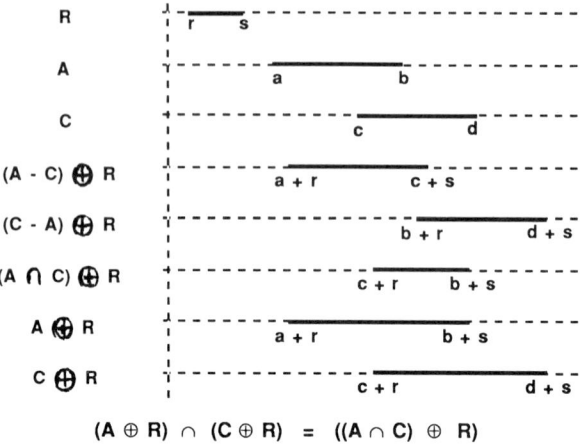

$$(A \oplus R) \cap (C \oplus R) = ((A \cap C) \oplus R)$$

Figure 5. Vector sum of overlapping 1D intervals

Thus if A contains C or visa versa, set intersection distributes over vector sum. If $A = \{a \dots b\}$ and $C\{c \dots d\}$ such that $a < c < b < d$ (Fig. 5) or $c < a < d < b$,

$$((A \cap B) \oplus R) \supset ((A - B) \oplus R)) \cap ((B - A) \oplus R)).$$

This is also true if $(A-B) \oplus R$ and $(B-A) \oplus R$ are disjoint because their intersection is then empty. Set intersection thus distributes over vector sum for arbitrary overlapping 1D intervals.

Since the ray classification parameter intervals M_i are disjoint, their complements M_i' all overlap. The vector sum distributes over set intersection and the complementary vector sum of two sets of parameter intervals becomes

$$M \oplus' N = \cup_{1 \le i \le I} \cap_{1 \le j \le J} \{M_i \oplus' N_j\}.$$

Therefore, as shown in Fig. 6, simple arithmetic on the interval limits can be used to compute the vector sum and complementary vector sum of two parameter intervals:

$$
\begin{aligned}
M_i \oplus N_j &= \{M_{i1} + N_{j1} \dots M_{i2} + N_{j2}\}, \\
\text{and} \quad M_i \oplus' N_j &= \{M_{i1} + N_{j2} \vdots \dots M_{i2} + N_{j1}\},
\end{aligned}
$$

where $M_i = \{M_{i1} \dots M_{i2}\}$ and $N_j = \{N_{j1} \dots N_{j2}\}$.

Unlike the normal vector sum, the complementary vector sum of two non-empty parameter intervals could be the empty set. This is represented

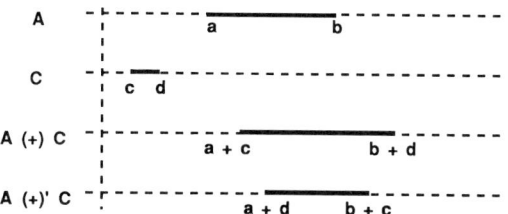

Figure 6. Combination of 1D intervals

by a start parameter larger than the end parameter in the previous equation.

Using the previous two equations and the expansions for the combination of the two sets of parameter intervals $M(\mathbf{q} - \mathbf{a})$ and $N(\mathbf{a})$, rays classified against volumes defined via the vector sum and complementary vector sum yield parameter interval sets defined respectively by

$$K = \cup_{\mathbf{a} \in A} \cup_{1 \leq i \leq I} \cup_{1 \leq j \leq J} \{M_{i1} + N_{j1} \ldots M_{i2} + N_{j2}\},$$

and

$$H = \cap_{\mathbf{a} \in A} \cup_{1 \leq i \leq I} \cap_{1 \leq j \leq J} \{M_{i1} + N_{j2} \ldots M_{i2} + N_{j1}\}.$$

These two equations cannot be programmed directly because they involve a continuous union and intersection respectively. This requires sub-expressions to be computed for an infinite number of vectors covering the plane A. In practice, as with ray casting against conventional set theoretic models, the plane is sampled over a finite region. A single ray classification against a volume defined as the vector sum or complementary vector sum of constituent volumes R and S may then be achieved using the algorithm on the next page.

Since the classification operations in this algorithm are recursive invocations and the volumes R and S may be defined by the conventional set operators, its implementation must include clauses for those operators. In the innermost loop, the set E of parameter intervals is combined with a single interval by repeated use of 1D merging algorithms used to implement the 1D union or intersection [19].

If the volumes R and S defined by the conventional set operators, the upper bound on the computation time complexity of the previous algorithm is $O(P^2/d^2)$, where d is the ray separation and P is the total number of primitive volumes. This complexity is quadratic in the reciprocal of the ray separation. However, the worst case occurs only if all rays intersect all primitives. It is more realistic for the density of the primitives to be constant and independent of their number. In this case, the size of the primitives is inversely proportional to their number, each ray will intersect a constant

```
Input :      Ray = q + vk ,
             Volume = R @ S, where @ is either ⊕ or ⊕'.

Output :     K = Set of disjoint parameter intervals for ray segments inside V.

BEGIN
    Initialise: K ← { }
    FOR  all integers r ∈ [0...R]
        FOR  all integers s ∈ [0...S]        # for all vectors on rectangular grid #
            a  ←  a₀  +  <r*d, 0, 0>  +  <0, s*d, 0>
            Classify the ray a+vn against the volume S, to generate parameter set N.
            Classify the ray q-a+vm against the volume R, to generate parameter set M.
            FOR  all I intervals Mᵢ in M
                E ← { }
                FOR  all J intervals Nⱼ in N
                    IF  @ = ⊕
                    THEN
                        E  ←  { Mᵢ₁ + Nⱼ₁ ... Mᵢ₂ + Nⱼ₂ }  ∪ E
                    ELSE IF  Mᵢ₁ + Nⱼ₂ < Mᵢ₂ + Nⱼ₁   # @ = ⊕' and interval is not empty #
                    THEN
                        E  ←  { Mᵢ₁ + Nⱼ₂ ... Mᵢ₂ + Nⱼ₁ }  ∩ E
                    ENDIF
                ENDFOR
                K ← K ∪ E
            ENDFOR
        ENDFOR
    ENDFOR
END
```

Vector sum of ray classifications

number of primitives and the complexity will be $O(P/d^2)$ [23]. For volumes defined entirely by vector sum operators, this becomes $O(P \log P/d^2)$ for primitive classification and $O(P/d^4)$ for classification combination! If the algorithm is modified to compute primitive and sub-tree ray classifications only once, the primitive classification complexity reduces to $O(P/d^2)$ and is independent of the set operator mix. Unfortunately, the complexity for tree traversal remains $O(P/d^4)$ and the storage requirement increases considerably.

It is well known that although these complexity bounds cannot be reduced, the expected performance can be improved significantly by exploiting bounding boxes and using other standard techniques described in [17] and [19]. Further speed improvements can be obtained by using more sophisticated techniques which exploit spatial locality, [20], [23] and [26]. An additional spatial locality peculiar to the vector sum can also be exploited. The sample points **a** on the plane A can be restricted to a region such that **a** is in the orthogonal projection of R and **q-a** is in the orthogonal projection of S. This could be easily incorporated in the related technique of Sears & Middleditch [20].

4 Continuous Tone Image Generation

The image of a solid object can be generated by casting a ray from the view point to that object through each pixel in the viewing plane. Pixel colour is determined from the lighting conditions and the properties of the surface at the point of ray penetration [4]. In particular, the colour depends on the surface normal at the penetration point. Penetration points are easily extracted from ray classifications and an approximation to the associated normals can be computed from the penetration points of neighbouring rays, e.g. by local surface fitting. Better images result from true normals. At surface points of volumes defined via the conventional set operators or the vector sum operator, the true normal is the normal at a surface point of one of the constituent volumes (Fig. 4). For volumes defined via the vector sum operator, surface normal selection requires additional knowledge of the volume in the neighbourhood of the associated penetration point. This section is concerned with the computation of such neighbourhoods.

When a ray is classified with respect to an analytic halfspace, the surface normals at the halfspace boundary are computed from the surface derivatives. Ray classification data can be extended to associate surface normals with parameter interval end points. The classification of a ray with respect to a conventional set combination of two objects consists of a set of parameter intervals whose end points are selected from the component classifications. These points correspond to ray/composite object penetration points and thus retain the surface normal associations of the same points in the component classifications.

In contrast, the algorithm of the previous section for the combination of classifications under the vector sum operator involves arithmetic operations on the parameter interval end points. In this case, it is not obvious which component object surface normal is equal to that of the composite object. It is necessary to retain knowledge of the surface geometry in the neighbourhood of the penetration point and combine such neighbourhoods during the combination of ray classifications.

The neighbourhood $Nb(\mathbf{v}, V)$ of a point **v** with respect to a volume V is that part of the volume within an arbitrarily small sphere B centred at that point. It is defined at the origin to simplify subsequent manipulation:

$$Nb(\mathbf{v}, V) = B \cap (V \oplus \{-\mathbf{v}\}).$$

For a volume V defined as the vector sum $R \oplus S$, it is shown in [11] that

$$Nb(\mathbf{v}, V) = \cup_{\mathbf{r} \in \Re^3} \{Nb(\mathbf{r}, R) \oplus Nb(\mathbf{v} - \mathbf{r}, S)\}.$$

Image generation is only concerned with surface points, i.e. points with neighbourhoods that are neither full nor empty. The neighbourhood $Nb(\mathbf{v}, V)$ of the previous equation is partially full only if there exists a point $\mathbf{r}$ such that both $Nb(\mathbf{r}, R)$ and $Nb(\mathbf{v} - \mathbf{r}, S)$ are partially full and if one is full only when the other is empty. This means that the neighbourhood of a point on the surface of a vector sum volume is derived from the vector sum of neighbourhoods of points on the surface of its constituent volumes. Although the neighbourhood $Nb(\mathbf{v}, V)$ is the union of multiple vector sum results, a surface point neighbourhood is usually generated entirely by one operand of that union because surface points usually result from the vector sum of only one pair of surface points from the constituent volumes. It is also possible for surface points of the constituent volumes to sum to an interior point of the result volume and the neighbourhood $Nb(\mathbf{v}, V)$ to be full.

It is shown in [11] that neighbourhoods of a volume W defined as the complementary vector sum $R \oplus' S$ are given by

$$Nb(\mathbf{v}, W) = \cap_{\mathbf{r} \in \Re^3} \{Nb(\mathbf{r}, R) \oplus' Nb(\mathbf{v} - \mathbf{r}, S)\}.$$

The infinite number of computations implied by the continuous union of the previous two equations are avoided by sampling, i.e. by computing neighbourhoods only for the surface penetration points of a finite number of rays cast on an orthogonal grid. These neighbourhoods are associated with the parameter interval end points in each ray classification. Interval end points thus provide appropriate point pairs $\mathbf{r}$ and $\mathbf{v}$-$\mathbf{r}$ to implement the previous two equations. It is merely necessary to extend the algorithm of the previous section to combine pairs of neighbourhoods whenever parameter intervals are combined. The same set operator is used for parameter intervals and the associated neighbourhoods because neighbourhoods combined under the vector sum or complementary vector sum involve the same continuous set operation as ray classifications similarly combined.

4.1   Sampling errors

Surface point neighbourhoods are curvilinear 'hemispheres' unless the corresponding point is a vertex or a point on an edge. In practice, a ray is never found to pass exactly through a vertex or edge. This has an insignificant effect on the calculation of properties which depend only on the ray segments within the object, e.g. the calculation of volume and moments of inertia, but it is intolerable for applications such as image generation which depend on neighbourhoods. Rays which intersect a volume face have hemispherical neighbourhoods, and the vector sum of two different hemispherical neighbourhoods is always full.

Fig. 7 shows the ray segments inside volumes R, S and $R \oplus S$ with the neighbourhoods associated with the top penetration points. Using an equation derived previously, the classification of ray k against volume $R \oplus S$ is given by

$$k \cap (R \oplus S) = ((a \cap R) \oplus (g \cap S)) \cup ((b \cap R) \oplus (f \cap S)) \cup ((c \cap R) \oplus (e \cap S)).$$

This is the union of only three ray classifications; the remaining classifications of the multiple union over the X axis are empty because one of the arguments of a constituent vector sum is empty. For example, ray d does not intersect volume S and $(d \cap R) \oplus (d \cap S)$ is empty because $d \cap S$ is empty. The union of the three ray segments in the previous equation gives an approximation to the classification of ray k against volume $R \oplus S$ (sampling errors cause the computed segment of k inside $R \oplus S$ to be slightly short because the segment of g inside S is shorter than the height of triangle S). Unfortunately, all three segments are the vector sum of different hemispherical neighbourhoods (Fig. 7) and are thus full. Their union is not the correct neighbourhood for ray k. This problem does not arise with volumes defined entirely via the conventional set operators because the neighbourhoods of ray segment end points are merely copies from those of ray segments inside component volumes.

Correct neighbourhoods of volumes defined via the vector sum operators are computed if vertices and edges of component volumes are reflected in the classifications of rays which pass nearby. Fig. 8 shows the appropriate neighbourhoods for the geometry of Fig. 7. Segments of rays which pass near a vertex or edge and 'clip' the associated volume (e.g. ray d are neglected. Those which pass near a vertex or edge and penetrate the volume have penetration point neighbourhoods which reflect that feature (e.g. ray g).

When a volume is clipped, the small ray segments are eliminated by ray classification combinations which inherently include 1D regularisation of ray segments. Such regularisation restricts parameter intervals to those which are the closure of their interior. In practice, to compensate for computation and sampling errors, a lower limit is imposed on the size of parameter intervals. This is achieved by coalescing close interval end points when two ray segments are combined under the conventional set operators. Under intersection or difference, this may eliminate the result interval. When the classification of ray d against halfspace $r2$ of volume R (Fig. 7) is set intersected with its classification against halfspace $r3$, the result is empty (Fig. 8). Computed classifications of rays e and h against

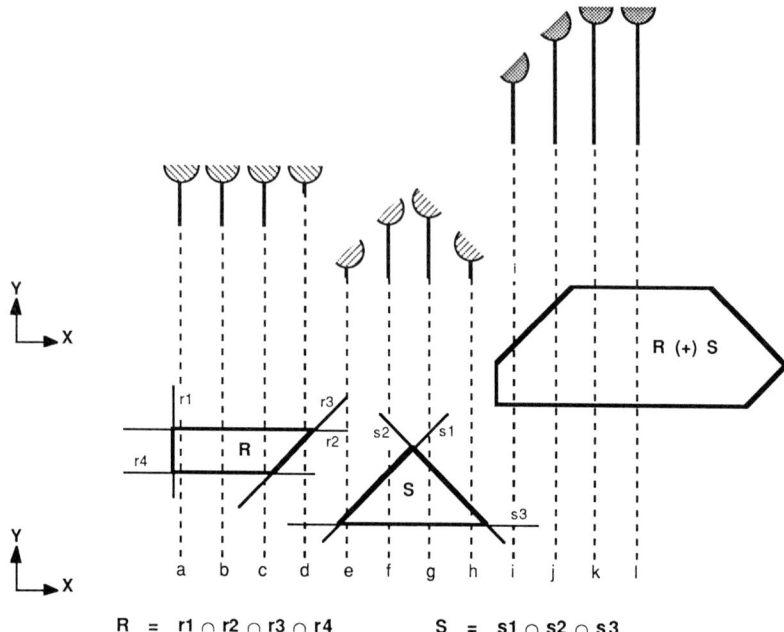

R = r1 ∩ r2 ∩ r3 ∩ r4   S = s1 ∩ s2 ∩ s3

Figure 7. True neighbourhoods

volume S are also empty. When two intervals are combined under union, coalesced end points may cause them to be combined to form a single larger interval. Neighbourhood manipulation is unnecessary because the neighbourhood of the coalesced point is full (it is empty when intervals are eliminated under intersection). If one segment contains the other, the coalesced point remains an interval end point under both intersection and union. In this case, the new neighbourhood is created by combining the neighbourhoods of the original points under the operator used to combine the segments. When the classification of ray f or g against half space $s1$ of volume S (Fig. 7) set intersected with its classification against halfspace $s2$, the volume penetration points are coalesced and the result neighbourhood is the quarter disc shown in Fig. 8.

The generation of correct neighbourhoods requires the regularisation tolerance to reflect the angle of ray/surface penetration in order to detect an edge or vertex which is closer than an adjacent ray. In the limit, when the surface lies parallel to the rays, it must be reflected in the neighbourhood of the adjacent ray segments within the volume. The neighbourhood of the segment of ray a in volume R (Fig. 7) reflects surface $r1$ (Fig. 8). All vertices and edges are thus represented by an appropriate set of rays with

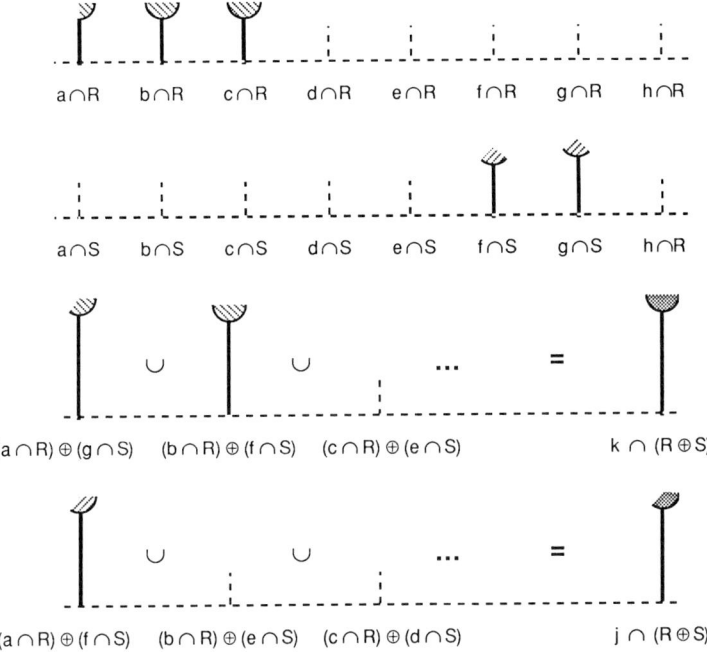

Figure 8. Computed neighbourhoods

vertex and edge penetration point neighbourhoods. The classification of ray j against the composite object $R \oplus S$ reflects the vertex between rays j and k but the classification of ray k does not (Figs. 7 and 8).

Inevitably, sampling also causes problems remote from edges and vertices because none of the sampled point pairs on constituent object surfaces sum exactly to a surface point of their vector sum (complementary vector sum). In Fig. 9, the classification of ray h with respect to volume $R \oplus S$ is the union of three classifications:

$$h \cap (R \oplus S) = ((a \cap R) \oplus (o \cap S)) \cup ((b \cap R) \oplus (n \cap S)) \cup ((c \cap R) \oplus (m \cap S)).$$

Figure 9 shows that all three consituent interval end points are within the volume $R \oplus S$ and that segments of rays b and n give rise to the longest segment and thus the surface point approximation. Since this point is slightly inside the object, its neighbourhood is full and of little use as an approximation to the surface point neighbourhood. A similar problem arises with the complementary vector sum and surface point neighbourhoods are found to be empty. Computed surface point neighbourhoods

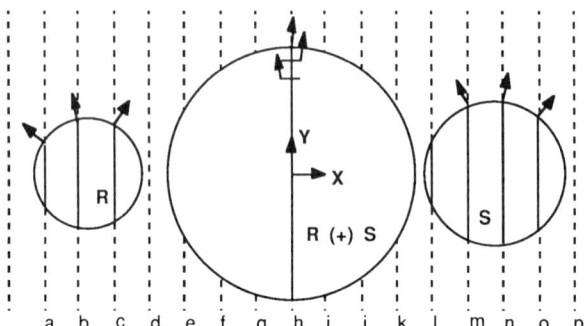

Figure 9. Computed surface normals

are full or empty because the component neighbourhoods are bounded by
smooth surfaces which cross at their centre. Neighbourhoods of component
object surface points which sum to a true surface point of the composite
object are bounded by surfaces which are tangential but do not cross.

Incorrect computed neighbourhoods of smooth surfaces are eliminated
by replacing all full (empty) neighbourhoods produced via the vector sum
(complementary vector sum) by a 'hemisphere' with a normal and signed
radius obtained as the sum (difference) of the component neighbourhood
normals and radii. Radii are positive for convex neighbourhoods and negative for concave neighbourhoods; reciprocal radii (curvature) are plus or
minus infinity at vertices and edges. If the replaced neighbourhood was correctly full or empty, the associated ray segment end point is not a surface
point and is eliminated by subsequent operations of the 'continuous' union
(intersection). Rays segments $((a \cap R) \oplus (o \cap S))$ and $((c \cap R) \oplus (m \cap S))$ in
Fig. 9 are subsequently eliminated. If it is not eliminated (segment of ray
$((b \cap R) \oplus (n \cap S))$ in Figure 9), the segment end point and its computed
neighbourhood are an approximation to a true surface point and neighbourhood. The computed neighbourhood of ray $((b \cap R) \oplus (n \cap S))$ has a
radius equal to the sum of the radii of the smaller circles and the slightly
incorrect surface normal shown in the diagram.

4.2 Surface normals

A technique for computing the vector sum of 3D neighbourhoods represented by two orthogonal 2D neighbourhoods is described in [11]. Such
neighbourhoods are useful for applications such as image generation with
anti-aliasing, but simple image generation only requires a single surface normal at ray penetration points. In this case, surface point neighbourhoods
can be represented by a surface normal, curvature and possibly higher

derivatives. Curvatures of infinite magnitude indicate that the normal is an average of adjacent face normals. Points coalesced during conventional set operations are associated with an average of their normals and, unless the normals are similar, infinite curvature to indicate that the point is near a vertex or edge. The curvature is negative for union to indicate local concavity and positive for intersection. If the normals are similar, the smaller curvature is selected for union and the larger for intersection.

When intervals are combined under a vector sum operator, their end points are summed to provide the end points of the result interval (inner loop of the ray casting algorithm described previously) and the associated surface normals and curvatures are computed from those associated with the point arguments. If both argument and curvatures are finite, the result normal and curvature are computed by summation as described in the previous section and shown in Fig. 9. Inevitably, the associated point is not a surface point of the composite volume; it may be eliminated by subsequent ray segment combinations. If it is not eliminated, the point is close to the surface, the component normals were similar (normals associated with rays b and n in Fig. 9) and their sum (the top normal of ray h) is a good approximation to the surface normal at a nearby point. If one argument has a curvature of infinite magnitude, its normal represents an edge or vertex and the other normal is selected. The normal associated with ray segment $(b \cap R)$ in Fig. 8 is selected for ray segment $(b \cap R) \oplus (f \cap S)$ and the normal associated with ray segment $(f \cap S)$ is rejected. If both argument curvatures are infinite, the result is infinite because the associated point is near a vertex or edge and an average of the normals is computed.

On completion of the 'continuous' set combination of ray classifications, a computed surface curvature which remains infinite is associated with a point near a vertex or edge. The associated normal is an average of adjacent face normals and may be used for image generation. Alternatively, a normal selected from a neighbouring ray could be used.

5 Summary and Conclusions

The methods of Middleditch & Sears [10] and Hoffman & Hopcroft [6] are unsuitable for rolling sphere blends. A new method based on the vector sum has been introduced to produce such blends on arbitrary volumes. This method can also be used for purposes other than blending and should be very useful in a CAD environment.

A simplified version of the proposed ray casting algorithm for continuous tone image generation has been implemented. This naive implementation does not exploit all possible efficiency enhancing techniques and uses a crude approximation for the computation of surface neighbourhoods. The resulting continuous tone images have some 'interesting' surface textures due to incorrect normals.

Middleditch [11] shows that the production of line drawings from volume models defined via the vector sum operators requires a complete boundary representation to be generated and that the computation time is excessive. The approach of Ghosh [5] for the vector sum of convex volume boundary representations should be more efficient, but it must be extended to accommodate the re-entrancy which may occur with concave volumes.

References

1. Braid, I.C. (1973). Designing with volumes. PhD Dissertation, University of Cambridge.

2. Braid, I.C. (1975). Six systems for shape design and representation. CAD Document No. 87, University of Cambridge.

3. Choi, B.K. and Ju, S.Y. (1989). Constant radius blending in surface modelling. *Computer Aided Design, 21*, 213–220.

4. Foley, J.D. and Van Dam, A. (1982). *Fundamentals of Interactive Graphics*. Addison Wesley.

5. Ghosh, P.K. (1986). A computation theoretic framework for shape representation and analysis using the Minkowski addition and decomposition operators. PhD Dissertation, University of Bombay.

6. Hoffman, C. and Hopcroft, J. (1985). Automatic surface generation in computer aided design. Department of Computer Science TR 85-661, Cornell University.

7. Lee, Y.T. and Requicha, A.A.G. (1982). Algorithms for computing the volume and other integral properties of solids: I. Known methods and open issues, II. A family of algorithms based on representation conversion and cellular approximation. *Communications of the ACM, 25*, 635–650.

8. Middleditch, A.E. (1984). Blend surfaces for set theoretic volume modelling systems. CAE Group ITM 84:4, Polytechnic of Central London.

9. Middleditch, A.E. (1984). The representation and manipulation of convex polygons. CAE Group ITM 84:6, Polytechnic of Central London.

10. Middleditch, A.E. and Sears, K.H. (1985). Blend surfaces for set theoretic volume modelling systems. *Computer Graphics, 19*, 161–170.

11. Middleditch, A.E. (1990). Set theoretic rolling sphere blends. CAE Group Report BRU/CAE/90:2. Brunel University.

12. Okino, N., Kakazu, Y. and Kubo, H. (1973). TIPS-1: Technical processing system for computer aided design, drawing and manufacturing. In *Computer Languages for Numerical Control*, J. Hatvany, editor, North Holland, 141–150.

13. Requicha, A.A.G. (1980). Representations for rigid solids: theory, methods and systems. *ACM Computer Surveys, 12*, 437–464.

14. Requicha, A.A.G. and Voelker, H.B. (1982). Solid modelling: a historical summary and contemporary assessment. *IEEE Computer Graphics and Applications, 2*, 9–24.

15. Requicha, A.A.G. and Voelker, H.B. (1983). Solid modelling: current status and research directions. *IEEE Computer Graphics and Applications, 3*, 25–37.

16. Rossignac, J.R. and Requicha, A.A.G. (1984). Constant radius blending in solid modelling. *Computers in Mechanical Engineering*.

17. Roth, S.D. (1982). Ray casting for modelling solids. *Computer Graphics and Image Processing, 18*, 109–144.

18. Salmon, G. (1914). *Analytic Geometry of Three Dimensions, Vol. II.* 5th edn., Chelsea.

19. Sears, K.H. and Middleditch, A.E. (1983). Tree structures and classification algorithms for set theoretic volume models. CAE Group ITM 83:3, Polytechnic of Central London.

20. Sears, K.H. and Middleditch, A.E. (1984). Set theoretic volume model evaluation and picture plane coherence. *IEEE Computer Graphics and Applications, 4*, 41–46.

21. Tilove, R.B. (1977). A study of geometric set membership classification. Production of Automation Project ITM-30, University of Rochester.

22. Tilove, R.B. (1980). Set membership classification: a unified approach to geometric intersection problems. *IEEE Transactions on Computers, C-9*, 874–883.

23. Tilove, R.B. (1981). Exploiting spatial and structural locality in geometric modelling. Tech. Memo 38, Production Automation Project, University of Rochester.

24. Varady T. and Pratt, M.J. (1984). Design techniques for the definition of solid objects with free form geometry. *Computer Aided Geometric Design*, *1*, 207–225.

25. Wilkinson, J.H. (1959). The evaluation of ill-conditioned polynomials, parts I and II. *Numerische Mathematik*, *1*, 150–180.

26. Woodwark, J.R. and Bowyer, A. (1986). Better and faster pictures from solid models. *Computer-aided Engineering Journal*, 17–24.

27. Woodwark, J.R. (1987). Blends in geometric modelling. In *The Mathematics of Surfaces*, R. R. Martin, editor, Oxford University Press, 255–297.

Generalised Distance Functions and Plastic Modelling over an Armature

M.H.E. Larcombe

Department of Computer Science, University of Warwick

Abstract This paper describes a technique of implicit surface definition which is designed to use primitives of great economy of storage. Blending of primitives is effected by using a generalisation of the concept of distance expressed in a manner similar to potential functions. Unlike simple potential function definitions of implicit surfaces which require complex evaluation in order to form "common" geometric forms, the generalised distance function approach defines useful primitive forms which are, in effect, clad with an adjustable implicit surface. Such implicit surfaces are then blended to form constructs of arbitrary complexity. The style of modelling is equivalent to armature and clay modelling where an initial coarse form is finished by the application of layers of modelling clay.

1 Introduction

The form of modelling to be described stems not from a graphics/cad context but from a class of problems in the mapping of rough terrain for mobile robot navigation and guidance [1]. While the provision of an economical form of topograhical model presents few problems the provision of a representation of surface features such as rocks, bushes, artifacts etc. in such a way that range or optical angle subtention queries could be tested economically presented a major problem. Even a small area of rough terrain can contain many thousands of such surface features, non of which oblige by having concise geometrical approximations. In practical terms it was clearly infeasible to model the surfaces of these thousands of items by such means as polynomial patches. Each item could well require many hundred patches and polynomial models do not lend themselves to economical range or angular subtention testing.

The earliest solutions to the problem attempted to use a simple potential modelling approach effected in conjunction with space subdivision techniques used to limit the search region (binary space division tree storage of simple point and line "sources"). Although this approach satisfied the requirements for small scale modelling the modelling of such features as hedgerows or riverbanks required large numbers of point and line primitives, with consequent excessive computational loads. The modelling process was also exceedingly time consuming requiring long periods of interactive design using graphical representations.

The requirement seemed to be for "source" distributions which lent themselves to more general modelling by representing more substantial primitives. Unfortunately when distributions were found which approximated primitives such as e.g. finite cylinders the evaluation of potential using such distributions became a major exercise in numerical spatial integration, and as such the surface evaluation problems became a greater load than was the case with the naive point source approximation.

The concept of some form of "potential" as a model of generalised distance from a surface instead of as the primary mechanism for defining the surface then arose. Could "armature" elements be defined which were in themselves reasonable shape primitives: could such primitives be defined such that range queries could be tested quickly in a single element condition: could such primitives be combined in such a manner that they could be combined, superimposed and merged?

This paper gives an initial set of such primitives and a means of combining them which not only satisfy the above conditions but also have extremely low storage and evaluation costs. It must be emphasised that the initial purpose of the modelling process was to approximate the forms of natural features and that precision of representation was definitely not a factor. The detailed representation of a particular clump of undergrowth would require vast resources, a "fuzzy" interpretation of an approximating surface is extremely useful in a navigation and guidance context.

Since the surfaces generated by the process are mathematically defined they are appropriate in the synthesis of forms such as may be requred in industrial design of "plastic" forms. In order to use these forms in a more conventional cad modelling process the resulting surfaces may well have to be themselves approximated by e.g. polynomial patches but this can clearly be done in a semi-automatic manner using surface fitting methods. It should be noted that the author has not pursued this aspect.

2   Merging using Generalised Distance Functions

Suppose there exists a set, the armature, of surface elements, the armature elements, and that for any armature element the minimum distance between the element surface and a general point in space can be computed. Given a finite set of such elements how can the individual element distances be merged to give a single "generalised distance" from the set of elements in such a way that a family of smooth implicit surfaces can be defined?

The surface will in general be disconnected and individual parts may not be simply connected.

Clearly if we take a simple set minimum the resulting surfaces will have slope discontinuities. While this may be appropriate for some surface modelling processes it is not truly merging, being very similar to the combination effects provided in constructive solid geometry (CSG). We can however get a measure in terms of the inverse of the minimum distance, the proximance, from a general point, $\underline{x}$, to an armature element surface. If we make $D(\underline{x})$ the inverse of the sum of the individual proximances we form a potential-like function which enjoys some of the properties of a simple scalar potential. In this interpretation the implicit surfaces become equipotential surfaces $D(\underline{x}) = k$, where k is a positive constant analogous to an offset distance.

These functions formed by the inverse of the sum of the individual proximances are the "generalised distance functions" of the title. Their value as we shall see can be defined everywhere but in computational terms we must be aware of the fact that individual proximances have infinite values at the surface of their corresponding armature element, requiring some care in evaluation in the neighbourhood of the armature.

The simple generalised distance function described above may be further extended by using weighted sums of the proximances where the weighting is associated with an armature element. In the general case $D(\underline{x}) = k$ increasing weight has the effect of expanding the spatial influence of the corresponding armature element. If we consider the constant k to be the "nominal" thickness of a "cladding" (as in clay and armature modelling) then the weighting can be considered as a magnification factor locally applied. The nearest physical concept would be the "fibrousness" of a surface in clay and armature modelling, the more fibrous or "hairy" the surface the more clay can be built up upon it. We can thus give to the modeller not only a choice of armature elements but also some variation in the nature of each element.

To extend the flexibility of the armature elements a further intrinsic cladding is allowed. This extends a "rind" of known thickness beyond the basic surface. In the examples given this element specific cladding is of uniform thickness but there is no reason why it should not have variation, provided the resulting extended surface continues to have a simple minimum distance algorithm.

3 General Requirements for Armature Elements

As is well known the finding of the minimum distance from a general point to an arbitrary mathematically defined surface generally involves non-trivial computation. The primary requirement for armature elements for use within a generalised distance function context is high computational efficiency in minimum distance computation. Accordingly the armature elements to be given will be primarily described algorithmically, and it being implied that we are working in three-dimensional Euclidean space the algorithms are essentially constrained quadratic minimisation processes. In the cases given in the Appendix all the armature element basic surfaces also have a relatively simple mathematical description, although this is usually piecewise rather than in the form of a single formula.

The primitives given comprise:

0) Simple Sphere (4)

1) Hemi-Spherically Capped Finite Cylinder or Bean (8)

2) Finite Cylinder or Rod (8)

3) Spherically Capped Finite Conical Frustrum or Top (10)

4) Right Cut Conical Frustrum (10)

5) Torus (8)

6) Cigaroid (Solid of Rotation of Circular Arc) (8)

7) Infinite Plane (6)

8) Rectangle (8)

9) Rectangular Box (12)

10) Disc (7)

11) Convex Lens (7)

12) Finite Cylindrical Tube (9)

where the numbers in brackets indicate the number of words required to define each armature element.

Given such a set of armature elements composite armature elements can be created and manipulated as though they were simple armature elements. This is simply a matter of data-structuring within the armature construction algorithms. In large systems the data structuring should allow for space partitioning to allow reduction in the search space in terms of both the functional evaluation and the mechanics of visualisation.

It is quite possible to define surfaces algorithmically which have an efficient minimal distance computation but which have no simple mathematical surface definition. For instance a recursively defined tree-like surface (natural tree, not a lattice representation), of the kind which if allowed to recurse indefinitely would be "fractal" but if terminated either by recursion depth or by physical dimension limit produces a finite two-dimensional surface, may be implemented as a very fast and simple minimisation algorithm but there appears to be no equivalent simple "mathematical" description of the surface - unless one defines the algorithm as the mathematical description of the surface, in which case it is not open to conventional differential geometry!

4 Rendering of Surfaces

While the original aim of the modelling process was to assist in a navigation process where graphical or even mechanical reproduction of the surfaces was not involved (indeed any graphical representation was simply to enable interactive cartography) the models lend themselves to most surface rendering techniques in graphics and to the forming of e.g. milling programs in CAD.

Given a chosen implicit surface, $D(\underline{x}) = k$, the gradient of the generalised distance function at the surface yields the surface normal (except at such singular points as saddle points where continuity of direction may have to be called upon, as is the case with many other surface models). This allows Gouraud or Phong forms of rendering or the application of ray tracing techniques.

The implicit surface produced by this technique is in general formed from disjoint sections. In order to produce the illustrations given in this paper a form of ray-tracing/z-buffer technique was used in conjunction with a variant of an area fill algorithm. The illustrations may use either perspective projection or orthogonal projections. In either case a ray is reverse projected and traced by a zero seeking algorithm over the function $D(\underline{x}) - k$. On encounter with the first zero (the nearest intercept) the gradient direction is used to control a Phong form of shading.

Area tracing over the projected parts of the over the surface is effected by using an area fill process controlled by the existence of a valid intercept in order to determine the extent of the projective area and the use of a neutral background colour in order to determine prior filling. The fill algorithm is restarted on the pixels which are projections of one or more points on each primitive element. This is only a semi-automatic process and the system used allowed manual restart of the rendering process. Since the process only renders where the surface projects on screen it is naturally more efficient than a full raster ray tracing.

The time taken to render depends on the number of minima of the function $D(\underline{x}) - k$ which are closer to the ray projection origin than the first zero. The generalised distance function along the ray traverse is almost linear at large distances from the aggregate of primitives but becomes a "roller coaster" form as the ray by-passes parts of the armature (Fig. 1).

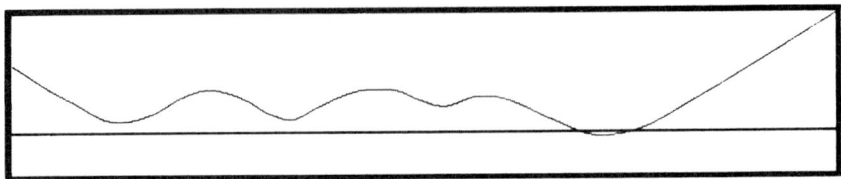

Figure 1. Generalised distance function along ray

Figure 2. Some armature elements

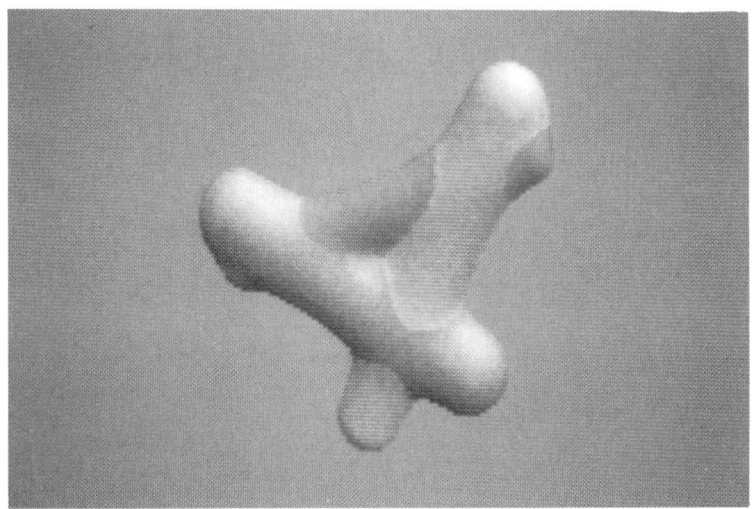

Figure 3. Intersecting beans

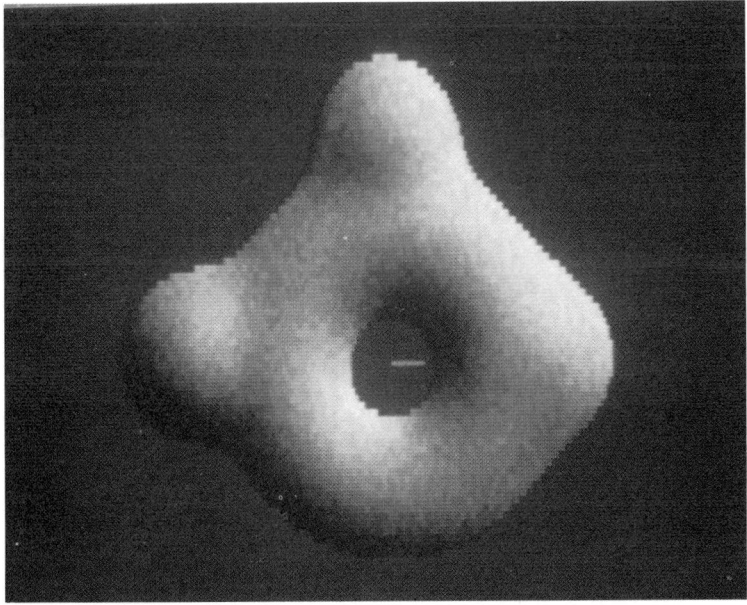

Figure 4. Torus merged with three spheres

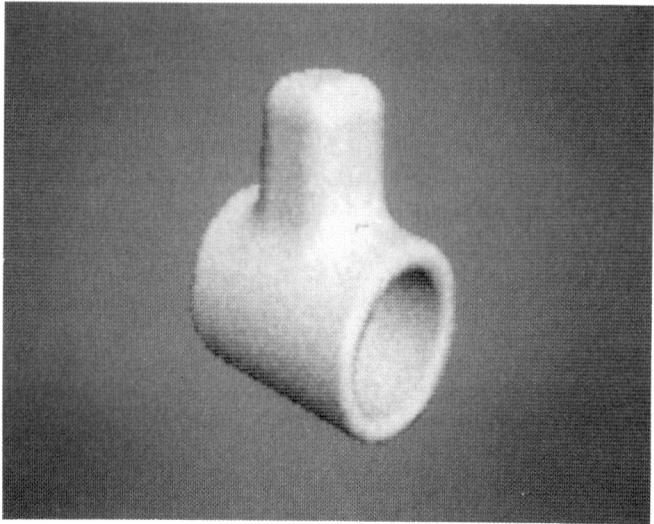

Figure 5. Tube and block

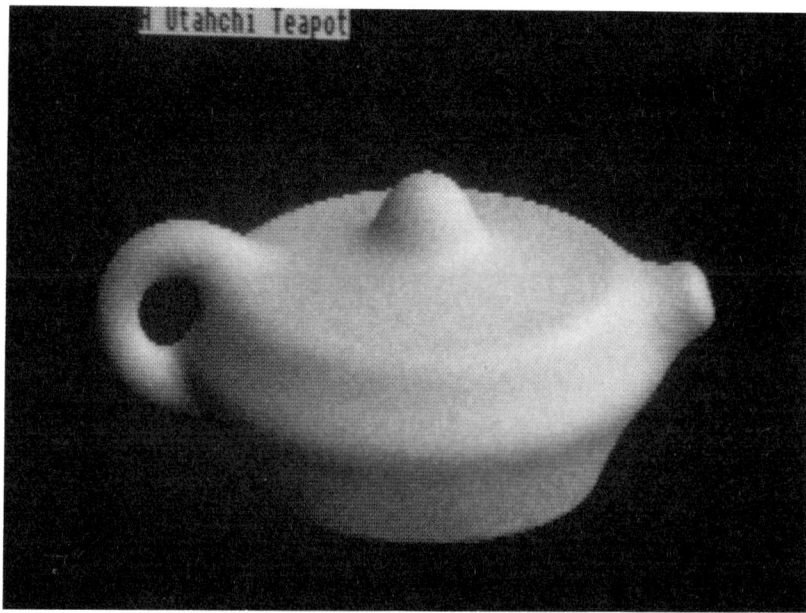

Figure 6. "Utahchi" teapot, composed of three cones, a tube, a sphere and a torus

Figure 7. Soft toy composed from six beans, a block, two spheres, and two discs

The algorithm used in this case is a modified Newton-Raphson process making use of the gradient but scaled to "undershoot" when the slope is negative. Where the slope becomes positive on any "roller coaster" part of the curve the system adopts a simple stepping mode.

More sophisticated zero seeking processes could be created by locally modelling successive minima and maxima by e.g. a polynomial model, and using this model to accelerate the iteration across the minima-maxima pair. The full zero seeking from the ray origin need not be applied for every ray. The images shown used an interpolatory process which takes sample rays at separations of several pixels and uses a second order estimation to start intermediate ray traces close to the surface.

In cases where mechanical reproduction is required the surface can be scanned in "slices" giving not only the profile in the slice plane but also the surface slope perpendicular to the slice plane, a method appropriate for a milling machine with cutter angle control.

5 Conclusion

Figs. 2-7 show individual shapes of primitives and combinations. The most complex form shown is the "teddy bear" which uses a combination of eleven armature elements. The "Utahchi Teapot", with apologies for the Japonification of the term, comprises only six armature elements (3 conical frustra, a tube, a sphere, and a torus) requiring only 118 bytes of storage compared with a typical two or three thousand for the classic Utah teapot represented in patch form. This exercise took by far the longest time to design (almost ten minutes), the teddy bear took about five minutes. The smaller examples took moments to set up. On the machine used (a 4 Mips RISC machine, the Acorn Archimedes 310) the rendering took 10 to 20 minutes but it should be stated that was not using compiled code but the interpreted high level BASIC used on the Archimedes machine.

The technique has certain features in common with other merging methods and with some other implicit surface definitions [2,5] but appears to be unusual in that it is designed for high economy of storage and in providing armature elements which are "natural" spatial modelling elements. It should also be very applicable in animation as the armature elements can be kinematically linked and the final generated implicit surface smoothly changes with the articulation, furthermore changes in the size and weighting of elements should allow the emulation of elastic or plastic deformations during motion. Unfortunately the author does not have immediate access to equipment of the power required to perform such animation in real time.

The technique is very simple in essence, the major effort comes in devising the primitives and their fundamental minimisation algorithms. Where very large scale modelling is required the range over which contributing elements are evaluated requires the use of space partitioning techniques in order to reduce the load of evaluation of the distance functions. By imposing a lower limit on the weighted proximance values suitable partitionings can usually be devised.

While the technique has not been exercised very greatly outside the context for which it was first created it appears to show some promise in the graphics and CAD modelling areas. It is therefore proferred as a possible addition to the armoury of techniques for surface representation.

References

1. Larcombe, M.H.E. (1987). Modeling of Rough Terrain for Navigation and Guidance. Report of NATO Advanced Research Workshop on Mobile Robot Implementation, Porto, Portugal.

2. Filip, D.J. (1989). Blending parametric surfaces. *A.C.M. Trans. on Graphics*, *8*, No.3.

3. Warren, J. (1989). Blending algebraic surfaces. *A.C.M. Trans. on Graphics, 8*, No 3.

4. Rockwood, A.P. (1989). The displacement method for implicit blending surfaces in solid models. *A.C.M. Trans. on Graphics, 8*, No.3.

5. Bloomenthal, J. and Wyvill, B. (1990). Interactive techniques for implicit modeling. *Computer Graphics (ACM), 24*, No.2.

6. Pentland, A. et al. (1990). The ThingWorld modeling system: virtual sculpting by modal forces. *Computer Graphics (ACM), 24*, No.23.

A1 A Selection of Algorithms for Generalised Distance Function Primitives

The following algorithms yield the minimum distance from a general space point, $\underline{x}$, to the surface of the particular primitive element. From a set of such distances, $d_i(\underline{x})$, $i = 1, \ldots, N$, and the corresponding weighted generalised distance function, with weights w_i, is given by:

$$D(\underline{x}) = \frac{1}{\sum_{i=1}^{N} \frac{w_i}{d_i(\underline{x})}}$$

Each algorithm is a constrained minimisation of $|\underline{x} - \underline{y}|$, where $\underline{y}$ can range over the surface of the element. The rind-like intrinsic cladding is represented by reducing the computed minimum distance, in the direction of the outward pointing distance vector, by the intrinsic cladding thickness, δ. The individual cladding is given by c.

In the following algorithmic definitions the function/procedure $d :=$ $unit(\underline{d})$ returns the length of $\underline{d}$ while simultaneously converting $\underline{d}$ to its unit form $\hat{d}$. The surface normal contribution is then derived from $\frac{-w_i \hat{d}_i}{d_i^2}$, while the weighted generalised distance function contribution is $\frac{w_i}{d_i}$.

0) Sphere

This is defined by a reference point $\underline{m}$ at the centre and a nominal radius ρ:

$$\underline{d} := \underline{x} - \underline{m};$$

$$d = unit(\underline{d}) - \rho - \delta;$$

1) Hemispherically Capped Cylinder or Bean

This is defined by a line segment of centre $\underline{m}$, in the direction $\hat{\underline{s}}$, and of semi-length, s, and of radius ρ.

$$\underline{d} := \underline{x} - \underline{m};$$

$$d_s := \hat{\underline{s}}.\underline{d};$$

$$\text{if } abs(d_s) > s \text{ then } d_s := sign(d_s)s;$$

$$\underline{d} := \underline{d} - d_s\hat{\underline{s}};$$

$$d := unit(\underline{d}) - \rho - \delta;$$

$$\text{if } d < 0 \text{ then } \{\underline{d} := -\underline{d}; \ d := -d; \};$$

2) Finite Cylinder or Rod

This is a similar definition to 1) above but the element is cut at the ends of the defining line segment. An additional parameter, the local cladding, c, is used which allows "rounding out" of the basic cylinder (this parameter is redundant in 0) and 1) above since it corresponds to an extension of the radius ρ).

$$\underline{s} := s\hat{\underline{s}};$$

$$\underline{d} := \underline{x} - \underline{m};$$

$$d_s := \hat{\underline{s}}.\underline{d};$$

$$\underline{t} := \underline{d} - d_s\hat{\underline{s}};$$

$$d_t := unit(\underline{t});$$

$$\text{if } s\,d_t > \rho\,abs(d_s) \text{ then}$$

$$\{\text{if } abs(d_s) > s \text{ then } d_s := sign(d_s)s;$$

$$\underline{d} := \underline{d} - d_s\hat{\underline{s}} - \rho\underline{t}; \}$$

$$\text{else}$$

$$\{ \text{ if } d_t > \rho \text{ then } d_t := \rho;$$

$$\underline{d} := \underline{d} - \ sign\,(d_s)\underline{s} - d_t\underline{t}; \}$$

$$\text{endif};$$

$$d := unit(\underline{d}) - c - \delta;$$

$$\text{if } d < 0 \text{ then } \{\underline{d} := -\underline{d}; d := -d; \};$$

Note that $\underline{t}$ has been reduced to a unit vector before the penultimate line.

3) Spherically Capped Cone or Top

This is again defined by a line segment of mid-point $\underline{m}$ in the direction $\hat{s}$, and of semi-length s but the form is defined by the radii ρ_a at $\underline{m} - s\hat{s}$ and ρ_b at $\underline{m} + s\hat{s}$ which describe the spherical end caps, the remainder of the form being the conical frustrum cotangent to these spheres.

$$\Delta := (\rho_a - \rho_b)/2; \quad \rho_m := \rho_b + \rho_a)/2 + c;$$

$$sin\,(\alpha) := \Delta/s; \quad as\,\alpha := \sqrt{1 - sin^2(\alpha)}; \quad tan\,(\alpha) := sin\,(\alpha)/cos\,(\alpha);$$

$$\underline{d} := \underline{x} - \underline{m};$$

$$d_s := \hat{s} \cdot \underline{d};$$

$$\underline{t} := \underline{d} - d_s \hat{s};$$

$$d_t := \text{unit}\,(\underline{t});$$

$$d_\rho := d_s + d_t\,tan\,(\alpha); \quad \text{if}\,(d_\rho > s)\,\text{then}\,d_\rho := sign\,(d_\rho)s;$$

$$\underline{d} := \underline{d} - d_\rho \hat{s}; \quad d := \text{unit}\,(\underline{d}) - \rho_m - d_\rho\,sin\,(\alpha);$$

$$\text{if}\,d < 0\,\text{then}\,\{\underline{d} := -\underline{d}; \quad d := -d\};$$

4) Conical Frustrum

This is defined in a similar manner to 3) above excepting that the cone is right cut by planes at the ends of the line segments with the given radii ρ_a and ρ_b. In this case a local cladding c is used to allow "rounding out" of the basic form.

$$\Delta := (\rho_b - \rho_a)/2; \quad \rho_m := (\rho_b + \rho_a)/2; \quad h := \sqrt{s^2 + \Delta^2};$$

$$cos\,(\alpha) = s/h; \quad sin\,(\alpha) := \Delta/h;$$

$$tan\,(\alpha/2) := \Delta/(h + s); \quad g := \rho_m - h; \quad \underline{s} := s\hat{s};$$

$$\underline{d} := \underline{x} - \underline{m}; \quad d_s i = \hat{s} \cdot \underline{d}; \quad \underline{t} := \underline{d} - d_s \hat{s}; \quad d_t := \text{unit}\,(\underline{t});$$

$$\text{if}\,d_t \leq g\,\text{or}\,d_t = 0\,\text{then}$$

$$\{d := d_s - sign\,(d_s)s; \quad \underline{d} := sign\,(d)\hat{s}; \quad d := abs(d);\}$$

else

$$\{\underline{d} := \underline{d} - g\underline{t}; \quad d_t := d_t - g;$$

$$d_q := d_s + \tan{(\alpha/2)}d_t; \quad d_p := \cos{(\alpha)}d_s + \sin{(\alpha)}d_t - \triangle;$$

$$C := abs(d_q) > (d_t - \tan{(\alpha/2)}d_s);$$

if C or (not C and $abs(d_p) > h$) then

$$\{\underline{d} := \underline{d} - \mathrm{sign}{(d_q)}\underline{s}; \quad d_{tt} := h + \mathrm{sign}{(d_q)}\triangle;$$

if $d_t > d_{tt}$ then $d_t := d_{tt}$;

$$\underline{d} := \underline{d} - d_t\underline{t}; \}$$

else

$$\{\underline{d} := \underline{d} - (d_p \cos{(\alpha)})\hat{\underline{s}} - (d_p \sin{(\alpha)} + h)\underline{t}; \}$$

endif;

$$d := unit\,(\underline{d}); \}$$

endif

$$d := d - c - \delta;$$

if $d < 0$ then $\{\underline{d} := -\underline{d};\; d := -d; \};$

5) Torus

This is defined in terms of its centre point, $\underline{m}$, its axis $\hat{\underline{s}}$, its major (centre line) radius, R, and its minor radius, r.

$$\underline{d} := \underline{x} - \underline{m};$$

$$d_s := \hat{\underline{s}}.\underline{d};$$

$$\underline{t} := \underline{d} - d_s\hat{\underline{s}};$$

$$d_t := unit(\underline{t});$$

$$\underline{d} := \underline{d} - R\underline{t};$$

$$d := unit(\underline{d}) - r - c - \delta;$$

$$\text{if } d < 0 \text{ then } \{\underline{d} := -\underline{d}; \ d := -d; \};$$

6) Cigaroid

This is the form generated by the rotation of a minor arc of a circle about its chord. Again the form is derived from a line segment of mid-point $\underline{m}$, direction $\hat{\underline{s}}$, and semi length s. The minor radius of curvature at the mid-point is given, r, and the major radius of curvature, R, is then determined by $R := (s^2/r + r)/2$.

$$\Delta := R - r;$$

$$\underline{s} := s\hat{\underline{s}};$$

$$\underline{d} := \underline{x} - \underline{m};$$

$$d_s := \hat{\underline{s}}.\underline{d};$$

$$\underline{t} := \underline{d} - d_s\hat{\underline{s}};$$

$$d_t := unit(\underline{t});$$

$$\text{if } abs(d_s) > s(1 + d_t/\Delta) \text{ then}$$

$$\{\underline{d} := \underline{d} - sign(d_s)\underline{s};$$

$$d := unit(\underline{d}); \}$$

else

$$\{\underline{d} := \underline{d} + \Delta\underline{t};$$

$$d := unit(\underline{d}) - R; \}$$

endif;

$$d := d - c - \delta;$$

$$\text{if } d < 0 \text{ then } \{\underline{d} := -\underline{d}; d := -d; \};$$

7) Infinite Plane

This is defined by some reference point, $\underline{m}$, in the plane, and a normal direction, $\hat{\underline{n}}$. Typically used as a ground or back plane. The local cladding is c.

$$\underline{d} := \underline{x} - \underline{m};$$

$$d_n := \hat{\underline{n}}.\underline{d};$$

$$d := sign(d_n)\hat{\underline{n}};$$

$$d := abs(d_n) - c - \delta;$$

$$\text{if } d < 0 \text{ then } \{\underline{d} := -\underline{d};\ d := d\};$$

8) Rectangle

This lamina is defined in terms of its mid-point, $\underline{m}$, two axis directions, $\hat{\underline{a}}$ and $\hat{\underline{b}}$, and their corresponding semi-dimensions, a and b. The local cladding is c.

$$\underline{d} := \underline{x} - \underline{m};$$

$$d_a := \hat{\underline{a}}.\underline{d};$$

$$d_b := \hat{\underline{b}}.\underline{d};$$

$$\text{if } abs(d_a) > a \text{ then } d_a := sign(d_a)a;$$

$$\text{if } abs(d_b) > b \text{ then } d_b := sign(d_b)b;$$

$$\underline{d} := \underline{d} - d_a\hat{\underline{a}} - d_b\hat{\underline{b}};$$

$$d := unit(\underline{d}) - c - \delta;$$

$$\text{if } d < 0 \text{ then } \{\underline{d} := -\underline{d};\ d := -d\};$$

9) Rectangular Box

The extension of 8) above to three dimensions. A further axis, $\hat{\underline{c}}$, and corresponding semi-dimension, c, is given. The local cladding here is taken as t, to avoid confusion.

$$\underline{d} := \underline{x} - \underline{m};$$

$$d_a := \hat{\underline{a}}.\underline{d};$$

$$d_b := \hat{\underline{b}}.\underline{d};$$

$$d_c := \hat{\underline{c}}.\underline{d};$$

if $abs(d_a) > a$ then $d_a := sign(d_a)a;$

if $abs(d_b) > b$ then $d_b := sign(d_b)b;$

if $abs(d_c) > c$ then $d_c := sign(d_c)c;$

$$\underline{d} := \underline{d} - d_a\hat{\underline{a}} - d_b\hat{\underline{b}} - d_c\hat{\underline{c}};$$

$$d := unit(\underline{d}) - t - \delta;$$

if $d < 0$ then $\{\underline{d} := -\underline{d};\ d := -d;\};$

10) Disc

A simple circular lamina, centre $\underline{m}$, axis $\hat{\underline{s}}$, radius ρ, and local cladding c.

$$\underline{d} := \underline{x} - \underline{m};$$

$$d_s := \hat{\underline{s}}.\underline{d};$$

$$\underline{t} := \underline{d} - d_s\hat{\underline{s}};$$

$$d_t := unit(\underline{t});$$

if $d_t > \rho$ then $d_t := \rho;$

$$\underline{d} := \underline{d} - d_t\underline{t};$$

$$d := unit(\underline{d}) - c - \delta;$$

if $d < 0$ then $\{\underline{d} := -\underline{d}; d := -d\};$

11) Convex Lens

This form is generated when a minor planar cut of a sphere is reflected in the cutting plane. It is defined in terms of a mid-point, $\underline{m}$, an axis $\hat{\underline{s}}$, with a semi-thickness s, and a lens radius ρ. The local cladding is c. The spherical radius, R, is derived from $R := (\rho^2/s + s)/2.$

$$\triangle := R - s;$$

$$\underline{d} := \underline{x} - \underline{m};$$

$$d_s := \hat{\underline{s}}.\underline{d};$$

$$\underline{t} := \underline{d} - d_s\hat{\underline{s}};$$

$$d_t := unit(\underline{t});$$

if $d_t > \rho(1 + abs(d_s)/\triangle)$ then

$$\{\underline{d} := \underline{d} - \rho\underline{t};$$

$$d := unit(\underline{d}); \}$$

else

$$\{\underline{d} := \underline{d} + sign(d_s);\triangle\hat{\underline{s}};$$

$$d := unit\,(\underline{d}) - R; \}$$

endif;

$$d := d - c - \delta;$$

if $d < 0$ then $\{\underline{d} := -\underline{d};\ d := -d; \};$

12) Tube

This is a hollow version of 2) The Finite Cylinder. The definition is identical but the algorithm interprets the interior distance in another manner. Local cladding is c.

$$\underline{d} := \underline{x} - \underline{m};$$

$$d_s := \hat{\underline{s}}.\underline{d};$$

$$\underline{t} := \underline{d} - d_s\hat{\underline{s}};$$

$$d_t := unit(\underline{t});$$

$$\underline{d} := \underline{d} - \rho\underline{t};$$

if $abs(d_s) > s$ then $d_s := sign(d_s)s;$

$$\underline{d} := \underline{d} - d_s\hat{\underline{s}};$$

$$d := unit(\underline{d}) - c - \delta;$$

if $d < 0$ then $\{\underline{d} := -\underline{d};\ d := -d; \};$

Texture Mapping Functions and Surface Flattening

Maria Nordgren and Charles Woodward

Helsinki University of Technology

1 Introduction

Surface flattening stands for spreading a curved surface onto a plane with minimal distortion. Applications of surface flattening include cartography, shipbuilding, shoe manufacturing, among many others. This paper considers in particular the use of surface flattening in *texture mapping* [1] for computer graphics.

Texture mapping is a technique to create a varying shading pattern onto the displayed surface. The texture is defined in the *source image* as an $m \times n$ array of pixels. Each pixel contains a value of the texture variable, $T : N^2 \to R$, *e.g.* $T(i,j) = colour$ (Fig. 1).

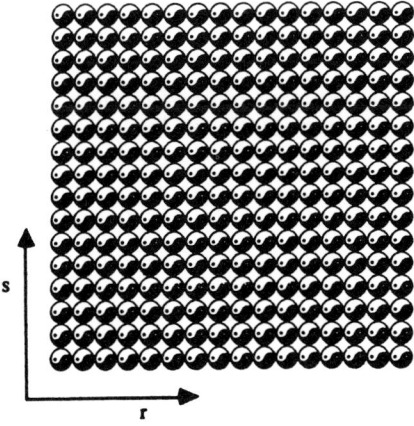

Figure 1. Source image

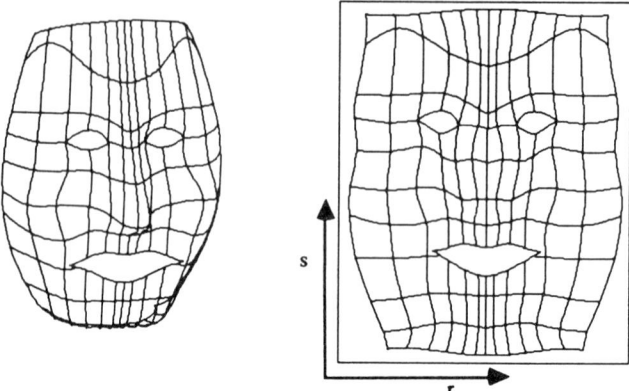

Figure 2. Texture mapping function

For a given surface

$$S : [u_0, u_M] \times [v_0, v_N] \rightarrow R^3, \tag{1.1}$$

the *texture mapping function,*

$$F : [u_0, u_M] \times [v_0, v_N] \rightarrow [0, m] \times [0, n], \tag{1.2}$$

decides which point in the source image should be assigned to each displayed surface point. For given surface parameter values (u, v) the texture value for $S(u, v)$ is thus equal to $T(F(u, v))$.

It is usually desirable that the texture is distorted as little as possible when mapped onto the 3D surface. Indeed, surface flattening can be considered a special case of texture mapping functions. $F(u, v)$ is then defined as the flattened copy of the surface, suitably transformed to the source image space (Fig. 2).

We will deal with both surface flattening and some simpler texture mapping functions in Sections 4 and 3, respectively. The formulation and some notation for our solution methods are explained in Section 2.

The discussion only considers flattening a single surface at a time. For flattening of composite surface models we do not have any new solutions to offer; texture mapping methods for this purpose must generally proceed along quite different lines than discussed here; see for instance [2] [3].

Due to space restrictions the discussion lacks practically all introductory material. For good background reading on the surface flattening problem, see [4]. Flattening applications specifically in texture mapping were first discussed in [5]. A general introduction to texture mapping is given in [1].

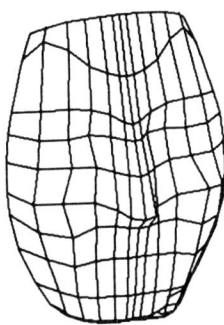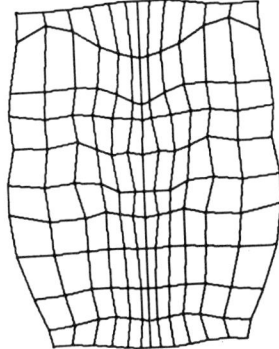

Figure 3. Discretized problem

2 Discretization

We assume the surfaces to be defined in some parametric tensor product form, as a rectangular $M \times N$ mesh of four-sided patches [6]. Instead of considering directly the parametric surface representation, however, we simplify the problem (and generalize the discussion) by approximating the surface with a similar set of bilinear patches (facets). The corner points of the facets, $s_{i,j}$, $i = 0, \ldots, M$, $j = 0, \ldots, N$, are thus equal to the surface knot points. In case of very curved surfaces consisting of only a few large patches we recommend subdividing the surface before discretizing.

Correspondingly, the flattened surface is first computed as a rectangular grid of 2D points, $f_{i,j}$, $i = 0, \ldots, M$, $j = 0, \ldots, N$. The points $f_{i,j}$ are afterwards interpolated with the original parametric representation, yielding $F(u, v)$ with an equal patch structure and parametrization as $S(u, v)$ (Fig. 3). (Setting the interpolation boundary conditions may require some heuristics depending on the particular parametric form, in particular if $S(u, v)$ is a closed surface.)

Finally, note that possible trim curves defined in the surface's parameter space (Fig. 4 left) do not have to be considered by the flattening algorithms at all, as the same parameter space is shared by both $S(u, v)$ and $F(u, v)$.

3 Texture Mapping Functions

The source images are customarily defined in a rectangular square. It is therefore desirable that the texturing function somehow obeys the rectangular shape, as well.

The simplest texture mapping function — we call it *uniform rectangle mapping* — is to scale the surface's parameter space linearly to the source

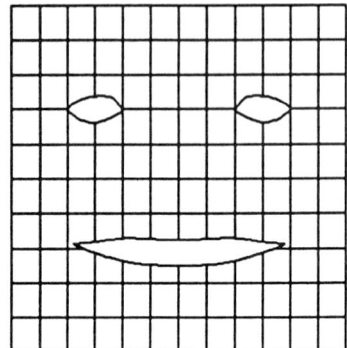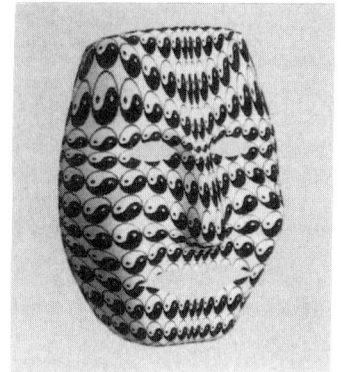

Figure 4. Uniform rectangle mapping

image (Fig. 4 left). Let us write it down in pseudocode for demonstration:

for $i = 0$ to M
    for $j = 0$ to N $\{f_{i,j}^r = i/M \times m;\ f_{i,j}^s = j/N \times n;\}$.

In the absence of other methods, uniform rectangle mapping is, in fact, the most frequently used method in practice [7]. However, it is highly dependent on the parametrization, distorting the texture according to the isoparametric lines of the surface. This is usually not desired, though playing with the parametrization can sometimes be used to produce special effects, like the angry forehead of the mask in Fig. 4.

The visual result is immediately improved by applying what we call the *adaptive rectangle* function. The idea is to deform the surface's parameter space according to the relative Euclidean distances between the surface knot points:

for $i = 0$ to M
    for $j = 0$ to 1 $\{f_{i,j \times N}^r = j \times m;\}$
for $j = 0$ to N
    for $i = 0$ to 1 $\{f_{i \times M,j}^s = i \times n;\}$
for $i = 0$ to M
    $u\_length = 0$; for $j = 1$ to N $\{u\_length\ +\ =\ d(s_{i,j-1}, s_{i,j});\}$
    for $j = 1$ to N $\{f_{i,j}^r = f_{i,j-1}^r + (d(s_{i,j-1}, s_{i,j})/u\_length) \times m;\}$
for $j = 0$ to N
    $v\_length = 0$; for $i = 1$ to M $\{v\_length\ +\ =\ d(s_{i-1,j}, s_{i,j});\}$
    for $i = 1$ to M $\{f_{i,j}^s = f_{i-1,j}^s + (d(s_{i-1,j}, s_{i,j})/v\_length) \times n;\}$

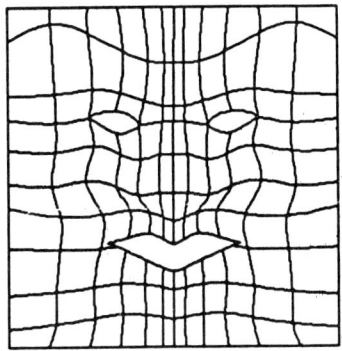 

Figure 5. Adaptive rectangle mapping

where $d(p, q) = \sqrt{(p^r - q^r)^2 + (p^s - q^s)^2}$. The adaptive rectangle function makes the texture mapping dependent mostly on the surface geometry but not the parametrization (Fig. 5). Generally stated, it is practically always superior to the uniform method, and, in fact, exactly what is required in many texture mapping cases.

Another similar solution, if parts of the texture image can be ignored, is called *adaptive parallel* mapping. It has two sides parallel to the source image boundaries, while the two other sides imitate the idea of a physically flattened surface (Fig. 6). The difference in computation to the rectangle function is that the relative distances are accounted for only in one parameter direction, and true distances in the other. The following pseudocode applies especially for surfaces with a central symmetry axis:

$$\text{for } i = 0 \text{ to } M$$
$$u\_length = 0; \text{ for } j = 1 \text{ to } N \ \{u\_length \mathrel{+}= d(s_{i,j-1}, s_{i,j});\}$$
$$f^r_{i,0} = -u\_length/2;$$
$$\text{for } j = 1 \text{ to } N \ \{f^r_{i,j} = f^r_{i,j-1} + d(s_{i,j-1}, s_{i,j});\}$$
$$\text{for } j = 0 \text{ to } N$$
$$v\_length = 0; \text{ for } i = 1 \text{ to } M \ \{v\_length \mathrel{+}= d(s_{i-1,j}, s_{i,j});\}$$
$$f^s_{0,j} = 0;$$
$$\text{for } i = 1 \text{ to } M \ \{f^s_{i,j} = f^s_{i-1,j} + (d(s_{i-1,j}, s_{i,j})/v\_length) \times n;\}.$$

Finally, following true distances in both surface directions in the same manner yields a simple flattening method which already resembles the intuition of actual flattening (Fig. 7). To distinguish this one-pass method from the more elaborated iterative one in the next Section, we call it the *rough flattening* function.

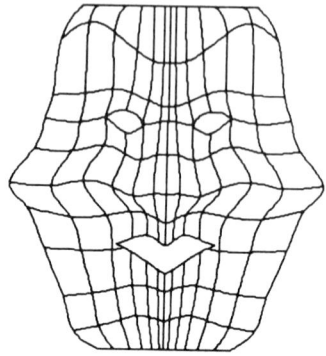

Figure 6. Adaptive parallel mapping

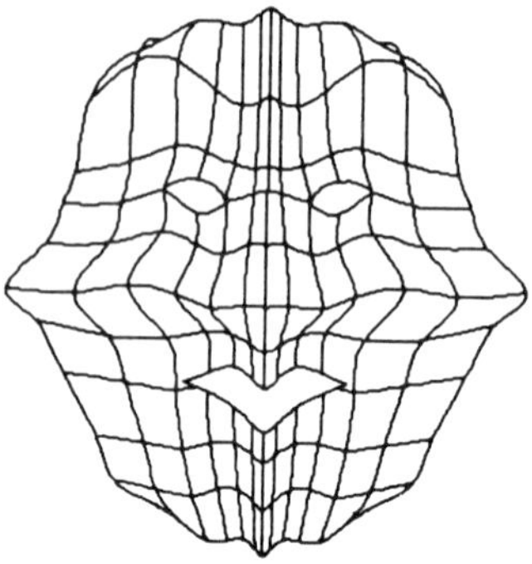

Figure 7. Rough flattening

4 Optimal Flattening

4.1 Isometry

A surface flattening F is called *isometric* if any curve defined on $S(u, v)$ has equal length on $F(u, v)$. In other words, it preserves all the surface areas and also the angles between any two intersecting curves on the surface [4].

An isometric flattening exists only in the case of so-called *developable surfaces* [8], such as cylinders and cones which can be rolled out onto a plane without any stretching or tearing. For non-developable surfaces the flattening can be done only approximately. A flattening is then said to be *optimal*, if it is as close to isometry as possible.

However, closeness to isometry can in practice only be measured approximately and in local terms. Two different examples of how to measure isometry are *conformal flattening* which attempts to preserve all the angles and shapes of small surface areas, and *equal area* flattening which preserves the areas but not distances [4]. Different mapping functions can thus result in widely varying planar shapes.

Besides preserving the local measures, the flattened surface should obviously not self-intersect, and in general it should be aesthetically appealing with smooth boundaries. As such things are difficult to compute, and as no clearly superior method exists to measure departure from isometry, the flattening algorithm should also provide flexibility to account for varying applications.

4.2 Flattening algorithm

In the next Section we define a *local norm* $n(f_{i,j})$ which measures the departure from isometry concerning point $f_{i,j}$ and its eight neighbours. The *global norm*, $N(F) = \sum_{i=0}^{M} \sum_{j=0}^{N} n(f_{i,j})$, correspondingly measures the isometry of the total surface.

The algorithm we use to minimize $N(F)$ follows the outline of [5]. First an initial guess is made for the flattening, which is then refined iteratively with the gradient method [9]. The overall algorithm looks like the following:

1. Discretize the surface $S(u, v)$ into the knot points $s_{i,j}$.

2. Compute the initial grid of points $f_{i,j}^{0}$. For this, choose one of the methods described in Section 3, *e.g.* rough flattening.

   Evaluate the global norm $N(F^{0})$. Set $k = 0$ and, say, $a = 0.1$.

3. Move each of the points to the negative gradient direction of $n(f)$,

$$f_{i,j}^{k+1} = f_{i,j}^{k} - a(\delta n(\{f^{k}\}_{i,j})/\delta f_{i,j}^{k}). \tag{4.1}$$

4. Evaluate the global norm $N(F^{k+1})$. If its value has increased from the previous one, decrease the step length by $a = a/2$ and go back to step 3; else set $a = a \times \sqrt{2}$ and $k = k + 1$;

5. Terminate the algorithm if no points $f_{i,j}$ were moved more than a given value, *e.g.* tenth of a pixel size in the source image. Otherwise go back to step 3.

6. Interpolate the obtained points $f_{i,j}$ into $F(u, v)$.

The method converges to a unique solution, provided the norm and the starting values make any sense. Compared with the original algorithm [5], note that the one above computes the flattening independently of the viewing direction. It also adjusts the step length automatically which we have found crucial for fast and reliable convergence. And finally, it defines the flattened surface with the original parametric representation instead of pixel-sized facets.

4.3 Flatness norm

We have tested two different norms from previous literature.

The norm in [4] is based on differential analysis, theoretically providing very close local measures to the actual departure from isometry. However, the analytic properties of the norm are essentially compromised when the problem is discretized. On the other hand, if a very dense point grid $f_{i,j}$ is used the method gets extremely slow due to the complex norm equation.

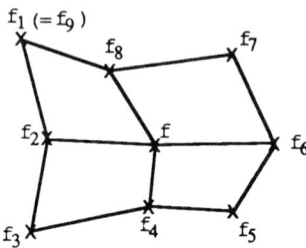

Figure 8. Notation of local variables

The norm is also difficult to differentiate; according to [4] the point moves are actually done by choosing one of the four N-E-S-W directions instead of evaluating the gradient.

According to our experience, a more promising approach is given by the easier local norm defined in [5]:

$$n_D(f) = \sum_{i=1}^{8}(d_i(s) - d_i(f))^2, \qquad (4.2)$$

where s and f are corresponding points of the grids S and F, and $d_i(p)$ is the distance between p and p_i (see Fig. 8). The gradient vector is correspondingly:

$$\delta n_D(f)/\delta f = \sum_{i=1}^{8} 2(\frac{d_i(s)}{d_i(f)} - 1)(f_i - f), \qquad (4.3)$$

The norm $n_D(f)$ seems to comply well with the definition of isometry, but, unfortunately, it does not contain sufficient information about the *directions* of the distance vectors $f_i - f$. This can very easily lead to self-intersecting results, like the nose in Fig. 9 (left) resulting in the locally distorted surface texture in Fig. 10. The self-intersection problem is also always encountered when there are large surface patches adjacent to much smaller ones. With highly curved surfaces, a torus for instance, the self-intersecting occurs even in global scale.

To fix the self-intersection problem we introduce a norm similar to $n_D(f)$, but involving areas instead of distances:

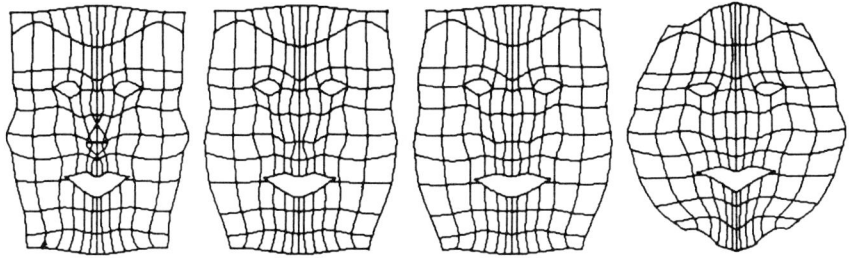

Figure 9. Flattenings with $n_{AD}(b; f), b = 0.0, 0.2, 0.5$ and 0.9

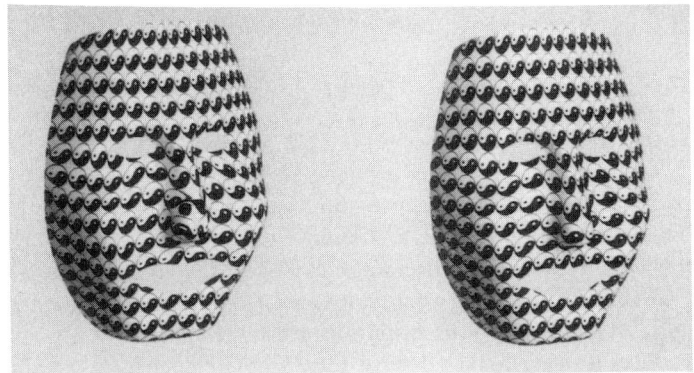

Figure 10. Mask textured using $n_{AD}(b; f), b = 0.0$ and $b = 0.5$

$$n_A(f) = \sum_{i=1}^{8} (\sqrt{|A_i(s)|} - \frac{A_i(f)}{\sqrt{|A_i(f))|}})^2, \qquad (4.4)$$

where $A_i(p)$ is the cross-product $((p_i - p) \times (p_{i+1} - p))/2$.

When computing the gradient vector the denominator can be taken as constant, leading to the simple expression:

$$\frac{\delta n_A(f)}{\delta f} = \pm \sum_{i=1}^{8} (\frac{\sqrt{|A_i(s)|}}{\sqrt{|A_i(f)|}} - \frac{A_i(f)}{|A_i(f)|})(f_{i+1} - f_i), \qquad (4.5)$$

where the sign is $+$ for f^r and $-$ for f^s.

The norm $n_A(f)$ prevents self-intersections by having negative areas $A(f)$, which make the norm values and gradient lengths correspondingly large. However, the norm $n_A(f)$ is too loose to be used as such — the equal area problem does generally not even have a unique solution to which the gradient method should converge.

The norm finally proposed is a linear combination of $n_D(f)$ and $n_A(f)$:

$$n_{AD}(b; f) = (1 - b)n_D(f) + bn_A(f), \qquad (4.6)$$

where b is a user defined constant. We have observed that the gradient method converges with any value of $b \in [0, 1)$, and also self-intersections tend to be avoided already with very small values of b. Value $b = 0.5$, for instance, usually gives aesthetically good results (Fig. 10).

When using the norm $n_{AD}(b; f)$ in applications other than texture mapping, it may be difficult to choose the optimal flattening among the variety produced with different b values. A solution is then to evaluate the flattening results with another global norm derived from the particular application, *e.g.* the norm [4] for shoe lace flattening. Computing the flattening by $n_{AD}(b; f)$ is fast enough to allow such optimization procedure even to be made automatic.

5 Conclusions

The texture mapping functions presented in Section 3 provide simple means to reduce undesired texture distortion in rendering parametric surfaces. To our knowledge, no previous literature exists on the subject. Instead of applying no texture mapping function at all (here called uniform rectangle mapping), as seems to be the general practice, the adaptive rectangle function provides a much better default for parametric surface visualization.

The surface flattening method described in Section 4 gives very good results for optimally minimizing the texturing distortion. The method is fast, flexible, and easy to use. The only required parameter is the area-distance weighting factor applied with the flatness norm, which allows the flattening to be tuned according to requirements of the application. The presented norm produces superior results compared to other norms we have tested. Finally, we believe the results are not restricted only to computer graphics but they apply equally well to many other surface flattening tasks.

References

1. Heckbert, P. (1986). A survey of texture mapping. *IEEE Computer Graphics and Applications, 6(11)*, 56–67.

2. Bier, E.A. and Sloan, K.R. (1986). Two-part texture mappings. *IEEE Computer Graphics and Applications, 6(9)*, 40–43.

3. Woodward, C., Rekola, P., Nordgren, M. and Slotte, T. (1989). An implementation of a CAD/CAM system in ceramic industry. *Proc. CAPE '89*, 163–170.

4. Manning, J.R. (1980). Optimal mappings onto a plane. PhD thesis, Loughborough University of Technology.

5. Ma, S. de and Lin, H. (1988). Optimal texture mapping. *Proc. Eurographics 1988*, 421–428.

6. Faux, I.D. and Pratt, M.J. (1979). *Computational Geometry for Design and Manufacture*, Wiley.

7. Upstill, S. (1990). *The Renderman Companion*, Addison–Wesley.

8. Redont, P. (1989). Representation and deformation of developable surfaces. *Computer Aided Design, 21(1)*, 13–20.

9. Blum, E.K. (1972). *Numerical Analysis and Computation: Theory and Practice*, Addison–Wesley.

Fourier Methods for Surface Smoothing in Computer Aided Design

P.A. Roach and R.R. Martin

University of Wales College of Cardiff

1 Introduction

Many modelling systems use cubic B-spline geometries to define the shape of surfaces. This method uses surface patches which are defined by a large number of control points, which *en masse* can be awkward and difficult to manipulate to achieve a desired result. This paper proposes a surface modelling system which performs smoothing operations using only a few selected parameters.

The system can be used to produce a smooth version of an object which has its surface crudely defined with polyhedra, or it can be used to edit existing shapes of more or less any geometric type produced by another CAD system. Samples of data are taken from the surface of the object to be altered, which are used to calculate Fourier transforms. Fourier smoothing techniques (similar to those used in smoothing signal noise) are then applied to this spatial frequency information. The system allows local control over shape, but does not easily lend itself to the imposition of engineering constraints. Thus the techniques we describe would be best used as a method of altering regions of objects to meet some aesthetic guidelines. To this end it is necessary that the few parameters used are given clear names and meanings intuitive to a designer, who will not know any details of Fourier analysis.

The system is not restricted to edge or vertex smoothing (as are some systems [3]) as the method used relies on sampling points from the whole of the surface region being smoothed, performing operations on spatial frequency distributions for these samples. This allows already curved surfaces to be smoothed, as well as ones with planar geometry, and, by suitably restricting the region, it is even possible to alter the shape of the interiors of patches.

This paper is intended as an analysis of the properties of Fourier transforms which are useful in a Fourier treatment of smoothing, and as a statement of the current position of the work. Section 2 establishes an outline of the way in which the system works. The basic ideas of Fourier smoothing are examined first in Section 3 in two dimensions where the bulk of the work has so far been directed, and some of these ideas are then extended into three dimensions in Section 4.

2 System Overview

This Section covers the method used by the system to accept user requests for smoothing objects, and how the system meets these requests.

2.1 Outline of steps taken by the system

Fig. 1 shows the order of the steps taken in smoothing objects. In this case an example of a simple L-shaped block is used. The part of the surface which is to be smoothed is identified by the user selecting edges or faces by labels assigned to them by the system, and specifying some proportions along edges or faces, if this information is needed, by keyboard input. Points are sampled from this surface region and Fourier transforms are used to calculate frequency distributions for the shape data. These Fourier transforms are filtered, or altered in some way to smooth the data, and their inverse Fourier transforms are calculated giving shape data for the smoothed surface region. A new surface may then be fitted to the shape data, and returned to the original object.

As can be seen from the example in Fig. 1, further adjustment of the overall object model may be needed. How this is dealt with is dependent on the method used to represent surface information. The method chosen for this system is Boundary Representation [11]. After smoothing a surface region, a method is needed to decide which other faces must be extended,

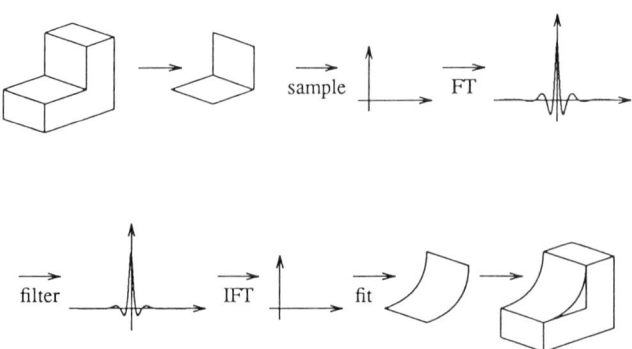

Figure 1. Steps taken by the system

or whether a new surface patch must be inserted into the object. Similar ideas have been discussed in [10].

Each face of the object surface is assumed to be stored as a series of parametric surface patches, with given boundary curves and vertices. The surface region to be altered may be made up of several surface patches or parts of patches. After smoothing, new patches are fitted to the smoothed shape data. By considering which of the surrounding patches originally had boundaries in common with the altered patches, action can be taken to extend faces or insert more patches, if needed.

2.2 User interface

The user interface must be kept simple, the user specifying the smoothing in a purely aesthetic way, leaving the system to determine exactly how to meet such requirements.

The input that the interface requires is: 1) the selection of the faces or parts of faces to be smoothed and those which are to remain fixed; 2) the specification of the degree of surface continuity desired at the boundaries of the shape (*i.e.* the continuity of the smoothed region with the rest of the object); 3) a small number of parameters which control the amount and nature of the smoothing.

2.3 Expectations of the system

Before proceeding with details, it is necessary to establish expectations of how the system should behave, and to determine whether Fourier methods can be used to meet these expectations. Some particular expectations are outlined below.

Firstly, smoothing performed on a particular configuration of faces and edges should produce the same results as the same operation performed on the complement of that configuration (*i.e.* the same faces and edges, but with the material on the other side). This is because smoothing is an operation which uses points sampled only from the surface of the object, not from the material.

Smoothing should be a locally defined operation. It is left up to the user to check that there are not any undesired global consequences such as the smoothed body having self intersections, or to request a global verification if unsure.

The system must be able to smooth an already smoothed surface region. As smoothing is an operation performed on arrays of points, it should be possible to perform an operation any number of times, giving different (but converging) results with each repetition of the operation.

Finally the system must allow an operation to be undone, should the results of that operation be unsatisfactory.

3   Fourier Smoothing in Two Dimensions

In order to examine exactly how Fourier methods can be used to meet the aims of the system, we will first consider Fourier smoothing methods in 2D. The system described in this section is a curve-smoothing system (the 2D analogue to the surface smoothing system).

3.1   Organisation of data

In order to produce a spatial frequency distribution for a given shape, data points must be sampled from the shape and organised in such a way that a suitable Fourier transform can be calculated. Two possible methods have been considered.

Firstly, the original shape could be represented by a two-dimensional array as $f(x, y)$ by assigning the value 1 to those (x, y) pairs which lie inside the shape, and 0 for all other points. After smoothing the result, $g(x, y)$, would include values between 0 and 1 and a threshold t could be used such that points with $g(x, y) > t$ lie inside the new shape. Varying t would provide a simple parameter to control the smoothing. However, such a method uses a large amount of storage space due to the large number of data points necessary to define the shape, most of which would correspond to points in the interior of the shape. It is only the boundary of the shape which is to be altered in smoothing. As well as being wasteful in storage, a two-dimensional Fourier transform would have to be used on data of this form, which would take time $O(N^2 \log N)$ for an $N \times N$ resolution grid. For engineering tolerances, N might be quite large.

Secondly, it is possible to represent the boundary of the shape as parametric functions of x and y, $x(s)$ and $y(s)$, where s is arc length or some other suitable parameter. These functions would each be sampled at N equally spaced values of s. This method requires two one-dimensional Fourier transforms, and takes time $O(N \log N)$ which is a significant reduction in computation. We have thus chosen to use this latter approach. The value of N chosen may depend on the properties of the shape being sampled. Data must be sampled sufficiently often so as not to lose important shape details and thus produce meaningless Fourier transforms, but not so often as to make N unnecessarily large. The question of undersampling is dealt with by Bracewell [2] and others.

3.2   Fourier transforms of simple shapes

To explain how Fourier components relate to shape, and to see how smoothing can be performed in Fourier space, we will illustrate the differences between the Fourier transforms of two simple (and distinct) shapes.

Firstly, consider a square, which is a simple example of an unsmooth shape:

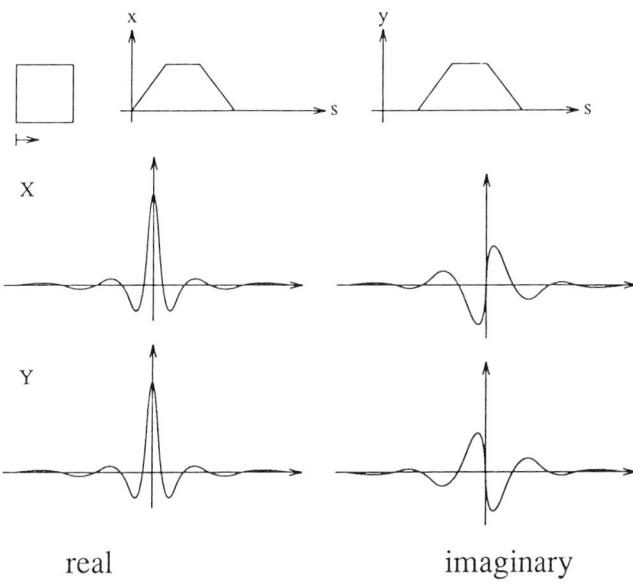

Figure 2. Fourier transforms of a parametrised square

In Fig. 2 the real and imaginary parts of the Fourier transforms have been shown separately. In this case the Fourier transforms for the x and y distributions are identical in real part, and reversed in imaginary part. This is because the parametric distributions themselves have the same shape and are shifted in s.

Let us compare these Fourier transforms with those of a circle (Fig. 3).

More generally, choosing N equidistant points on the boundary of the circle to be $(x_c + r\cos(\theta + 2\pi j/N), y_c + r\sin(\theta + 2\pi j/N))$ where (x_c, y_c) is the centre of the circle, θ is the angle between the starting point and the x-axis, and j takes values from 0 to $N-1$, the FT data for these sequences are, using the notation $X = FT(x)$, $Y = FT(y)$:

$$X \text{ real } = 0\ 0 \ldots 0 \quad r\cos\theta \quad x_c \quad r\cos\theta \quad 0 \ldots 0\ 0$$
$$ \text{imag } = 0\ 0 \ldots 0 \quad -r\sin\theta \quad 0 \quad r\sin\theta \quad 0 \ldots 0\ 0$$

$$Y \text{ real } = 0\ 0 \ldots 0 \quad r\sin\theta \quad y_c \quad r\sin\theta \quad 0 \ldots 0\ 0$$
$$ \text{imag } = 0\ 0 \ldots 0 \quad r\cos\theta \quad 0 \quad -r\cos\theta \quad 0 \ldots 0\ 0$$

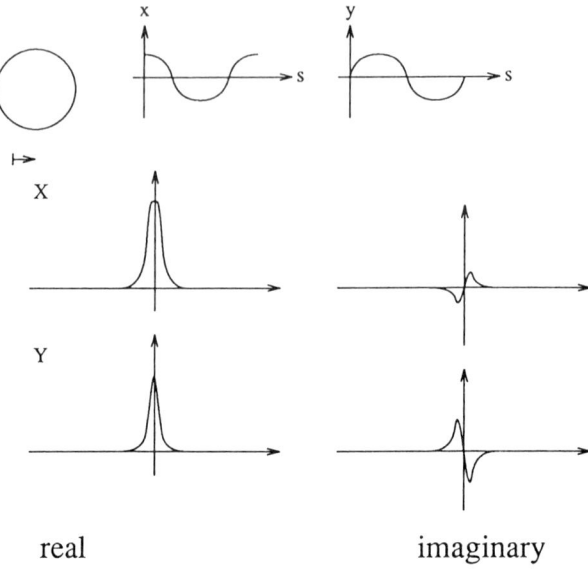

Figure 3. Fourier transforms of a parametrised circle

From Fig. 3 and the above sequences it is clear that the Fourier transforms of equidistant points sampled from a circle have zero values for the
higher frequency components, in direct contrast to the Fourier transforms
for a square. This ties in with the observation that places of rapid change
(the corners of the square) are represented by high spatial frequencies in
the Fourier transform of a shape (these are absent for the circle). The
more rounded a shape is (*i.e.* the more it resembles a circle) the smaller its
higher frequency components, which immediately suggests that removing
or reducing these components offers a means of smoothing.

3.3   Symmetry of Fourier transform data

Both sets of Fourier transforms seen in Figs. 2 and 3 display symmetry;
more specifically the real parts are symmetric and the imaginary parts are
anti-symmetric. This occurs as a result of the original sampled shape data
being purely real.

$$X_k = \frac{1}{N} \sum_{j=0}^{N-1} x_j \mathrm{e}^{-2\pi i j k/N}$$

where x_j is a complex data point, *i.e.* $x_j = Re(x_j) + iIm(x_j)$. But since shape data is being used, $Im(x_j) = 0$. So

$$x_j e^{-2\pi ijk/N} = x_j \left(\cos(-2\pi jk/N) + i\sin(-2\pi jk/N)\right)$$

Now, cos is a symmetric function, and the multiplication by x_j will maintain this symmetry, so $Re(X_k)$ is symmetric. Similarly, since sin is an anti-symmetric function, $Im(X_k)$ is antisymmetric.

Conversely we should note that any method involving the alteration of Fourier components must ensure that these symmetries are maintained in order to produce purely real data as output.

3.4 Interpolation between FTs of different shapes

One possible method of using the observation in Section 3.2 is to linearly interpolate between the Fourier transforms of different shapes and produce a new Fourier transform of a shape which is an interpolation of the two original shapes. One of the shapes is envisioned as being the original shape, the other as being some smooth template such as a circle, both having data sequences of equal length to ensure Fourier transform sequences of equal length. However, such interpolation gives the same result as interpolating in real space and so there is no point in using Fourier methods in this manner.

This follows since the Fourier transform is a linear operator. Treating x and y parametric data separately, linearly interpolating the Fourier components of the data sequences with a given weight has the effect of interpolating the parametric data by the same weight. This is equivalent to a real space linear interpolation of the shape data.

3.5 Simple filtering

A well documented method of decreasing high frequency information of a Fourier transform while leaving lower frequencies relatively unchanged is to use a Butterworth low pass filter [5]. This is often used in image enhancement, to lessen the effects of sharp transitions (such as noise) in the grey levels of an image [5]. The filter (shown in Fig. 4) has values H_k, where k is the frequency, between 0 and 1. If some parametric function $f(s)$ has sampled data points f_j and corresponding Fourier components F_k, then the data for the smoothed shape g_j produced by filtering has Fourier components G_k defined by

$$G_k = F_k H_k$$

H_k is given by

$$H_k = \frac{1}{1 + [k^2/\omega_0{}^2]^N}$$

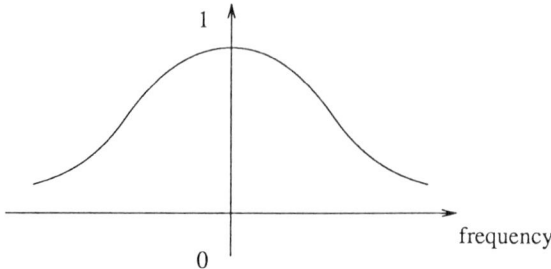

Figure 4. Butterworth low pass filter

Changing the value of N allows control over the rate at which the filter falls towards zero, and changing ω_0 determines how many frequencies are significantly reduced. The filter is symmetric, and so will preserve the symmetry of the Fourier data, ensuring that purely real data are output. It should be noted here that the zero[th] frequency of the Fourier transform of a sequence is equal to the mean of that sequence [2]. The value of the Butterworth filter at $u = 0$ is always 1, and so the mean of the data sequence, and hence sum of data points, is unchanged. Furthermore taking a limiting case of the Butterworth filter $w_0 \to 0$, G_k tends to:

$$
\begin{aligned}
X \text{ real} &= 0\ 0\ \ldots\ 0\ 0\ \bar{x}\ 0\ 0\ \ldots\ 0\ 0 \\
\text{imag} &= 0\ 0\ \ldots\ 0\ 0\ 0\ 0\ 0\ \ldots\ 0\ 0
\end{aligned}
$$

$$
\begin{aligned}
Y \text{ real} &= 0\ 0\ \ldots\ 0\ 0\ \bar{y}\ 0\ 0\ \ldots\ 0\ 0 \\
\text{imag} &= 0\ 0\ \ldots\ 0\ 0\ 0\ 0\ 0\ \ldots\ 0\ 0
\end{aligned}
$$

where $\bar{x}$ and $\bar{y}$ are the means of the original data sequences. And since

$$
x_j = \sum_{k=0}^{N-1} X_k \mathrm{e}^{(2\pi i j k/N)} = X_0 = \bar{x}
$$

and similarly $y_j = \bar{y}$ the shape data sequences tend to:

$$
\begin{aligned}
X \text{ data} &= \bar{x}\ \bar{x}\ \ldots\ \bar{x}\ \bar{x}\ \bar{x}\ \bar{x}\ \bar{x}\ \ldots\ \bar{x}\ \bar{x} \\
Y \text{ data} &= \bar{y}\ \bar{y}\ \ldots\ \bar{y}\ \bar{y}\ \bar{y}\ \bar{y}\ \bar{y}\ \ldots\ \bar{y}\ \bar{y}
\end{aligned}
$$

Thus in this extreme example the two-dimensional data collapses to a single point, the centre of gravity of the shape. More generally, for other Butterworth filters, this effect results in a new shape after filtering which is smoother, but has also shrunk. For very small values of ω_0, a large number of frequency values are reduced greatly, producing pronounced shrinkage. Larger values of ω_0 produce less shrinkage (as reduction in lower frequency

Figure 5. Closed shapes smoothed using Butterworth low pass filters

values is slight) but also affect the higher frequency components less, and produce less smoothing of the shape.

Results of this method are shown in Fig. 5, (with each original shape followed by the smoothed version and both versions superimposed), in which the final real data output have been joined by line segments. In principle a new curve may be fitted to this data; methods of doing this are covered extensively in [1, 4, 7].

The problem of shrinkage can be resolved by scaling points about the centre of gravity of the shape after smoothing, in order to ensure that the smoothed shape is roughly the size of the original shape. Size here might mean area, or it might be desired that the smoothed shape fits roughly within the boundary of the original shape.

This method is of limited usefulness as, although control over the amount of smoothing of closed shapes can be achieved, the method does not extend well to open shapes, as we shall show in the next Section.

3.6 Filtering of open shapes

The overall system works by isolating parts of the original object, smoothing them separately and replacing them in the original shape afterwards. Thus it is important to examine the smoothing of open shapes. These behave in a different manner to closed shapes. Endpoints which are not coincident have the same effect as corners, *i.e.* they contribute to the high frequency information of the Fourier transform.

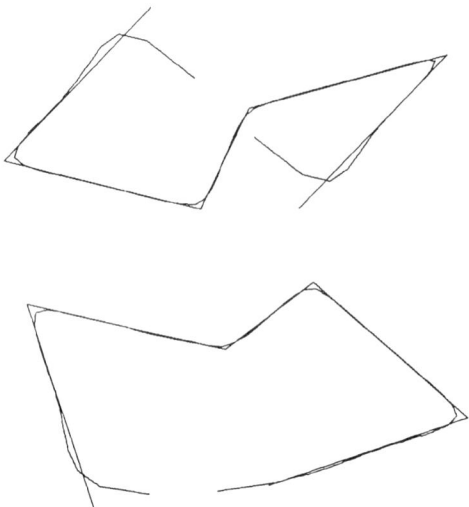

Figure 6. Open shapes smoothed using Butterworth low pass filters

If a filter, such as a Butterworth low pass filter, is used to reduce high frequency components, then as before the shape will tend in the limit to a single point.

In particular, as well as rounding off corners, the endpoints move closer together as a result of the shape tending to collapse to a point, (as can be seen in Fig. 6, in which the smoothed versions are superimposed on the original shape). If an open shape has been isolated from a larger shape and it must be returned to the rest of the shape after smoothing, then the endpoints must either remain fixed during smoothing or be returned to their original positions after smoothing. Also, tangent continuity at the endpoints, or higher order continuity, will often be desired. A method which fixes endpoints *after* smoothing would involve also altering many other points on the smoothed shape. This may introduce undesirable effects, and ultimately lessen the control over smoothing which the filtering allowed. Fixing endpoints during, and after, smoothing will be dealt with separately.

3.7   Fixing endpoints after smoothing

As this is of limited usefulness, it will only be covered briefly. An open shape can be scaled after smoothing (by scaling x and y data sequences separately about their mean values) in order to fix the endpoints of the smoothed shape at their positions in the original shape. (No shifting of the shape is necessary as the centre of gravity of the shape is unaltered by this

Figure 7. Smoothed open shape after scaling

method of smoothing). However, this leads to a great deal of distortion of the smoothed shape, an example of which is shown in Figure 7 (in which the smoothed shape, scaled to fix the endpoints, is shown below the original shape).

One approach to avoid this problem is to weight the endpoints by repeating their values in the data sequences, *i.e.* to replace the original sampled data

$$x_0 \; x_1 \; \ldots \ldots \; x_{N-2} \; x_{N-1}$$
$$y_0 \; y_1 \; \ldots \ldots \; y_{N-2} \; y_{N-1}$$

by

$$x_0 \; \ldots \; x_0 \; x_0 \; x_1 \; \ldots \ldots \; x_{N-2} \; x_{N-1} \; x_{N-1} \; \ldots \; x_{N-1}$$
$$y_0 \; \ldots \; y_0 \; y_0 \; y_1 \; \ldots \ldots \; y_{N-2} \; y_{N-1} \; y_{N-1} \; \ldots \; y_{N-1}.$$

In smoothing, the new endpoints are moved closer together, the adjacent points are moved less, and so on. In this way it is the additional points which are moved the most, and after smoothing these points can be discarded, leaving endpoints which have been moved by a smaller amount. These endpoints can be fixed to their corresponding positions in the original shape by scaling the data points about the centre of gravity of the shape. The problem of distortion of the shape due to scaling still occurs, but to a

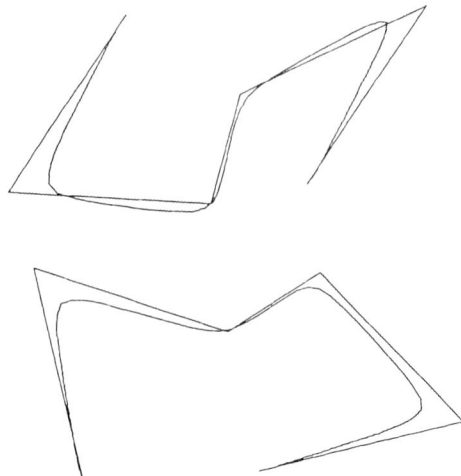

Figure 8. Smoothing open shapes defined with augmented sequences, after scaling

lesser extent, (as seen in Fig. 8, with the smoothed versions superimposed on the original shapes).

It should also be noted that if too many augmented points are used they will dominate the shape data, and after smoothing the shape will tend in the limit to a straight line between the endpoints.

3.8   Fixing endpoints during smoothing

In order to see how this may be done we will consider the concept of a cyclic convolution. The convolution of two functions can be thought of as a (weighted) running mean, and describes the action of an observing instrument when it takes a weighted mean of some physical quantity over a narrow range of some variable [2]. Using discrete data, the convolution of two sequences having N_1 and N_2 values respectively will have $N_1 + N_2 - 1$ elements. If convolution ideas are used to smooth shapes then the task is simplified if the number of points produced by the convolution is the same as the number of points used to define the original shape. For this a cyclic convolution may be used [2]. For some parametric function $f(s)$ with data sequence f_j, its convolution with a shape $h(s)$ is denoted $g(s)$ which is defined as

$$g(s) = f(s) * h(s)$$

$$g_j = \sum_{j'=0}^{N-1} f_{j'} h_{j-j'}$$

Since a cyclic convolution is used, in terms of the Fourier transforms of data sequences f_j, g_j and h_j, written F_k, G_k and H_k respectively,

$$G_k = F_k H_k$$

Thus, performing the cyclic convolution of $f(s)$ and $h(s)$ is equivalent to filtering the Fourier transform of the data points f_j using the Fourier transform of the data points h_j.

A mask (a shape which is convolved with the original shape data) with data h_j has size M if $h_j \neq 0$ only for M points. Now, if we have a mask of size M (typically $M \ll N$) then calculating a convolution on a sequence of N values in real space using such a mask takes time $O(MN)$. In order to perform a convolution in Fourier space the Fourier sequences must be of equal length and so the convolution shape must first be padded out with zeros to have N values. The convolution is then performed by multiplying the corresponding components of the Fourier transforms of the sequences (calculated using a Fast Fourier transform algorithm), taking time $O(N \log N)$ [2]. If $M < \log N$ then calculation of the convolution is computationally more efficient in real space. Using a small mask produces only local change in the original shape data, such as rounding of a corner. Larger masks are more useful for gross shape changes of a more global nature.

Let us now consider possible shapes with which to convolve parametric shape data. Firstly, consider a top hat function (shown in Fig. 9 (i)), having the following discrete sequence:

$$\frac{1}{2M+1} [0, \ldots, 0, 1, \ldots, 1, 1, 1, 1, 1, \ldots, 1, 0, \ldots, 0]$$

It is defined by $2M + 1$ points having the value 1 (where $M < N/2$) with all other points being 0, and so has width $2M$. Considering the real space consequences of convolution of a data sequence f_j with this sequence, the resultant data sequence, g_j has the values:

$$g_j = \frac{1}{2M+1} [f_{j-M} + \ldots + f_{j-1} + f_j + f_{j+1} + \ldots + f_{j+M}]$$

where f_r is defined cyclically (*i.e.* $f_r = f_{r+N} = f_{r-N}$). This just averages each point with the M points on either side of it. This will smooth the

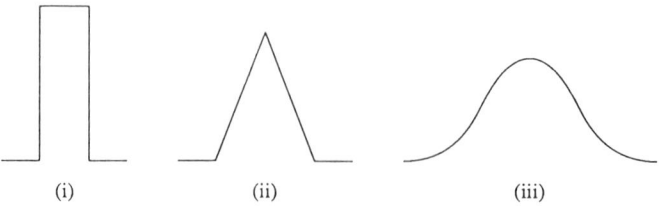

(i)   (ii)   (iii)

Figure 9. Convolution masks

shape, but somewhat crudely for large values of M as a point will be affected as much by distant points as by near neighbours. In particular corners will have a large effect on many of the surrounding points.

An improvement on the former can be made by considering convolution using shapes which give greater weight to the original point and its near neighbours, such as a triangular function (shown in Fig. 9 (ii)). Such a function (with width $2M - 2$) has a discrete sequence:

$$\frac{1}{M}\left[0,\ldots,0,\frac{1}{M},\ldots,\frac{M-2}{M},\frac{M-1}{M},1,\frac{M-1}{M},\frac{M-2}{M},\ldots,\frac{1}{M},0,\ldots,0\right]$$

Convolution here would produce data points g_j given by:

$$\frac{1}{M}\left[\frac{1}{M}f_{j-M+1}+\ldots+\frac{M-1}{M}f_{j-1}+f_j+\frac{M-1}{M}f_{j+1}+\ldots+\frac{1}{M}f_{j+M-1}\right]$$

Due to the weighting towards near neighbours the previous problems of shape distortions do not occur to the same extent. In fact, this triangular shape can be obtained by convolving a top hat function (of half the width) with itself (*i.e.* it is the auto-correlation of the top hat function [2]). This can be improved further by using the convolution of the triangular function with a top hat function, which is a piecewise quadratic curve (shown in Fig. 9 (iii)). Here the weighting has been moved more towards the near neighbours at the expense of more distant points. This causes a point to be more affected by its near neighbours than was the case with the triangular function, and less by more distant points than occurs with either of the previous shapes considered.

The Fourier transform of a triangle (Fig. 10) can be thought of as a filter which falls away, but not steadily. Convolving a shape with this triangular mask effectively makes it smoother in a similar way to the Butterworth filter by decreasing the high frequency components generally more than the low frequency ones. Where this filter differs from the Butterworth filter is that some points may be left unchanged, if the convolution shape is chosen to be narrower than the shape data sequence, and this property can be exploited in order to fix the endpoints of the shape.

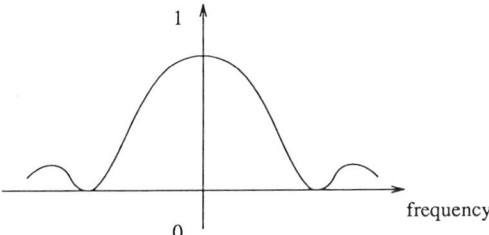

Figure 10. Fourier transform of a triangle

In general, for any open shape, the endpoints of the shape will be moved along with the $M - 1$ points adjacent to each endpoint (if there are that many points). If a point is preceded and succeeded by $M - 1$ points of the same value, then it will be unaffected by the convolution. Thus repeated endpoints can be placed in the sequence to surround the endpoints, *i.e.* the data sequence

$$[x_0, x_1, x_2, \ldots\ldots\ldots, x_{N-3}, x_{N-2}, x_{N-1}]$$

can be replaced by

$$[x_0, \ldots, x_0, x_0, x_0, \ldots, x_0, x_1, x_2, \ldots\ldots\ldots, x_{N-3}, x_{N-2}, x_{N-1}, \ldots$$

$$\ldots, x_{N-1}, x_{N-1}, x_{N-1}, \ldots, x_{N-1}]$$

A similar sequence is constructed for the y coordinates. After convolution, the augmented points are discarded, leaving the real endpoints unaltered.

This process is very similar to the method of construction of B-spline curves [6,9], and the exact relationships are still being considered. The narrowest width cases of the masks described (top hat, triangle, quadratic) correspond to a progression of B-spline basis functions (having 2, 3 and 4 knots respectively [6]). Note however that this convolution method uses masks of *varying* widths, and can be extended to other masks. Given the similarity of the methods it should be possible to use known properties of B-splines to obtain desired continuity. Turning the argument on its head, Fourier methods might even provide an efficient way of evaluating B-splines.

This convolution method produces satisfactory results (examples are shown in Fig. 11, with smoothed versions superimposed on the original shapes), and so provides a basic approach to the construction of filters. It yields a sequence of data points (joined in the example above by line segments) to which a curve can be fitted. If curve continuity of higher order

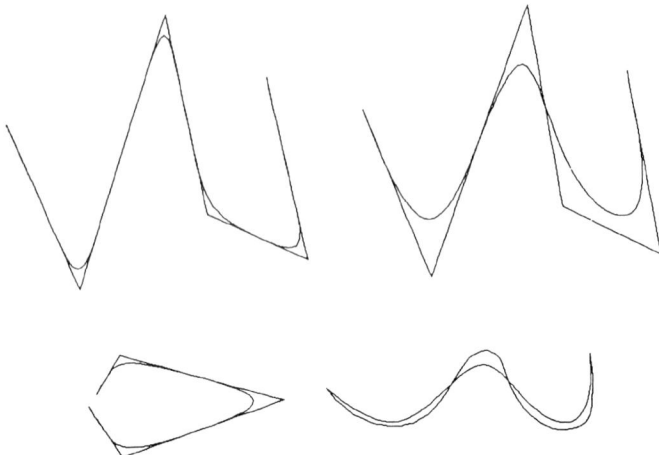

Figure 11. Results of convolutions using triangular masks of different widths, using augmented points

than just point continuity is desired at the endpoints, then the positions of the points adjacent to the endpoints could be fixed in some way. Exactly how these points are fixed is dependent in part on the curve fitting method used. B-spline properties might provide a method of ensuring desired continuity, but we will consider here an extension of the above idea which will fix (x_1, y_1) and (x_{N-2}, y_{N-2}) to have particular values after smoothing (relative to their original positions).

Placing copies of the endpoints on either side of the endpoints may be thought of in shape terms as extending the original shape by placing straight lines of zero gradient through these endpoints. The convolution of a straight line of zero gradient with any mask produces another straight line of zero gradient. This idea can be furthered by using straight lines of non-zero gradient. If the lines have width greater than that of the mask then the centres of the lines (the original endpoints) will be unaffected by convolution of the shape data with the mask. The data sequence may now be written

$$[x_0 - (M-1)a, \ldots, x_0 - 2a, x_0 - a, x_0, x_0 + a, x_0 + 2a, \ldots,$$

$$x_0 + (M-1)a, x_1, x_2, \ldots\ldots, x_{N-3}, x_{N-2}, x_{N-1} + (M-1)b, \ldots,$$

$$x_{N-1} + 2b, x_{N-1} + b, x_{N-1}, x_{N-1} - b, x_{N-1} - 2b, \ldots, x_{N-1} - (M-1)b]$$

where a and b are the gradients of the lines through the endpoints.

Now, the values of the data points placed between each endpoint and its adjacent points will affect the values of the adjacent points after smoothing. Thus the gradients of the lines can be altered in order to move these points. In particular they can be chosen to fix (x_1, y_1) and (x_{N-2}, y_{N-2}) to specified positions. For the x data:

If the value of x_1 after smoothing, written x'_1 is defined by $x'_1 = x_1 + \alpha$, and the convolution shape is defined as

$$\frac{1}{\sum_{j=0}^{N-1} h_j} [h_{M-1}, \ldots, h_1, h_0, h_1, \ldots, h_{M-1}]$$

then x_1 can also be defined as

$$x'_1 = \frac{1}{\sum_{j=0}^{N-1} h_j} ([h_0 x_1 + h_1 x_2 + \ldots + h_{M-1} x_M] +$$

$$[h_1(x_0 + (M-1)a) + h_2(x_0 + (M-2)a) + \ldots + h_{M-1}(x_0 + a)])$$

then putting $A = \frac{1}{\sum_{j=0}^{N-1} h_j} [h_0 x_1 + h_1 x_2 + \ldots + h_{M-1} x_M]$ and equating gives:

$$a = \left[x_1 + \alpha - A - \frac{x_0 \sum_{j=1}^{M-1} h_j}{\sum_{j=0}^{N-1} h_j} \right] \frac{\sum_{j=0}^{N-1} h_j}{\sum_{j=1}^{M-1} h_j(M-j)}$$

A similar condition holds for the point x_{N-2}:

$$b = \left[B + \frac{x_{N-1} \sum_{j=1}^{M-1} h_j}{\sum_{j=0}^{N-1} h_j} - x_{N-2} - \beta \right] \frac{\sum_{j=0}^{N-1} h_j}{\sum_{j=1}^{M-1} h_j(M-j)}$$

where $x'_{N-2} = x_{N-2} + \beta$ and

$$B = \frac{1}{\sum_{j=0}^{N-1} h_j} [h_0 x_{N-2} + h_1 x_{N-3} + \ldots + h_{N-M-1} x_M]$$

Similar conditions can also be derived for the y data sequences. Computation of these equations takes time $O(N)$, and so would have little effect (for large N) on the overall time to perform convolutions using Fourier transforms.

Throughout this section it has been assumed that N is a power of two. If this is not the case then an algorithm could be designed to take advantage of such factors as N possesses (which may take time worse than $O(N \log N)$), or the sequences used could be padded to the nearest power of 2. If the sequence is padded with zeros, this may introduce undesired steps in the data, possibly producing unacceptable distortions of the shape. The sequence could instead be padded out by more meaningful data. This method of fixing the positions of the points adjacent to the endpoints provides both a way of fixing the tangents at the endpoints and of padding out sequences (to the nearest power of 2) with meaningful data.

Convolution ideas are used to construct filters in order to produce predictable smoothing. The system offers users a range of convolution shapes (producing known results) from which filters for the Fourier transforms can be calculated. Augmented data sequences are used to ensure continuity constraints set by users.

4 Fourier Smoothing in Three Dimensions

Having outlined how smoothing can be performed in 2D, the extension of these ideas into 3D will be outlined.

4.1 Organisation of data

A similar situation to the 2D case is faced here. Organising the data in a three-dimensional array as $f(x, y, z)$ by assigning non-zero values to those (x, y, z) points which lie inside the shape would be costly in terms of storage, and would necessitate the use of a three-dimensional Fourier transform on a very large number of data points. This would be impracticably slow.

As in the two-dimensional case, parametrisation can be adopted, using two parameters in this case to define points on the surface at regular intervals of each parameter. This gives three parametric functions; $x(u, v)$, $y(u, v)$, $z(u, v)$. Three 2D Fourier transforms are required, using far fewer data points than the above method.

4.2 Sampling points

The translation of the previous ideas into 3D involves consideration of the topology of the objects being smoothed. In 2D only two line segments meet at each vertex and so parametrisation follows simply along the boundary of a shape. The relationships between adjacent data points is preserved when this data is placed in a one-dimensional sequence. In 3D, however, three or more faces may meet, as in Figure 12. The parametrisation does not run naturally from one face to another in such a case, and since the data is to be stored in a 2D array it becomes more difficult to preserve the relationships between adjacent points. Without these relationships the Fourier transform information becomes less meaningful. A subset of surfaces (made up of a series of surface patches) for which data definitely can be sampled meaningfully is identified below.

If a centre (in some topological sense) of an object can be identified then 'loops' can be formed around it such that a line drawn from the centre meets the loops at equal (parametric) distances. Equal numbers of data points

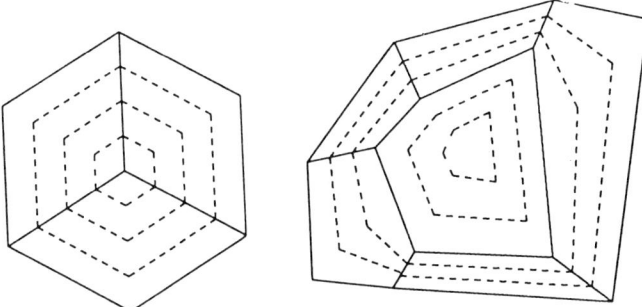

Figure 12. Vertex and face looping centres

may then be sampled at regular intervals along each loop. Examples of choosing a centre and the creation of loops can be seen in Fig. 12.

The centre of looping of an object might be a vertex, an edge or a face, and in many object cases there are simple rules governing the applicability of each. Firstly, certain properties have to be established about the object. The object may be said to be n faces deep about the centre if in jumping from any face towards the centre, in the shortest possible path, at most n steps are needed. In the left hand diagram the object is one face deep about the vertex common to all faces, and in the right hand diagram the object is one face deep about the central face. A similar property can be used to describe the boundary of the object. Taking each face which has an edge on the boundary, the boundary can be said to be m faces deep if from *all* faces on the boundary m steps are required to reach the chosen centre in the shortest possible path. In both diagrams, the objects have boundaries which are one face deep. If a vertex or a face is chosen as a centre of looping, and $n=m$ about it, then looping is possible.

The case of choosing an edge as a centre of looping is slightly different. Possible cases are two faces joined along one edge, and any four faces of a cube. More complicated cases are shown in Fig. 13. Here depth must be measured from the edge in two directions: from the vertices of the edge (in the direction of the edge), and from the edge itself (against the direction of the edge). In Fig. 13 (i), the depth of the object against the direction of the central edge is three faces, (as the faces containing the central edge are counted), and the depth in the direction of the edge is zero. The objects in Fig. 13 are shown roughly in profile. In each case a central line of six faces is shown. In (ii), two other faces join these, and a similar two faces on the other side are imagined to be occluded by these. With (iii), six extra faces are shown, with a similar pattern imagined to be on the other side.

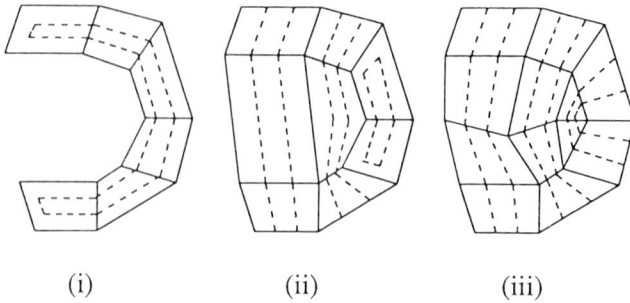

(i)   (ii)   (iii)

Figure 13. Edge looping centres

Four measures of depth are used here, shape and boundary depth in the direction of the central edge (denoted n_e and m_e), and shape and boundary depth against the direction of the central edge (denoted n_a and m_a). In order to be sure that looping is possible, $n_e = m_e$ and $n_a = m_a$, and $m_e \leq m_a$. Within these conditions there are three possible cases to be considered, an example of each case being shown in the figure: (i) $m_e(= n_e) = 0$. Looping treats a row of faces as if it were all a central face; (ii) $m_e(= n_e) < m_a(= n_a)$ and $m_e > 0$. A depth difference d is determined as $d = m_a - m_e$. The faces against edge direction to a depth of d from the central edge are treated as a central face, leaving an equal depth of faces in all directions around them, which are looped in the same way as faces surrounding a central face; (iii) $m_e(= n_e) = m_a(= n_a)$. Looping is formed around the central edge to cover the shape to a depth of one face (as if this is all one central face), and the rest of the object is looped in the same way as the faces surrounding a central face.

Data sampled from these loops may be organised in three coordinate matrices as $x_{u,v}$, $y_{u,v}$ and $z_{u,v}$ where u is the loop number, taking values from 0 to $N_u - 1$, and v is the number of the point taken around each loop, taking values between 0 and $N_v - 1$.

If a smoothing operation is to be performed on a surface region which does not conform to the above conditions, it might still be possible to loop the region in some way, or to break it down into parts, each of which can be looped separately. Operations might then be performed on these parts separately to produce the desired smoothing. Exactly how this might be made to work within the system, and what effects this would have on smoothing are still being considered.

4.3 Filters in three dimensions

Butterworth low pass filtering ideas extend simply into 3D. If this filter is used on the Fourier data of an object, the object will tend to collapse to a single point in space, that point being the centre of gravity of the object, $(\bar{x}, \bar{y}, \bar{z})$.

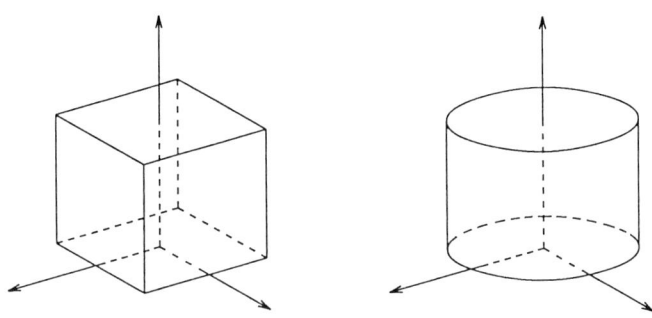

Figure 14. 3D extensions of the top hat mask

Convolution ideas are also straightforward, but the extensions of the convolution masks used in 2D are not unique. For example, the top hat function might extend to either of the 3D masks shown in Fig. 14. Using the right hand mask would seem to be the most sensible, since in real space this would involve averaging each point with all points within equal Euclidean distance from that point. However, the normal tensor product B-spline surface method uses a mask shaped like the left hand one.

Similarly, the triangular mask might extend to either of the masks in Fig. 15, these diagrams being the autocorrelations of those shown in

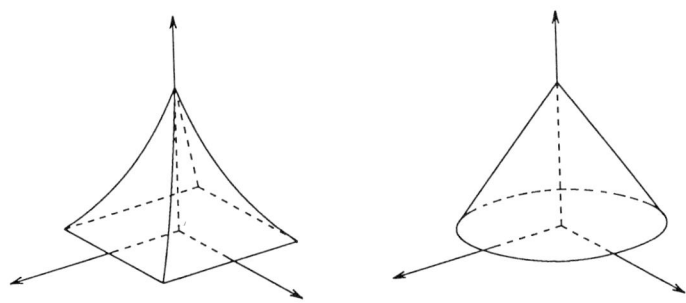

Figure 15. 3D extensions of the triangular mask

Fig. 14. For the same reason given above the second mask is possibly preferable, but it should be noted that the lazy-pyramid on the left equates to the normal tensor product B-spline surface produced by conventional methods.

5 Conclusions

This paper has established that it is possible to smooth shapes (in both 2D and 3D) by changing the components of their Fourier transforms. Further investigation is necessary (particularly in smoothing three-dimensional objects) in order to alter smoothness predictably to meet æsthetic criteria.

The width of the mask used in convolution with the original shape data determines the locality of the smoothing operation. For the reasons given in Section 3.8 local operations are less efficiently performed in Fourier space than in real space, and so real space local averaging would best be used. However, for more global operations (*i.e.* gross shape smoothing) a significant reduction in computation can be achieved using the Fourier methods described. Furthermore, there is a fundamental connection between such methods and B-splines, which is still to be fully explored.

It should be noted that the number of arithmetic operations needed to perform the calculations described may be high (as was mentioned in Sections 3.1 and 4.1). The development of digital signal processing chips may provide the necessary computing power, and indeed such chips are starting to appear as standard hardware in some recent workstations.

References

1. Bartels, R.H., Beatty, J.C. and Barsky, B.A. (1987). *An Introduction to Splines for use in Computer Graphics and Geometric Modelling*, Morgan Kaufmann, California.

2. Bracewell, R.N. (1986). *The Fourier Transform and its Applications*, McGraw Hill, Singapore.

3. Dickens, N.A. (1989). A survey of literature on blending in solid modelling. CAD Development Unit, Siemens Research Group Internal Report, Cambridge.

4. Farin, G. (1988). *Curves and Surfaces for Computer Aided Geometric Design*, Academic Press, California.

5. Gonzalez, R.C. and Wintz, P. (1987). *Digital Image Processing*, Addison-Wesley.

6. Gordon, W.J. and Riesenfeld, R.F. (1974). B Spline curves and surfaces. *Computer Aided Geometric Design*, Ed. R.E. Barnhill, R.F. Riesenfeld, Academic Press, New York.

7. Lancaster, P. and Salkauskas, K. (1986). *Curve and Surface Fitting, An Introduction*, Academic Press, London.

8. Nutbourne, A.W. and Martin, R.R. (1988). *Differential Geometry Applied to Curve and Surface Design*, Ellis Horwood, Chichester.

9. Rogers, D.F. and Adams, J.A. (1990). *Mathematical Elements for Computer Graphics*, 2nd edn., McGraw Hill, New York.

10. Varady, T., Vida, J. and Martin, R.R. (1989). Parametric blending in a boundary representation solid modeller. *Geometric Modelling Studies, 1*, MTA SZTAKI, Hungary, 35–62.

11. Woodwark, J. (1986). *Computing Shape*, Butterworths, England.

Properties of the B-spline Representations of PDE Surfaces that are Generated Using the Finite Element Method

J.M. Brown,[1] **M.I.G. Bloor,**[1] **M.S. Bloor,**[1]
M.J. Wilson[1] **and H. Nowacki**[2]

[1]*University of Leeds and* [2]*Technische Universitat Berlin*

1 Introduction

This paper examines the B-spline approximations to surfaces produced by the PDE method ([5] and [6]). The approximations were introduced because of the ease of data exchange of B-spline surfaces between different CAD/CAM systems. Earlier work examines the accuracy of the approximations . In this paper attention will be turned to the fairness.

A PDE surface is the solution of a partial differential equation (PDE) that fulfils suitable boundary conditions (see [3] and [4]). To obtain the approximate B-spline surface control vertices, the PDE is solved using the finite element method with a basis consisting of B-spline functions, boundary conditions being applied using Lagrange multipliers [5]. The accuracy of this B-spline approximation is affected by both the degree of the B-spline functions and the size of the finite element mesh. Two examples of PDE surfaces are considered; firstly, a blend between a cylinder and a plane, and secondly a yacht.

When designing and manufacturing objects, the free form surfaces involved need to be analysed for a variety of reasons; for example, for the sake of the aesthetics or the method of production, as discussed by Beck et al. [2]. The 'fairness' or 'smoothness' of a surface has been of concern for both aesthetic surfaces (such as a blend or a car body) and for functional surfaces (such as ship hulls). Everyday, using our eyes to receive reflected light, we automatically assess the smoothness of surfaces around us, some objects being more pleasing to look at than others. The fairness of a surface is subjective and as a result, fairness measures are difficult to formulate. Farin and Sapidis [10] remark that the definition of the concept of surface fairness is 'still an open question'. To gain an insight into the fairness of a surface different analyses can be carried out, such as the observation of reflection lines introduced by Klass [12], and the examination of Gaussian curvature used by numerous authors [8], [9], [13] and [14]. In this paper plots of Gaussian curvature have been obtained for both the PDE surfaces and their B-spline approximations. The integral over the surface of the sum of squares of the principal curvatures have also been calculated. This integral is an engineering fairness criterion (see Nowacki and Reese [15] and Hagen and Schulze [11]).

2 Background

In this section, the PDE method, the B-spline approximation to PDE surfaces and relevant methods for analysing surfaces will be examined.

2.1 The PDE method

A PDE surface, $\underline{X} = (x(u,v), y(u,v), z(u,v))$ (a function of parameters u and v), is obtained by solving a partial differential equation in u, v-space, subject to conditions on $\underline{X}$ and its derivatives with respect to u and v. The PDE can be expressed as

$$D_{u,v}^m(\underline{X}) = \underline{F}(u,v), \qquad (2.1)$$

where $D_{u,v}^m()$ is a partial differential operator of order m in the independent variables u and v, and $\underline{F}(u,v)$ is a vector valued function of u and v. Up to this point in time, only elliptic PDEs have been considered. The order of the PDE is governed by the number of specified conditions on $\underline{X}$ and its derivatives. For example, if function and tangency conditions on the boundary are required, then a fourth order PDE is necessary.

The conditions to which $\underline{X}$ and its derivatives are subjected are of importance to the form the surface takes. The generation of the blend and yacht surfaces (see Fig. 1) examined in this paper, have been considered by Bloor and Wilson in [3] and [4], respectively. Function and tangency conditions are specified on the boundary, so a fourth order PDE is required. The same PDE,

$$\left(\frac{\partial}{\partial u^2} + a^2 \frac{\partial}{\partial v^2}\right)^2 \underline{X} = 0, \qquad (2.2)$$

where a is constant, is used to produce both the blend and the yacht, the difference in form results from the different boundary conditions.

For reference later, Fig. 1 shows points on the lines of constant u and v. The blend is periodic in v (with $0 \le v < 2\pi$) and the lines of constant u (with $0 \le u < 1$) run from the cylinder (where $u = 0$) to the plane (where $u = 1$). The yacht is also periodic in v (with $-\frac{\pi}{2} \le v < \frac{\pi}{2}$) and the lines of constant u (with $0 \le u < 1$) lie between the deck line (where $u = 0$) and the base of the keel (where $u = 1$).

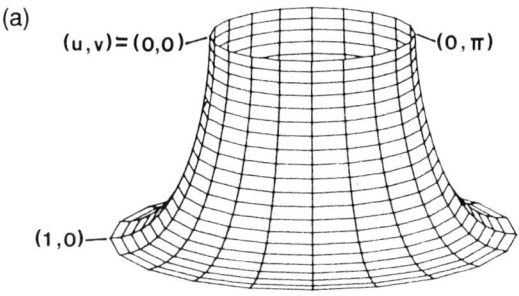

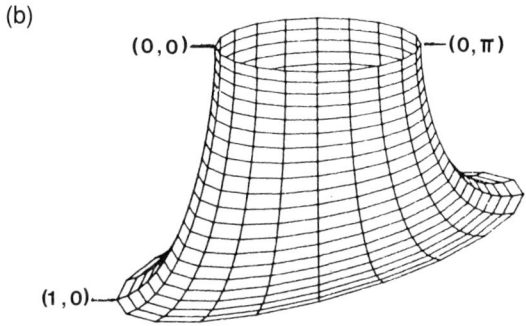

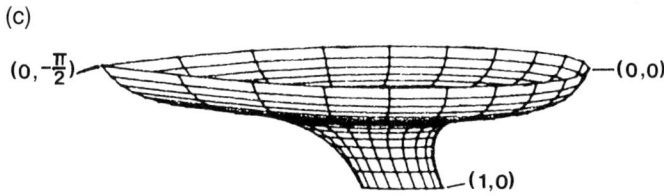

Figure 1. PDE surfaces: (a) a blend between a cylinder and horizontal plane, (b) a blend between a cylinder and an inclined plane, and (c) a yacht

2.2   The B-spline approximation to the PDE surface

A PDE surface can be approximated by a B-spline surface,

$$\underline{X}(u,v) = \sum_{i=0}^{m} \sum_{j=0}^{n} \underline{p}_{ij} B_{i,k}(u) B_{j,l}(v), \tag{2.3}$$

with non-decreasing knots in the u-direction being $\{u_0, u_1, \ldots, u_{m+k}\}$, and in the v-direction $\{v_0, v_1, \ldots, v_{n+l}\}$, where $\underline{p}_{ij} = (r_{ij}, s_{ij}, t_{ij})$ are a topologically rectangular set of control vertices, and $B_{i,k}(u)$ and $B_{j,l}(v)$ are B-spline basis functions of degree $k-1$ and $l-1$, respectively. For a given surface the values of k and l are fixed, and provided that the knots are distinct in the range u_{k-1} to u_{m-1} and in the range v_{l-1} to v_{n+1}, the surface will have C^{k-2} continuity in the u-direction and C^{l-2} continuity in the v-direction. Further details of B-spline surfaces are given by Bartels et al. [1].

A B-spline approximation can be obtained by solving the PDE using the finite element method, with the product of the B-spline basis functions $B_{i,k}(u)B_{j,l}(v)$ as the basis, and using Lagrange multipliers to apply the boundary conditions. The solution is in the form of a $(m+1) \times (n+1)$ mesh of control vertices, $\underline{p}_{ij} (i = 0, \ldots, n, j = 0, \ldots, m)$, which gives a B-spline surface consisting of $(m-k+2) \times (n-l+2)$ patches. The method of approximation is examined in further detail by Brown et al. in [5] and [6]. The effect on the B-spline surface approximations' fairness of the degree of the basis functions ($k-1$ and $l-1$), and of the size of the finite element mesh is examined in this paper.

2.3   Methods for analysing the surfaces

To examine a surface, classical measures based on curvature are widely used. For a point on a nonsingular surface, a family of normal section curves is produced by the intersection of planes, containing the normal to the surface, with the surface. From all the possible normal section curves, one curve has the greatest curvature, κ_1, and one has the least curvature, κ_2, (with the exception of surfaces that are locally planar or spherical). The Gaussian curvature, K, at the point is given by

$$K = \kappa_1 \kappa_2, \tag{2.4}$$

and mean curvature, H, is given by

$$H = \frac{1}{2}(\kappa_1 + \kappa_2). \tag{2.5}$$

For a given surface, $\underline{X}(u,v)$, the curvatures were computed using the following formulae (where $\underline{X}_u$ means the differentiation of $\underline{X}$ with respect to u). Let

$$
\begin{aligned}
E &= \underline{X}_u \cdot \underline{X}_u, \\
F &= \underline{X}_u \cdot \underline{X}_v, \\
G &= \underline{X}_v \cdot \underline{X}_v.
\end{aligned}
\tag{2.6}
$$

$\underline{N}$ is the unit normal to the surface, given by

$$
\underline{N} = \frac{\underline{X}_u \times \underline{X}_v}{|\underline{X}_u \times \underline{X}_v|}.
\tag{2.7}
$$

Let

$$
\begin{aligned}
L &= \underline{X}_{uu} \cdot \underline{N}, \\
M &= \underline{X}_{uv} \cdot \underline{N}, \\
N &= \underline{X}_{vv} \cdot \underline{N}.
\end{aligned}
\tag{2.8}
$$

Then K and H can be calculated as follows:

$$
K = \frac{LN - M^2}{EG - F^2},
\tag{2.9}
$$

$$
H = \frac{EN - 2FM + GL}{2(EG - F^2)}.
\tag{2.10}
$$

These formulae can be found in books concerning differential geometry such as [7]. Using (2.4) and (2.5)

$$
\int_s (\kappa_1^2 + \kappa_2^2) ds = \int_s (4H^2 - 2K) ds.
\tag{2.11}
$$

This integral being the fairness criterion considered in [11] and [15].

Considering Gaussian curvature at a point, if $K > 0$ the surface is known as elliptic (full or hollow), if $K < 0$ it is hyperbolic (saddle shaped), and if $K = 0$ it is parabolic (developable). Mean curvature indicates whether a region is full ($H > 0$) or hollow ($H < 0$). For previous discussions of Gaussian curvature and fairness see [8], [9], [13] and [14].

Although it is common practice to display Gaussian curvatures using a colour coded map, as done by Dill in [8] and [9], for this paper it was necessary to use contour lines of constant Gaussian curvature; the method used by Munchmeyer in [13]. The contours are plotted over the u, v-plane, with equal intervals between contours of the solid lines (but not for the dashed lines). For the solid lines $\{1, 2, 3, ..., 10, 11, 12, ..., 21\}$ corresponds to $\{-0.150, -0.135, -0.120, ..., -0.015, 0.0, 0.015, ..., 0.150\}$ and for the dotted lines $\{a, b, c, d\}$ is $\{-125, -25, -5, -1\}$ and $\{p, q, r\}$ is $\{1, 5, 25\}$. It should be noted that 11 corresponds to $K = 0$.

3   Results

Fig. 1 shows the PDE surfaces that have been considered: (a) a blend between a cylinder and horizontal plane, (b) a blend between a cylinder

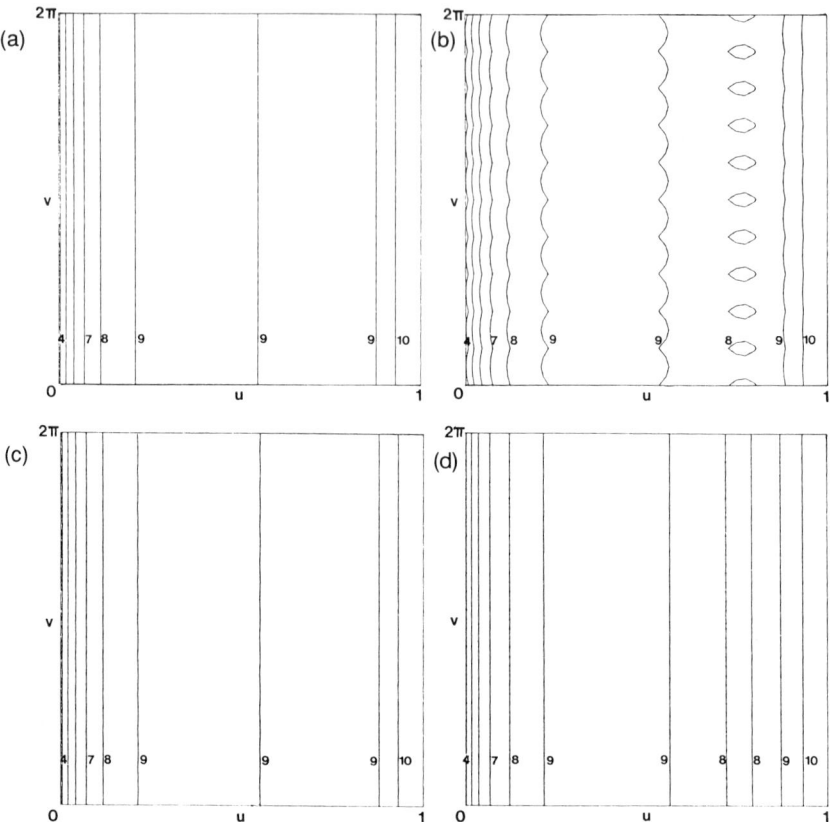

Figure 2. Contour plots of Gaussian curvature for the blend between the cylinder and horizontal plane for (a) a PDE surface, and uniform B-spline surfaces with: (b) $k = l = 4$ and a 5×10 finite element mesh, (c) $k = l = 5$ and a 5×10 mesh, and (d) $k = l = 6$ and a 5×10 mesh

and an inclined plane, and (c) a yacht. Figs. 2, 3, 4 and 5 show contour plots of Gaussian curvature. In Fig. 2 are the plots for the blend between the cylinder and horizontal plane, in Figs. 3 and 4 are those for the blend between the cylinder and inclined plane and in Fig. 5 are those obtained for the yacht. A constant mesh size of 5×10 ($u \times v$), with different degrees of basis polynomials, was used for the B-spline surfaces examined in Figs. 2 and 3, whereas the different mesh sizes of 5×5 and 5×15 were used for Fig. 4.

When comparing bi-harmonic like PDE surfaces with their B-spline approximations, for a constant mesh, the minimum error occurs when the

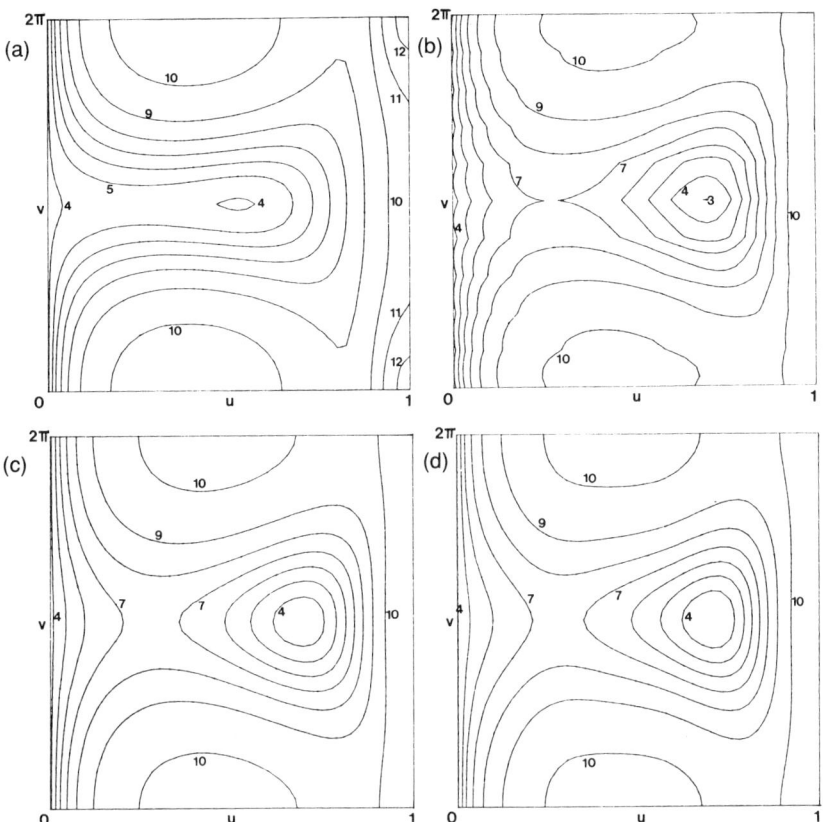

Figure 3. Contour plots of Gaussian curvature for the blend between the cylinder and inclined plane for (a) a PDE surface, and uniform B-spline surfaces with: (b) $k = l = 4$ and a 5×10 finite element mesh, (c) $k = l = 5$ and a 5×10 mesh, and (d) $k = l = 6$ and a 5×10 mesh

surface is biquartic ($k = l = 5$). For example, for the horizontal plane blend of height 2, using a 5×10 finite element mesh, the maximum errors were $O(10^{-3}), O(10^{-4})$ and $O(10^{-2})$, and the mean square errors were $O(10^{-6}), O(10^{-8})$ and $O(10^{-5})$ for k and l equal to 4, 5 and 6, respectively.

The plots of Gaussian curvature (Figs. 2, 3, 4 and 5) give an indication of the fairness of the PDE surfaces and their B-spline approximations, and they also enable a comparison to be made between the surfaces. It may be expected that the formation of a surface by the solution of an elliptic PDE would produce a fair surface, owing to the smoothing nature of the differential operator, unlike some surfaces that are produced using interpo-

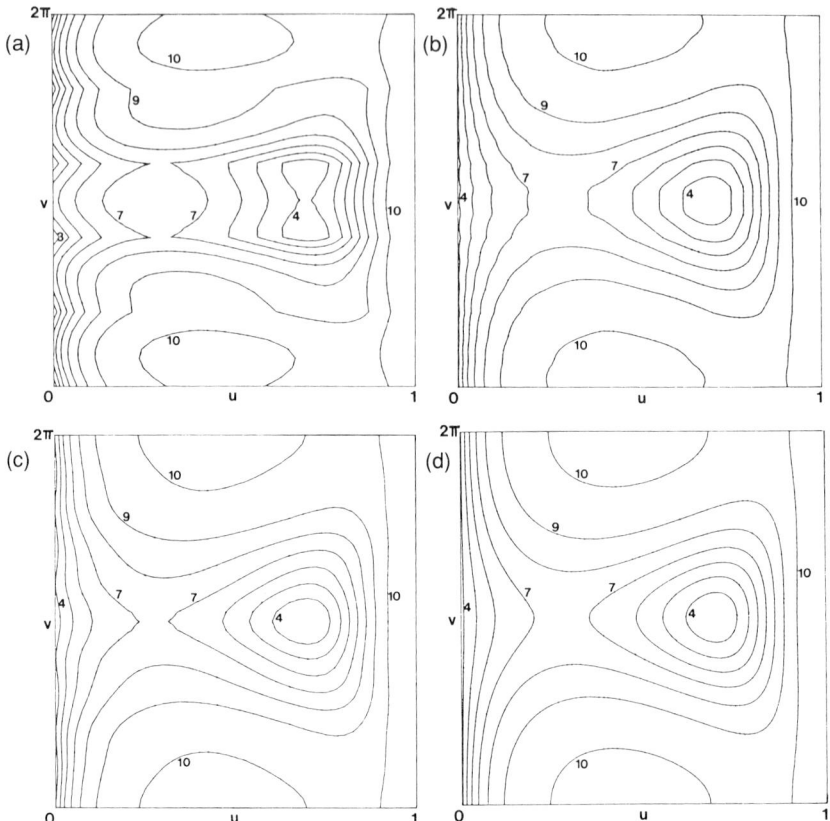

Figure 4. Contour plots of Gaussian curvature for the blend between the cylinder and inclined plane for uniform B-spline surfaces with: (a) $k=l=4$ and a 5×5 finite element mesh, (b) $k = l = 4$ and a 5×15 mesh, (c) $k = l = 5$ and a 5×5 mesh, and (d) $k = l = 5$ and a 5×15 mesh

latory schemes. The fairness of the PDE surfaces considered is confirmed by the plots of Gaussian curvature (Figs. 2(a), 3(a) and 5(a)) which do not possess any irregularities.

When $k = l = 4$ for horizontal and inclined plane blends (Figs. 2(b), 3(b) and 4(a) and (b)) and when $k = l = 5$ with a 5×5 mesh for the inclined plane blend (Fig. 4(c)), there are wiggles present which repeat from patch to patch along the contours. 'Aberrations' such as these were observed by Munchmeyer [13] who later said that if a surface possesses 'true wiggles' then it is not considered fair [14] thus, it is better if they can be avoided. In the case considered here, of the B-spline surface approximation, the

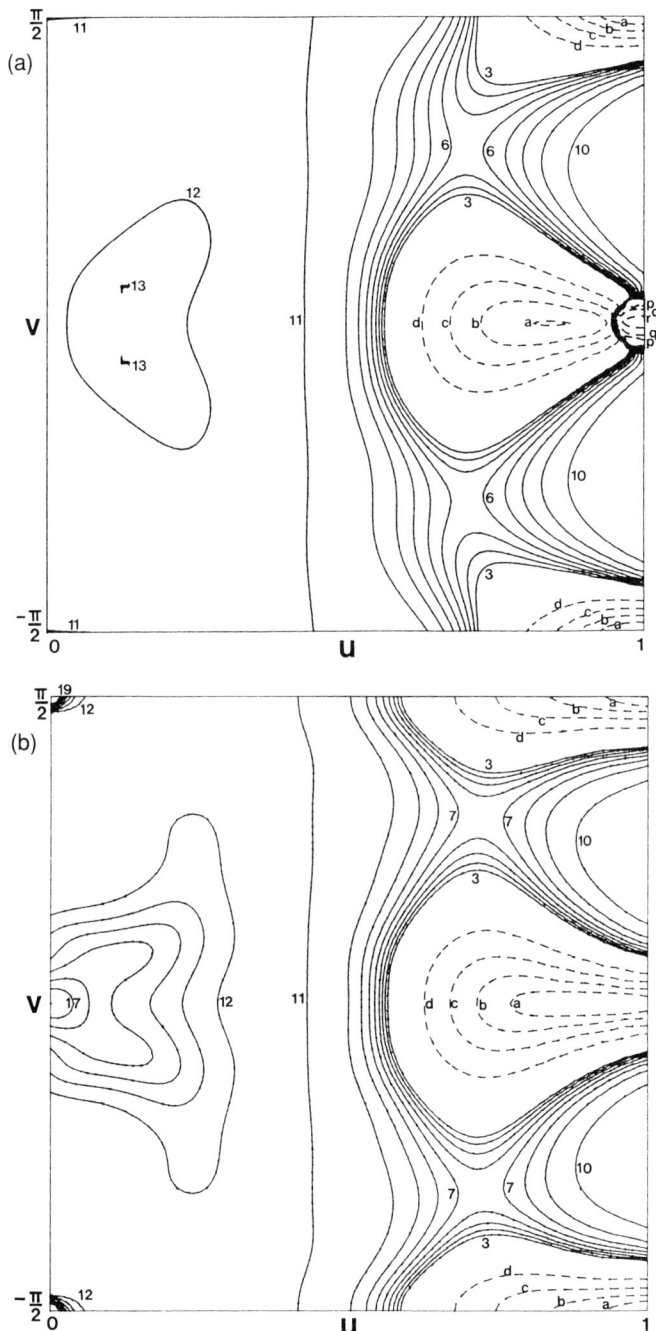

Figure 5. Contour plots of Gaussian curvature for the yacht for (a) a PDE surface, (b) a uniform B-spline surface with $k = l = 6$ and a 5×10 finite element mesh

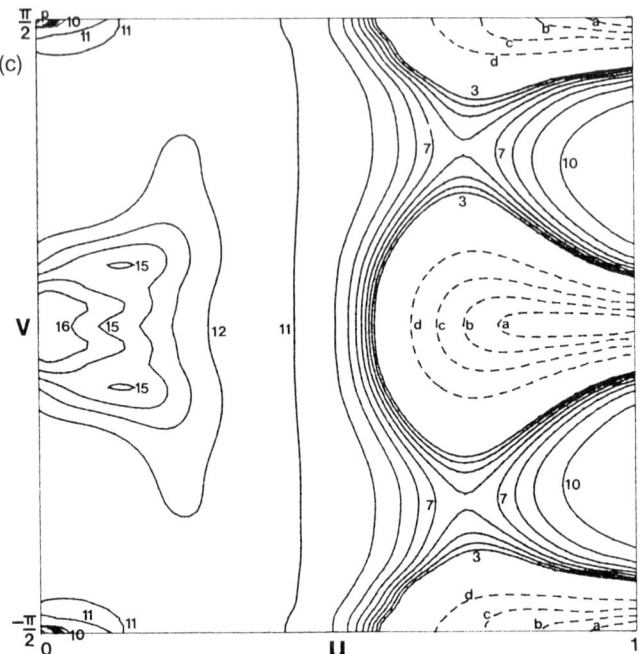

Figure 5. *cont.*(c) and a non-uniform B-spline surface with $k = l = 6$ and a 5×18 mesh

plots give an indication of the factors that affect the wiggles. There is a sharp contrast between the plots for the bicubic surfaces ($k = l = 4$) and higher degree surfaces ($k, l \geq 5$); for the bicubic case the lines of constant curvature have wiggles and are also continuous but not continuously differentiable. The later feature is not surprising as bicubic B-spline surfaces have discontinuous third derivatives, and the curvatures are calculated using second derivatives (see section 2.3). The number of patches used also has a rôle to play in the elimination of wiggles; if there are an inadequate number (as in Fig. 4(c) with $k = l = 5$) wiggles will appear. These results suggest that to avoid unwanted wiggles the degree of the polynomial and the number of patches must both be of sufficient size.

From the plots it can also be seen that, for the mesh sizes used, some of the characteristics of the Gaussian curvature plots of the bi-harmonic like PDE surface are lost when the surface is approximated by a B-spline surface. From Fig. 3(a) the Gaussian curvature for the inclined plane blend is seen to be negative over most of the region, but for a small region near $(u, v) = (1, 0)$ (and $(1, 2\pi)$, due to periodicity) it is positive. This positive

Table 1.

		Gaussian curvature minimum	maximum	$\int_s (\kappa_1^2 + \kappa_2^2)ds$
Horizontal plane blend				
	PDE surface	-0.108	0.000	0.924
$k = 1 =$	Finite element mesh			
4	5×10	-0.112	0.000	0.927
5	5×10	-0.109	0.000	0.924
6	5×10	-0.106	0.000	0.946
Inclined plane blend				
	PDE surface	-0.108	0.020	1.017
$k = 1 =$	Finite element mesh			
4	5×10	-0.120	0.000	0.863
5	5×10	-0.116	0.000	0.861
6	5×10	-0.119	0.000	0.883

region does not manifest itself in the B-spline approximations considered (Figs. 3(b), (c), (d) and 4). This can also be seen by observing the maximum values of Gaussian curvature for the inclined plane blend tabulated in Table 1. The positive region of Gaussian curvature for the yacht near the rear end of the base of the keel where $(u, v) = (1,0)$ (see Fig. 5(a)), also does not appear in the plots of Gaussian curvature for the B-spline approximation (Figs. 5(b), (c)). Due to the importance of second derivatives in the calculations of Gaussian curvature (see section 2.3), the similarity of curvature plots will probably increase if second derivatives on the boundary of the surface are specified as boundary conditions (along with functional and tangency conditions). The PDE generating the surface would then have to be sixth order.

For the yacht, a non-uniform B-spline approximation was examined (Fig. 5(c)). Extra knots were inserted in the regions near the front ($v = -\frac{\pi}{2}$ and $v = \frac{\pi}{2}$) and rear ($v = 0$) of the yacht. When the approximation was compared with the PDE surface, there was a reduction in the mean square error, but the plot of Gaussian curvature seemed to be more irregular. Increased accuracy and fairness for the yacht, and surfaces like it (with a discontinuity in the tangency conditions on the boundary), will probably result from the use of multiple knots and control vertices. More work needs to be done on their use and also on the effect of non-uniform knots on the fairness.

The integral over the surface of the sum of the squares of the principal curvatures (2.11) was calculated for the blends (see Table 1). It was thought that this integral may have been of value for the comparison of fairness of the PDE surfaces and their approximations. For the horizontal plane blend the integral for the biquartic B-spline (which was earlier found to be the closest approximation) is the nearest to that of the PDE surface. For the inclined plane though, where the curvature features present in the PDE surface are inadequately represented in the B-spline approximations, the values of the integrals arising from the B-spline approximations are markedly different from the value of the integral for the PDE surface. This difference, due to the curvature features, can be seen in plots of the sums of squares of the principal curvatures over the u, v-plane. For all the cases examined, the most accurate B-spline approximation (for a constant mesh) has the minimum integral (although this integral is not necessarily the closest to that for the PDE surface). This fact may be of use when looking at the accuracy of B-spline surface approximations to PDE surfaces for which there are no analytic solutions.

4   Concluding Remarks

From observing plots of Gaussian curvature it can be deduced that the PDE surfaces considered are fair because these plots do not show any small length scale disturbance. The uniform B-spline approximations retain this smoothness provided the number of surface patches (therefore the number of elements in the finite element mesh) is large enough, and the degree of the B-spline polynomials is sufficient, otherwise wiggles may occur in the Gaussian curvature plots, suggesting an unfairness. Curvature plots of the B-spline surfaces will probably be closer in form to those of the PDE surface if the PDE is also subject to second derivative boundary conditions. A sixth order PDE would then be necessary. This will be investigated further along with the effect of non-uniform knots on the fairness of the B-spline approximation.

The integral of the sum of squares of the principal curvatures may sometimes be useful in the comparison of the fairness, and possibly the accuracy of the different B-spline approximations of the same PDE surface. However, since it does not provide a measure of the criteria normally used to judge fairness, it must be used with some caution.

For the bi-harmonic like PDE surfaces the most accurate and the closest approximations from fairness considerations was found to be given by the biquartic B-spline surface. It is hoped that a general result can be obtained for n^{th} order PDE surfaces, where it is conjectured that the closest B-spline approximation will be given using basis functions of degree n.

Acknowledgements

Joanna M. Brown is supported by a studentship from the Science and Engineering Research Council. The authors are indebted to the British Council for support from the British German Academic Research Collaboration Program. J.M. Brown is grateful to P.D. Kaklis for some helpful discussions during her visit to Technische Universität Berlin.

References

1. Bartels, R.H., Beatty, J.C. and Barsky, B.A. (1987). *An Introduction to Splines for Use in Computer Graphics and Geometric Modelling*, Morgan Kaufman.

2. Beck, J., Farouki, R., and Hinds, J. (1986). Surface analysis and methods. *Computer Graphics and Applications, 6*, (12), 18–36.

3. Bloor, M.I.G. and Wilson, M.J. (1989). Generating blend surfaces using partial differential equations. *Computer-Aided Design, 21*, (3), 165–171.

4. Bloor, M.I.G. and Wilson, M.J. (1990). Using partial differential equations to generate free form surfaces. *Computer-Aided Design, 22*, (4), 202–212.

5. Brown, J.M., Bloor, M.I.G., Bloor, M.S. and Wilson, M.J. Generating B-spline representations of PDE surfaces. Submitted to *Mathematical Engineering in Industry*.

6. Brown, J.M., Bloor, M.I.G., Bloor, M.S. and Wilson, M.J. (1990). Generation and modification of non-uniform B-spline surface approximations to PDE surfaces using the finite element method. *Proceedings of ASME 16th Design Automation Conference*, Chicago.

7. Carmo, M.P. do. (1976). *Differential Geometry of Curves and Surfaces*, Prentice-Hall.

8. Dill, J.C. (1981). An application of color graphics to the display of surface curvature. *Computer Graphics*, (Proc. SIGGRAPH) *15*, (3), 153–161.

9. Dill, J.C., and Rogers, D.F. (1982). Color graphics and ship hull curvature. In: Rogers, D. F., Nehrling, B. C., and Kuo, C., eds., *Computer Applications in the Automation of Shipyard Operation and Ship Design IV*, North-Holland.

10. Farin, G. and Sapidis, N. (1989). Curvature and fairness of curves and surfaces. *Computer Graphics and Applications, 9,* 52–57.

11. Hagen, H. and Schulze, G. (1987). Automatic smoothing with geometric surface patches. *Computer Aided Geometric Design, 4,* 231–236.

12. Klass, R. (1980). Correction of local surface irregularities using reflection lines. *Computer-Aided Design, 12,* (2), 73–76.

13. Munchmeyer, F. (1980). The Gaussian curvature of Coons biquintic patches, advances in computer technology 1980 1. *Proceedings of ASME Century 2 International Computer Technology Conference,* pp. 383–387.

14. Munchmeyer, F. (1987). Shape interrogation: a case study. In: Farin, G., ed., *Geometric Modelling,* SIAM, Philadelphia, pp. 291–301.

15. Nowacki, H. and Reese, D. (1983). Design and fairing of ship surfaces. In: Barnhill, R. E. and Boehm, W., eds., *Surfaces in CAGD,* North-Holland, Amsterdam, pp. 121–134.

Generalized Cyclides for Use in CAGD

W.L.F. Degen

Universität Stuttgart, Germany

1 Introduction

In Computer Aided Geometric Design (CAGD) most of the free-form surfaces are composed by a network of tensor product B-spline or Bézier surface patches. In some cases, however, it is desirable to have a global representation of a larger part, e.g. as a blending surface. In general, quadrics are not flexible enough, though having favourable geometric properties. On the other hand, the order of, say, a Bézier patch considered as an algebraic surface, increases rapidly with the degrees of the polynomials in its representation (order $=2mn$). Thus it takes considerable efforts to calculate its equation or to derive those of sectional curves or to solve similar geometric tasks.

Therefore it seems to be worthwhile looking for a class of surfaces with more degrees of freedom than those of quadrics but being of low order and having a fine intrinsic geometry.

In an earlier period of research [1],[5],[6] we studied certain surfaces (called Blutel surfaces [2]) with a family of conics on it and the additional property that the tangent planes along each generating conic envelope a quadratic cone. There is a second family of curves on such a Blutel surface the tangents of which are the generators of these cones (so-called *conjugate curves*) and the problem was to determine all Blutel surfaces such that the conjugate curves are conics too. This was a very hard problem of differential geometry but it turned out to have an elegant solution: besides some special cases (which we omit here) the surfaces in question are the projective images of the cyclides of Dupin, hence of fourth order (obviously, vice versa, the cyclides of Dupin have the properties described above). Therefore we call these surfaces "generalized cyclides" for shortness (though sometimes the vaster class of fourth order algebraic surfaces having a singular conic is referred to by this name).

The purpose of this paper is to show how these generalized cyclides (including Dupin's ones) can be used in CAGD, especially as blending surfaces. Before that, we will derive its main geometric properties from their general parametric representation. Furthermore, we will see that they may be considered as special biquadratic Bézier surfaces in homogeneous coordinates and we will determine the conditions for its control points to belong to that class. At the end we will give some explicit examples.

2 The Universal Representation of Generalized Cyclides and their Basic Properties

In the following we use *homogeneous coordinates* in Euclidean 3-space E^3, i.e. each point $P \in E^3$ is associated with a vector from $\mathbb{R}^4$ $X = (1, x, y, z)$ and multiples ρX $(\rho \neq 0)$ represent the *same* point (x, y, x being Cartesian coordinates in E^3); directional vectors $\vec{v} = (v_x, v_y, v_z)$ are represented as $V = (0, v_x, v_y, v_z)$ and the set of all its multiples ρV is a direction (or a point "at infinity"). Recall that lines (planes) are represented by the set of all linear combinations of its base points (2 resp. 3 in number, lin. in dep.).

From [6] we take the following universal representation of a generalized cyclide (double Blutel surface) with skew axes:

$$X(s,t) = g(t)[a(s)A + b(s)B] + f(s)[c(t)C + d(t)D] \qquad (2.1)$$

whereby A, B, C, D designate the homogeneous coordinate vectors of four space points (not in a plane) and $f(s), a(s), b(s)$ respectively $g(t), c(t), d(t)$ two triples of quadratic forms (with respect to its parameter s, resp. t) being linearly independent, each. The lines AB and CD are called the first and second axis of our surface.

From (2.1) we conclude immediately

Theorem 1:

The generalized cyclides have the following properties:

(a) The parametric lines of both families are *conics* (except at the zeros of $g(t)$ resp. $f(s)$).

(b) The planes of these conics belong to a *pencil* for each of both families with AB as the axis of the first family (s varying) and CD as the axis of the second one (t varying).

(c) If s_1 is a zero of $f(s)$ then all conics of the first family pass through the point

$$S_1 = a(s_1)A + b(s_1)B$$

on the first axis (and analogously for another zero s_2 of $f(s)$ or the zeros t_1, t_2 of $g(t)$). Such a point S_1 is a *singular point* of the surface.

(d) Every pair of conics from the same family are in perspective position i.e., they are projected to each other from a certain center such that corresponding points have equal parameter values. Furthermore this center is incident with the other axis (different from the intersection line of its two planes).

Proof:

Recall that a conic is representable by

$$X(s) = B_0 + sB_1 + s^2 B_2 \quad s \in \mathbf{R} \tag{2.2}$$

whereby B_0, B_2 are two different points on it and B_1 is the pole of the line $B_0 B_2$; if the base points are changed by a regular (3,3)-matrix then the coefficients will transform into a triple of linearly independent quadratic forms and vice versa. For t fixed and $g(t) \neq 0$ we get $X(s, t)$ proportional (same point as X) to

$$\tilde{X}(s) = a(s)A + b(s)B + f(s)E(t) \tag{2.3}$$

with

$$E(t) = g^{-1}(t)(c(t)C + d(t)D) \tag{2.4}$$

Hence $\tilde{X}(s)$ represents a conic in the plane ABE. Now (c) is obvious. Taking two different (but not singular) values t_1, t_2 for t in (2.3) we get two conics C_1, C_2 of the same family with *equal* coordinate functions $a(s), b(s), g(s)$ with respect to the bases $A, B, E(t_1)$ and $A, B, E(t_2)$, respectively; hence they are perspective with axis AB and *a* center on $E(t_1)E(t_2)$ (which is b). ∎

Furthermore, differentiating (2.1) with respect to s, we get

$$\frac{\partial X}{\partial s} - \frac{\dot{f}}{f}X = g\left[(\dot{a} - \frac{\dot{f}}{f}a)A + (\dot{b} - \frac{\dot{f}}{f}b)B\right] \tag{2.5}$$

(arguments omitted; dots designating derivatives with respect to s). The vector on the right being proportional to

$$P(s) = (\dot{a}f - \dot{f}a)A + (\dot{b}f - \dot{f}b)B \tag{2.6}$$

which is *independent* of t (and *linear* with respect to s), we conclude

Theorem 2:

All the tangents to conics of the first family at the points along a fixed conic C of the second family pass through a fixed point P on the first axis, i.e. the surface is enveloped along C by the *quadratic cone* C^* with vertex P. While s is varying, P runs on the first axis and the correspondence $C \to P$ is one to one.

The analogous properties hold if the two families of conics (and corresponding cones) are interchanged. Hence the generalized cyclides are, along their parameter lines (generating conics), enveloped by two families of quadratic cones with vertices running on the axes.

Remarks:

1. By theorem 2, the generalized cyclides are completely self-dual.

2. Verify these properties for a Dupin cyclide e.g. the torus (second axis at infinity).

3. A generalized cyclide can have at most two singular points on each axis (including the conjugate complex and coincident cases).

4. If the corresponding root (of $g(t)$ resp. $f(s)$) is real and simple, then the singularity is a vertex like that of a quadratic cone. This property will enable us to construct (parts of) surfaces looking like curved horns.

Further results

For completeness, we add a brief survey on further results about generalized cyclides omitting, however, the proofs, which can be found in the original papers [1] and [6]. For simplicity we use the following notations: C designates some generating conic from any of the two families the cyclide is consisting of; p denotes the plane in which C is contained and a stands for that one of the two axes which is contained in p (recall that the planes of the generating conics belong to the pencils); finally E denotes the intersection point of the second axis (different from a) with p.

Now we can summarize these results as follows:

Theorem 3:

(a) The order of a generalized cyclide (as an algebraic surface) is at most four.

(b) The order reduces to three exactly in the case if each conic C passes through E (for both families).

(c) The order reduces to two (the cyclide degenerates into a quadric) if E is the pole of the axis for any C from each of both families.

(d) For any C, the intersection of p with the cyclide consists of C and another conic C' of the same family.

(e) C' is the image of C by a perspectivity with center E and axis a and the cross-ratio κ of this perspectivity is constant within each one of both families.

(f) The two numeric values κ_1, κ_2 of these cross-ratios are the only invariants, i.e., they determine the generalized cyclide up to a projective transformation of the space.

(g) The remaining singularities of the generalized cyclide (besides that on the axes) are the intersection points of C and C' (see (d)). These singularities build up a further fixed conic C_0 (in general not contained in any family) which, however, may be complex or degenerated. There are exactly two exceptions, where this property does not hold: first, for cyclides of order three, and second for cyclides where the condition of (b) is valid only for one family.

The classical cyclides of Dupin are usually defined as surfaces in Euclidean space that can be enveloped by two families of spheres. The circles of contact play the role of the generating conics in this case. Now the plane at infinity contains a complex conic in which it is intersected by each sphere (absolute sphere circle) and this one is C_0 for Dupin's cyclides.

In Euclidean differential geometry the cyclides of Dupin are important because the circles C are at the same time *curvature lines* but this property is lost for generalized cyclides. However it remains true that the generating conics form a *conjugate net*.

3   Generalized Cyclides as Rational Biquadratic Bézier Surfaces

It is well known [3],[7] that *rational* Bézier curves as well as surfaces can be written in exactly the same formula as ordinary ones if *homogeneous coordinates* (see section 2) for the control points are used; the factors ρ with which the vector $X = (1, x, y, z) \in \mathbf{R}^4$ is to be multiplied are called the "weights". If they are all positive the convex hull property will remain true otherwise poles and changes of orientation can occur.

Now, the Bernstein polynomials $B_k^n(t)(k = 0, \ldots n)$ of some degree $n \in \mathbf{N}$ are a base for the set (vector space) of all polynomials of this degree, hence each polynomial of degree n can be written as a linear combination of the Bernstein polynomials.

Applicating this principle to the representation (2.1) of the generalized cyclides (with $n = 2$) we conclude at once that this representation transforms into a biquadratic Bézier form.

Theorem 4:
  Each generalized cyclide can be represented as a rational biquadratic Bézier parametric surface (using homogeneous coordinates).

However, the inverse is not true. As mentioned in the introduction, the order of a biquadratic Bézier surface equals *eight* in general but the cy-

clides are of order *four* at most. Therefore there must be special conditions
to be fulfilled by the control points in order for the represented surface to
be a generalized cyclide.

We want to derive these conditions, now. One way to do this would be
to express the quadratic forms $a(s), b(s), c(t), d(t), f(s), g(t)$ explicitly by
the Bernstein polynomials, to insert these expressions into (2.1) and then
to calculate the control points as linear combinations of the base points
A, B, C, D. However this way is not only tedious but gives no further
insight; thus we prefer a geometric way.

We start from a biquadratic Bézier representation

$$X(s,t) = \sum_{k,j=0}^{2} B_k(s)B_j(t)C_{kj} \qquad (3.1)$$

(omitting superscripts for quadratic Bernstein polynomials and using homogeneous coordinates for the control points C_{kj}) and assume that the
surface given by it is a generalized cyclide. In particular we look at that
part which is the image of (s, t) in unit square of $\mathbf{R}^2$. Then the boundary
curves are parts of four conics $\mathcal{C}_0, \mathcal{C}_1, \mathcal{C}_0,' \mathcal{C}_1'$ the first two belonging to the
first family, the remaining ones to the second. Therefore, the planes p_0, p_1
of $\mathcal{C}_0$, $\mathcal{C}_1$ intersect in a (the first axis) and, analogously, p_0', p_1' (the planes
of $\mathcal{C}_0', \mathcal{C}_1'$) intersect in b (the second axis).

On the other hand, a rational quadratic Bézier curve

$$X(s) = \sum_{k=0}^{2} B_k^2(s)C_k \qquad (3.2)$$

is a conic $\mathcal{C}$ and the control points C_k form a tangent triangle (C_0, C_2 are
on $\mathcal{C}$ and C_1 is the pole of C_0C_2).

We apply this to the four boundary curves getting

$$\begin{array}{ll} p_0 = C_{00}C_{10}C_{20} & p_1 = C_{02}C_{12}C_{22} \\ p_0' = C_{00}C_{01}C_{02} & p' = C_{20}C_{21}C_{22} \end{array} \qquad (3.3)$$

Furthermore, by theorem 3(e) $\mathcal{C}_0, \mathcal{C}_1$ are in perspective position with corresponding points having equal parameter value s and with center V on b
(see Fig. 1).

Since the control points C_{00} and C_{02} have equal parameter values $s = 0$
they are projected to each other from V (i.e. C_{00}, C_{02}, V are colinear) and
the same is valid for C_{20} and C_{22} (value $s = 1$). *Therefore the four basic
control points $C_{00}, C_{02}, C_{20}, C_{22}$ are coplanar.* Furthermore, the tangents to
$\mathcal{C}_0$ and $\mathcal{C}_1$ at the endpoints C_{00}, C_{20} and C_{20}, C_{22} respectively correspond
to each other, too, in this perspectivity, hence they meet each other in
points of a. But these tangents are the joining lines of control points with
an index 1. It follows, in particular, that $C_{00}C_{10}$ and $C_{02}C_{12}$ meet at some

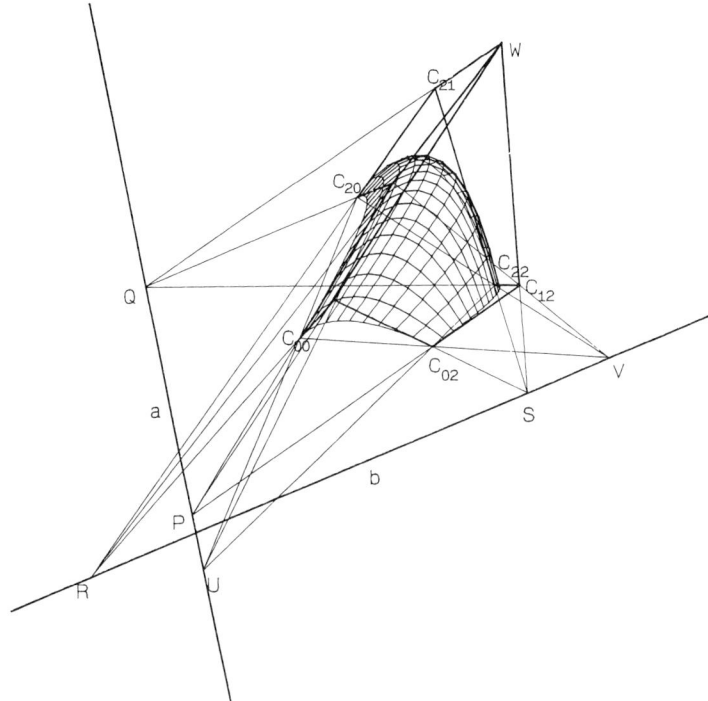

Figure 1. A generalized cyclide as a Bézier surface and its control points

point P on a as well as $C_{20}C_{10}$ and $C_{22}C_{12}$ meet at some point Q on a. In addition C_{10}, C_{12} and V are colinear, C_{10} and C_{12} being intersection points of corresponding tangents.

Analogously, interchanging the two families of conics, we conclude: $C_{00}C_{01}$ and $C_{20}C_{21}$ meet at some point R on b as well as $C_{02}C_{01}$ and $C_{22}C_{21}$ meet at some point S on b, too. Furthermore the three lines $C_{00}C_{20}, C_{01}C_{21}, C_{02}C_{22}$ meet at some point U on a (U being the center of perspectivity).

Now, we consider the tangent planes of the cyclide at the basic control points $C_{00}, C_{02}, C_{20}, C_{22}$. We already know that they are given by $C_{10}C_{00}C_{01}, C_{01}C_{02}C_{12}, C_{10}C_{20}C_{21}$ and $C_{21}C_{22}C_{12}$ since these planes contain the corresponding point of contact and the two tangents to the parameter lines passing through it. Now, defining W as the intersection point of PC_{01} with QC_{21} (we already know, that these four points are coplanar since PQ and $C_{01}C_{21}$ meet at U), we conclude that W lies on the intersection line of the planes $C_{10}C_{00}C_{01}$ and $C_{10}C_{20}C_{21}$; but this is true also for R and C_{10}, hence R, C_{10} and W are colinear. Similarly we find that

S, Q_{12} and W are colinear. Thus each of the four lines $PC_{01}, QC_{21}, RC_{10}$ and SC_{12} meet at W and this point is contained in each of the four tangent planes under consideration. Now, it follows immediately that W must coincide with the remaining control point C_{11}. With this in mind, we recognize at once that the configuration of the control points together with the axes a and b and the projection centers U, P, Q on a and V, R, S on b is completely symmetric: each triple of control points with a common index is projected onto both the others in the same direction.

If we are thinking of the control points as a scheme of a matrix corresponding to the indices, we can summarize the results as follows:

Theorem 5:

If a rational biquadratic Bézier surface represents a generalized cyclide then the following conditions hold:

(i) The three planes spanned by the control points from a column (row resp.) belong to *pencil* with the axis a (b resp.) and these two axes do not meet each other.

(ii) For each pair of triples from two columns (rows resp.), there exists a *perspectivity* with center on the axis b (a resp.) mapping these triples of control points onto each other.

Next we want to proof the converse

Theorem 6:

The conditions of theorem 5 are sufficient (hence characterizing) for a rational biquadratic Bézier surface to be a generalized cyclide.

Proof:

The axes being skew, by (i), there is, for each point C in space a line $l(C)$ passing through C and meeting both axes. Applying this to the control points, we can write

$$C_{kj} = \lambda_{kj} E_k + \mu_{kj} F_j \qquad (3.4)$$

with some coefficients λ_{kj}, μ_{kj} and E_k, F_j denoting the points where the above lines meet the axes (E_k on a, F_j on b). Writing only *one* index at E and F uses the fact that triples of control points from columns or rows span a plane containing the axis a or b respectively, hence their lines $l(C)$ have the intersection point of that plane with the other axis in common.

Now, we use (ii) and can derive by duality from this condition that λ_{kj} is independent of k and μ_{kj} independent of j. Thus we can rewrite (3.4) in the form

$$C_{kj} = \psi_j E_k + \varphi_k F_j \qquad (3.5)$$

The points E_k being on a and F_j on b implies

$$E_k = \alpha_k A + \beta_k B, \quad F_j = \gamma_j C + \delta_j D \qquad (3.6)$$

with certain coefficients $\alpha_k, \beta_k, \gamma_j, \delta_j$ and pairs of base points A, B for the axis a and C, D for b. Inserting (3.6) in (3.5) and the result in (3.1) we obtain indeed (2.1) where the quadratic forms are given by

$$
\begin{array}{ll}
f(s) = \sum \varphi_k B_k(s) & g(t) = \sum \psi_j B_j(t) \\
a(s) = \sum \alpha_k B_k(s) & c(t) = \sum \gamma_j B_j(t) \\
b(s) = \sum \beta_k B_k(s) & d(t) = \sum \delta_j B_j(t)
\end{array}
\tag{3.7}
$$

This completes the proof. ∎

4 Applications

4.1 Preliminary remarks

We are proposing to use the generalized cyclides mainly in *solid modelling* either as blending surfaces or to construct larger parts of a solid's surface rather than as a general surface patch to build up a free-form surface.

Adding the generalized cyclides, the classical arsenal—consisting of planes, spheres, cylinders, cones and other *quadratic* surfaces—will be considerably enlarged. Since their parameter lines are conics, especially ellipses, we will be able to construct plenty of tube surfaces (ducts, pipes etc.) with a great richness of shapes. Furthermore, the classical cyclides of Dupin, in particular the most familiar one of its examples, the torus, will be found at our disposal (compare [4],[9] for more details).

4.2 General conditions for a blending of a given surface with a generalized cyclide

We consider the problem to construct the surface of two (or more) joined solids. It is clear that only in the case when the common boundary (where the solids come together) is a *conic* $\mathcal{C}$ we have the chance to do it with a cyclide using one of its parameter lines for that boundary.

However, in most cases of applications, the two surfaces should join with G^1-continuity. This is equivalent to the coincidence of its tangent planes along that common conic. But since a cyclide, by theorem 2, is enveloped by a *cone* $\mathcal{C}^*$ along $\mathcal{C}$ this one must be a tangent cone for the other surface, too.

Vice versa, it is obvious that this condition is sufficient. Let us point out the result as a

Rule:
For a tangent continous (G^1 continuous) blending of any surface with a generalized cyclide along one of its parameter lines the following conditions are necessary and sufficient:

(i) The common boundary curve must be (a part of) a conic.

(ii) Along $\mathcal{C}$ both surfaces must have a common tangent cone $\mathcal{C}^*$.

Of course, beyond $\mathcal{C}$, only the vertex P of $\mathcal{C}^*$ must be known. Note that, all over the paper we assume *parallel lines* having an intersection point at infinity (its direction vector). Thus *cylinders* are included into the notion of a cone.

We see that most of our cyclide's shape parameters remain arbitrary. Only one of its generating conics and the unique point P of the corresponding tangent cone is fixed still now. However, by theorem 2, the following is crucial for the further construction:

(iii) The vertex P of the tangent cone $\mathcal{C}^*$ along $\mathcal{C}$ must lie on the second axis b while the first axis a of our generalized cyclide must be contained in the plane p of $\mathcal{C}$.

4.3   The profile construction

Now we assume to be asked for a blending on both sides. To fix the ideas let us think of a *duct*, but the cross section need not be a circle, it may be any ellipse as well. These two ellipses should become two generating conics $\mathcal{C}_0, \mathcal{C}_1$ of our cyclide belonging to the same family. We see at once that they can be prescribed with a great range of freedom; by theorem 1, (b) and (c), and the rules of the previous subsection the following conditions must be satisfied:

(i) The two planes p_0, p_1 of $\mathcal{C}_0, \mathcal{C}_1$ resp. must be different. Its intersection line a becomes the first axis of the cyclide.

(ii) The intersection points of $\mathcal{C}_0$ with the axis a must be the same as those of $\mathcal{C}_1$ (including the case of conjugate complex ones). If the cyclide is not allowed to have singularities, then these points must be conjugate complex.

(iii) The second axis b is determined by $b = PQ$ (assuming $P \neq Q$). Both vertices P and Q must not be incident with the first axis a. Therefore a and b are skew.

Of course, for reasons of *numeric stability*, P and Q must be neither very close to each other nor very close to the axis a.

In the next step we pick out some plane of the pencil with axis b, calling this one the "profile plane" $\bar{p}$. (See Fig. 2.)

It is clear that the profile plane contains a generating conic $\mathcal{C}_0'$ of the second family passing through R and S (corresponding to the same parameter value, say $s = 0$, on $\mathcal{C}_0$ and $\mathcal{C}_1$). By theorem 2 its tangents must pass through P and Q respectively.

Thus we know already a tangent triangle R, S, T of $\mathcal{C}_0'$ hence $\mathcal{C}_0'$ is determined up to one further parameter. This can be either an additional point to determine the weights of the base points B_k in a representation

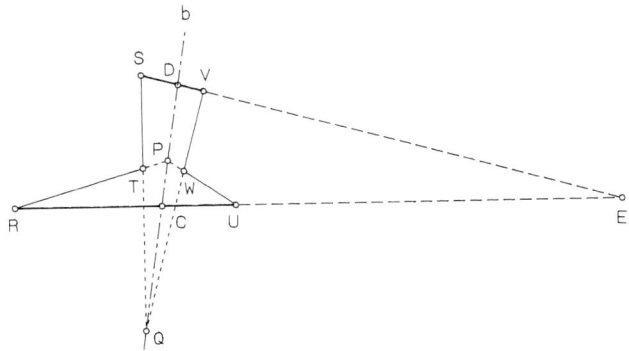

Figure 2. The configuration in the profile plane. $E = a \cap \bar{p}$, $\{R,U\} = \mathcal{C}_o \cap p$, $\{S,v\} = \mathcal{C}_1 \cap \bar{p}$, $C = p_0 \cap \bar{p} \cap b$, $D = p_1 \cap \bar{p} \cap b$

like (2.2) or simply the weight of T when using a Bézier representation for $\mathcal{C}_0'$ like (3.2) with the control points $C_0 = R, C_1 = T, C_2 = S$. So we have:

(iv) The generating conic $\mathcal{C}_0'$ (of the second family) contained in the profile plane is determined up to one parameter; choosing the weight of T in a suitable way (shape parameter!) one immediately gets $\mathcal{C}_0'$ in a Bézier representation (3.2) with the control points R, T, S.

At this stage of the construction, however, we must recognize, that the configuration is overdetermined! The second conic $\mathcal{C}'$ in which the profile plane intersects the cyclide—and for which the triangle U, W, V must play the same role as R, T, S for $\mathcal{C}$—by theorem 3(e), has to be the image of $\mathcal{C}$ by the perspectivity given by its center E and axis b (=PQ). So we have the final condition:

(v) For the existence of a blending generalized cyclide (as a duct with given ellipses and tangent cones at both ends) the cross ratios of E, C, R, U and E, D, S, V must be equal.

Geometrically, this means that E, T, W are colinear and RS, UW meet at a point on b (each of these properties implies both the others).

After having constructed the second conic $\mathcal{C}'$ in this way all is done in the profile plane. In the next subsection we will extend this into the space.

4.4 Completion into the space

The main idea is that in the profile plane we have a representation like (2.3), (2.4) but with interchanged axes:

$$\tilde{X}(s_0,t) = g(t)E + c(t)C + d(t)D \tag{4.1}$$

with

$$E = \frac{a(s_0)}{f(s_0)} A + \frac{b(s_0)}{f(s_0)} B \qquad (4.2)$$

We can start with the Bézier representation of C_0' with control points R, T, S obtained in the previous subsection. Expressing R, T, S as linear combinations of E, C, D (all points being in the profile plane) we immediately obtain the three quadratic forms $g(t), c(t), d(t)$.

Now we switch from the profile plane to a plane from the first pencil, say to the plane p_0 of C_0. Within this plane we can see $C + \bar{p} \cap p_0 \cap b$, the first axis a and, of course, C_0. E being a point in both planes $\bar{p}$ and p_0 and, additionally, on the axis a we can choose $A := E$. We still need a second point B on a. In principle we could choose any point on b different from A but as we will see later on it is favourable to choose the polar point of A with respect to C_0.

Now we can express the representation of C_0 in whatever a basis (e.g. in Bézier form) and substitute that basis by C, A, B; thus we obtain as before in the profile plane the coefficients of the quadratic forms $f(s), a(s), b(s)$. By the way, the only data coming into these calculations is the cross-ratio of A, C, R, U. Now all quantities of (2.1) are determined and the obtained cyclide will have all the desired properties.

Thus we have reached our goal to completely construct the both-side blending cyclide and no further conditions were necessary.

Theorem 7:

The conditions (i),(ii),(iii) and (v) are necessary and sufficient for the existence of a regular blending tube with given conics and tangent cones at its ends.

4.5 Using trigonometric functions in the parametrisation

The parametrisations of the ellipses from the first family of our cyclide is not suitable for practical reasons because the distribution of points with equi-distant parameter values is very unequal. This can easily be avoided by introducing trigonometric functions. Note first that $f(s)$ is positive definite because $C_0 \cap a = \emptyset$ and second that our choice of B (polar to A) implies $f(s)$ to be without a linear term. Then, by a suitable factor, we can achieve

$$f(s) = 1 + s^2.$$

With this, the quotients $a(s)/f(s)$ and $b(s)/f(s)$ can be written as linear combinations of $1, (1 - s^2)/(1 + s^2), 2s(1 + s^2)$. Introducing φ by

$$\varphi = 2 \arctan s$$

we get $\cos \varphi$ and $\sin \varphi$ instead of the latter two rational functions of s. The result will be a much more uniform distribution of the parameter lines.

4.6   An example: the nozzle of a can

Finally we will demonstrate the constructions by a realistic example. We imagine the design of a can's nozzle. The can is given as a flattened rotational ellipsoid. On one side of it an oblique plane p_0 cuts away a certain part and at the hole a nozzle must be inserted.

The nozzle should be a curved duct with decreasing cross-section and must have a smooth join to the can. Upon these requirements all classical types of surfaces fail. But a generalized cyclide will do it very well!

The can being a quadric, the conditions (i),(ii) of subsection 4.2 are satisfied. We take the data of the can and its hole as well as that of the point P which is the vertex of the tangent cone along the hole. At the top end where the nozzle is open we prescribe the analogous data as if there were an other surface to be joined smoothly.

Of course we choose the symmetry plane (containing the can's axis of rotation) as a profile plane. Then we proceed exacly as described in subsection 4.3 and perform the profile construction. The result is shown in Fig. 3.

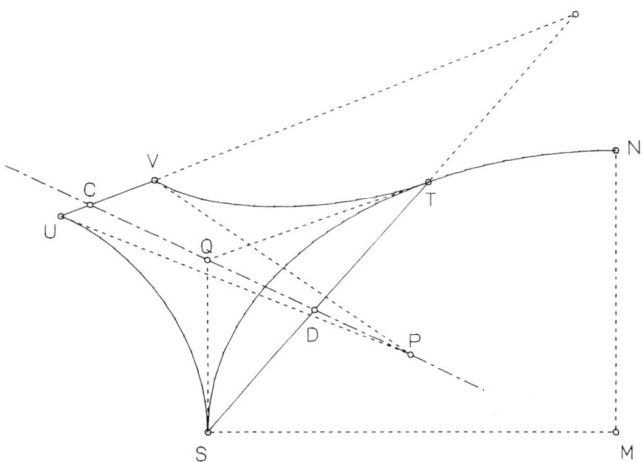

Figure 3. The profile construction of the nozzle

Furthermore we take the profile plane as x-z-plane of a cartesian coordinate system with origin at the center of the can; then the first axis becomes parallel to the y-axis (the second one is contained in the profile plane). Because of the symmetry the polar point B of A is at infinity, so we take $B = (0, 0, \beta, 0)$. The parameter β has to be adjusted in such a way that the second semiaxis of the hole (which is parallel to the y-axis) becomes right. For the real calculation in our computers we used the parameter transformation according to subsection 4.5. The final result is shown in Fig. 4.

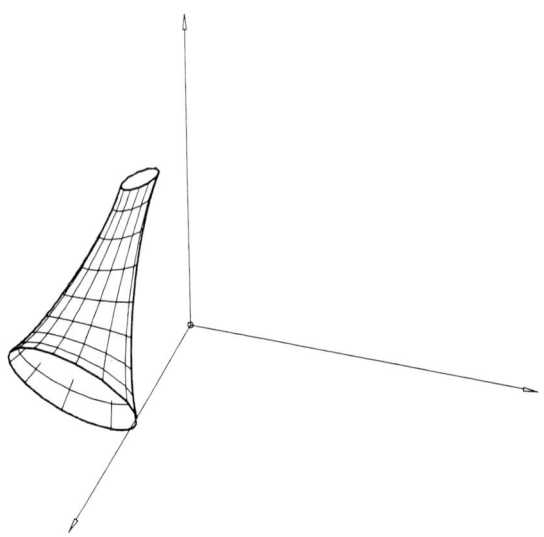

Figure 4. The nozzle of a can as an example of a generalized cyclide

References

1. Barner, M. (1987) Eine differentialgeometrische Kennzeichnung der allgemeinen Dupinschen Zykliden. *Aequationes Mathematicae, 34,* 277–286.

2. Blutel, E. (1890) Recherches sur les surfaces qui sont en même temps lieux de coniques et enveloppes de cônes du second degré. *Ann. sci. ecole norm. super.,* (3) *7,* 155–216.

3. Boehm, W. (1987) Rational geometric splines. *Comp. Aided Geom Design, 4,* 67–77.

4. Boehm, W. (1990) On cyclides in geometric modeling. *Comp. Aided Geom. Design, 7,* 243–255.

5. Degen, W.L.F. (1982) Surfaces with a conjugate net of conics in projective space. *Tensor* (n.s.), *39*, 167–172.

6. Degen, W.L.F. (1986) Die zweifachen Blutelschen Kegelschnittflächen. *Manuscripta math.*, *55*, 9–38.

7. Degen, W.L.F. (1988) Some remarks on Bézier curves. *Comp. Aided Geom. Design*, *5*, 259–268.

8. Farin, G.E. (1983) Algorithms for rational Bézier curves. *Comp. Aided Design*, *15*, 73–77.

9. Pratt, M.J. (1990) Cyclides in computer aided geometric design. *Comp. Aided Geom. Design*, *7*, 221–242.

A Consistent Zero-Thresholding Inequality for the Gaussian and Mean Curvatures

Li-Dong Cai

University of Sheffield

1 Introduction

The Gaussian curvature K and the mean curvature H [6], as the fundamental properties of surfaces, play an important role in surface perception, segmentation and recognition. By using the signs of curvatures K and H, surfaces (or their points) can be classified exactly into eight types [1].

However, the situation of surface classification in practice is not as good as in theory. As the surface curvatures are related to the second order derivatives, they are sensitive to noise, especially for those curvatures around zero. How to treat the zero curvatures therefore becomes a critical problem. So far, the zero thresholding of curvatures has been discussed by many papers, where the thresholds are empirically imposed in the processing [1, 9].

An empirical imposition might be feasible for single scale processing. However, it is no longer appropriate for multi-scale processing [8, 2, 3, 4, 5] due to the scale effects (including the reduction of noise, the distortion of surface and the changes of significant features at different scale levels). This suggests the complicated situation of the K and H zero thresholding in scale space [8].

The thresholding problem could be addressed in two aspects. One aspect is how to give surface features an explicit scale-based thresholding, rather than an empirical imposition at individual scale level. Ponce and Brady [7] have shown how to formulate the scale space behaviour of the curve primitives, such as, step, roof, smooth join, shoulder and bar, and use these results to set thresholds in edge detection. The other aspect is how to find possible inter-relationship between zero thresholds of different surface features, rather than to impose some isolated therefore perhaps casual values on different thresholds. This aspect becomes increasingly significant since the *combinations* of the signs of the Gaussian and the mean curvatures has been more and more frequently used in the surface segmentation. In this paper, we try to estimate and formulate the relationship between zero thresholds of the Gaussian and the mean curvatures so that once a threshold is fixed the other can be automatically produced by this formula.

2 An Example of Improper Zero Thresholding of K and H

There are nine combinations of the signs of K and H as shown in Table 1.

Table 1. Surface shapes from the signs of K, H

K \ H	$-$	o	$+$
$-$	saddle ridge	minimal	saddle valley
o	ridge	flat	valley
$+$	peak	(none)	pit

However, only eight of them can be used to classify surfaces. The rest one '$K > 0, H = 0$' is an impossible case for surface type. From the definition of K and H in differential geometry, we have

$$K = C_1 \cdot C_2 \tag{2.1}$$

$$H = \frac{1}{2}(C_1 + C_2) \tag{2.2}$$

where C_1 and C_2 are principal curvatures of surface. So, $H = 0$ implies $C_1 = -C_2$, which leads to $K < 0$. Thus, '$K > 0, H = 0$' is a 'phantom' type in theory.

Note that a strict zero is rarely met in computation. Small numbers will be seen as zero in computation if they are less than a given zero threshold. It is thus possible that an improper thresholding may lead to the 'phantom' case. When this happens, the surface patch is usually nearly flat. We now give a simulated example.

Given a plastic basin as shown in Fig. 1(a), the surface is composed of a conic wall patch W and a planar bottom patch B.

Suppose the principal curvatures of the wall patch W are 0.0 and 5.0, and the principal curvatures of the bottom patch B are both zero. According the KH sign classification in Table 1, the shape of the wall patch W will be 'valley' type and the shape of the bottom patch B will be 'flat' type. So it is easy to classify the basin surface by setting small zero thresholds for K and H, say,

$$\epsilon_K = \epsilon_H = 10^{-3}. \tag{2.3}$$

But when the bottom is under perturbation, its curvatures may change slightly as illustrated in Fig. 1(b). For example, the principal curvatures now become $C_1 = C_2 = 9 \times 10^{-4}$. As both curvatures are close to zero,

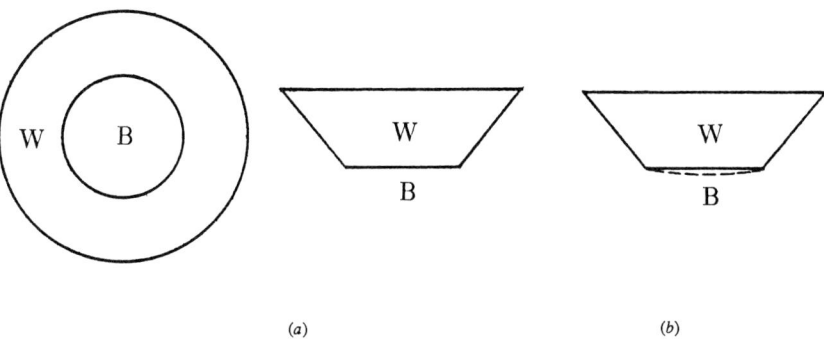

(a)   (b)

Figure 1. A plastic basin and its shape under a weak perturbation

the bottom patch can still be classified as the 'flat' type by a proper zero thresholding. From (2.1) and (2.2), the Gaussian curvature K of the bottom patch will be much smaller than its mean curvature H, so it seems reasonable to get a planar patch by setting a zero K threshold which is much smaller than the zero H threshold, say,

$$\epsilon_K = 10^{-7}, \qquad \epsilon_H = 10^{-3}. \tag{2.4}$$

However, this will be an improper zero thresholding. Although the shape of the wall patch can be correctly classified as a 'valley' type because the curvature of the wall patch satisfies $H = 2.5 > \epsilon_H$ and $K = 0.0 < \epsilon_K$, the shape type of the bottom patch we shall get is not a 'flat' type '$K = 0$, $H = 0$' but the 'phantom' type '$H = 0, K > 0$' because the curvature of the wall patch satisfies

$$H = 9 \times 10^{-4} < \epsilon_H, \qquad K = 8.1 \times 10^{-7} > \epsilon_K. \tag{2.5}$$

In fact, we have the following inequality:

$$K = C_1 \cdot C_2 = |C_1| \cdot |C_2| \le [\tfrac{1}{2}(|C_1| + |C_2|)]^2 = [\tfrac{1}{2}(C_1 + C_2)]^2 = H^2. \tag{2.6}$$

Hence, once a zero threshold ϵ_H is set up, the thresholds ϵ_K and ϵ_H should satisfy the inequality below to guard against the noisy perturbation around $K = 0$ and $H = 0$.

$$\epsilon_K \ge \epsilon_H^2. \tag{2.7}$$

Setting $\epsilon_K \ge 10^{-6}$ corresponding to $\epsilon_H = 10^{-3}$ may produce the right result as is expected.

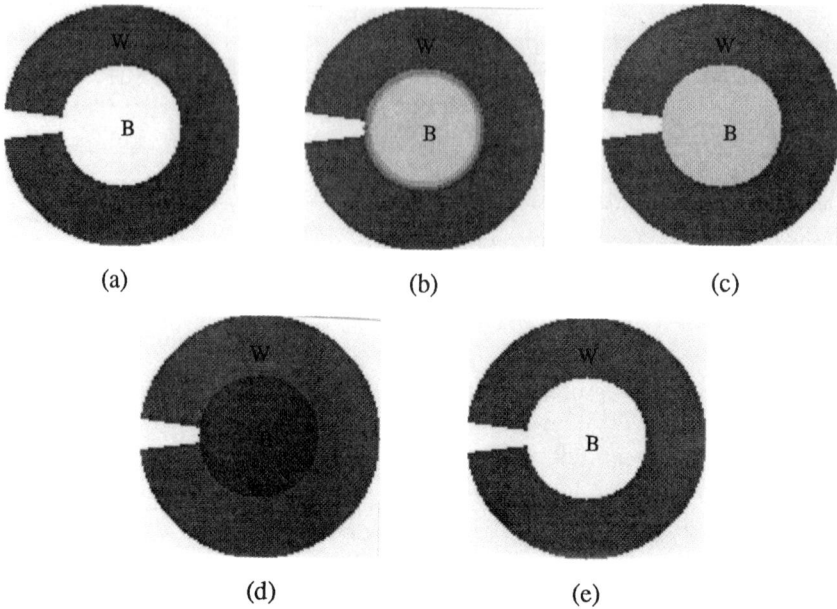

(a)   (b)   (c)

(d)   (e)

Figure 2. Experimental results of the zero thresholding for the basin surface classification

The experimental results of different zero thresholdings are shown in Fig. 2 using the KH sign images (except Fig. 2(b) using the reflection image).

Where the whole background is a plane, the wall of the basin is a conic patch and the bottom of the basin is either a plane under no perturbation or a shallow 'pit' shape under a weak perturbation; and a portion of the wall has been cut off for convenience to compare the shape type of the background with that of the basin bottom so that they are 'merged' into one in the KH sign image when the basin bottom is classified as the same type as the background.

Fig. 2(a) shows the correct classification result of the basin surface under no perturbation, where the shape type of the basin bottom and the background is classified as the same by using the thresholding (2.3). This is also the result we want to get when the basin is under a weak perturbation. Fig. 2(b) shows the reflection (cosine-shading) image of the basin surface under a weak perturbation, where the basin bottom is different from the background due to the shallow 'pit' shape resulting from the perturbation. Fig. 2(c) shows the shallow 'pit' shape can be detected by using the thresholds:

$$\epsilon_K = 8.0 \times 10^{-7}, \qquad \epsilon_H = 10^{-4}. \qquad (2.8)$$

Fig. 2(d) shows a wrong classification result of the basin surface under a weak perturbation. Due to the improper thresholding (2.4) the basin bottom has the 'phantom' shape type, whose colour shown in Fig. 2(d) is not only different from the colours of the background and the wall but also different from that of the 'pit' shape type shown in Fig. 2(c). Fig. 2(e) shows the desired classification result of the basin surface under a weak perturbation, where by using the proper thresholding (2.7) the basin bottom returns to the same shape type to the background as shown in Fig. 2(a).

It might be attractive to use thresholds as in (2.8) because they can partition surfaces and detect weak perturbations as well. Unfortunately, this is only possible for some very simple or for simulated data as in the above example. In the presence of noise, the situation is more complicated for the real data of unknown objects. A meaningful solution is to find a proper thresholding which is feasible to work and compatible with (2.7) as well. This leads to the following discussion.

3 Estimation to the Relationship Between Zero Thresholds ϵ_K and ϵ_H

The above inequality (2.7) is a brief formula, but it only considers the planar surface case and ignores any effect of scales. Recall that Besl and Jain [1] once used both

$$\epsilon_K = 0.015, \quad \epsilon_H = 0.03 \qquad (3.1)$$

and

$$\epsilon_K = 0.015, \quad \epsilon_H = 0.06 \qquad (3.2)$$

as zero thresholding in their experiment. The inequality (2.7) actually cannot give a satisfactory explanation for these thresholdings, where not only $\epsilon_K \gg \epsilon_H^2$ but also $\epsilon_K = O(\epsilon_H)$. The zero thresholding problem thus looks too flexible to handle.

Fortunately, a more certain and appropriate relationship can be found between the zero thresholds ϵ_H and ϵ_K for all eight surface types. We obtain it by simply introducing a small noise perturbation to principal curvatures then observing the behaviour of the Gaussian and the mean curvatures in this circumstance.

Suppose that the principal curvatures have small perturbations ξ_1 in C_1 and ξ_2 in C_2. They introduce perturbations E_H in H and E_K in K respectively:

$$E_H = \frac{(C_1 + \xi_1) + (C_2 + \xi_2)}{2} - \frac{C_1 + C_2}{2} = \frac{\xi_1 + \xi_2}{2} \qquad (3.3)$$

$$E_K = (C_1 + \xi_1)(C_2 + \xi_2) - C_1 C_2 = (C_1\xi_2 + C_2\xi_1) + \xi_1\xi_2. \quad (3.4)$$

Note that the noise perturbation is from the same source and what we are really concerned with here is the magnitude of the perturbation. (3.3) and (3.4) can be further simplified. Let $\xi_1 = \xi_2 = \xi$, therefore

$$E_H = \xi \qquad (3.5)$$
$$E_K = (C_1 + C_2)\xi + \xi^2. \qquad (3.6)$$

Now we can see that E_K and E_H are related via H as below:

$$E_K = 2H \cdot E_H + E_H^2. \qquad (3.7)$$

Set $|E_H| = \epsilon_H$, then

$$|E_K| \leq 2|H| \cdot \epsilon_H + \epsilon_H^2. \qquad (3.8)$$

Hence, to guard against the perturbation, the zero threshold of the Gaussian curvature K should satisfy $\epsilon_K \geq \sup |E_K|$. Otherwise, for instance, an exact $K = 0$ plus the perturbation E_K will be beyond the zero band $[-\epsilon_K, \epsilon_K]$, thus leading to an incorrect surface classification. Therefore we obtain the following inequality, namely, the *consistent zero thresholding* inequality:

$$\epsilon_K \geq 2|H| \cdot \epsilon_H + \epsilon_H^2. \qquad (3.9)$$

Comparing with (2.7), a term containing the mean curvature H appears in (3.9). This term shows the zero thresholds of K and H are related via the mean curvature H, which also brings the scale effects into the zero thresholding. By (3.9), the value of ϵ_K can be automatically derived from ϵ_H. When $H \equiv 0$, (3.9) degenerates to $\epsilon_K \geq \epsilon_H^2$, i.e. (2.6). And when $|H| = \epsilon_H$, (3.9) becomes

$$\epsilon_K \geq 3\epsilon_H^2 \qquad (3.10)$$

So (3.9) provides a more general formula than (2.7).

In the case of ϵ_K being set for the whole surface S, we may have

$$\epsilon_K = O(\epsilon_H) \quad \text{when} \quad \max_S |H| > 0.5 \qquad (3.11)$$

$$\epsilon_K > \epsilon_H \quad \text{when} \quad \max_S |H| > 0.5 \qquad (3.12)$$

and

$$\epsilon_K \gg \epsilon_H \quad \text{when} \quad \max_S |H| > 5.0 \qquad (3.13)$$

In this way, we can explain why a zero threshold imposed on the Gaussian curvature is usually close to and sometimes even larger than the zero

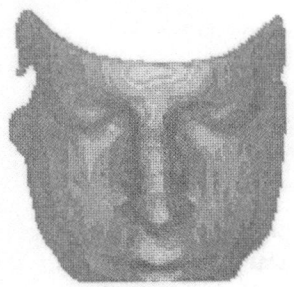 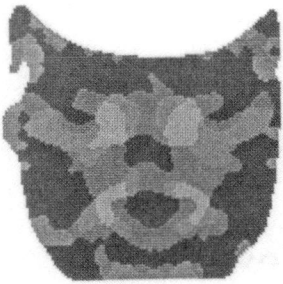

Figure 3. A human face and its KH sign image. Each grey colour denotes a surface shape shown in Table 1

threshold of the mean curvature. It also gives a better interpretation to Besl and Jain's zero thresholding mentioned at the beginning of this section.

In scale space, when the smoothing scale increases, the surface becomes more and more smooth. Note that the high curvature points on a smooth surface are always in a small number even though including some false high curvature points produced by noise. Also note that the aim of setting ϵ_K and ϵ_H is to properly treat those zero curvature points. Setting H an average value over the whole surface may be feasible in the scale space processing. The following formula is thus suggested:

$$\epsilon_K = \epsilon_H^2 + 2 \underset{S}{Average} |H| \cdot \epsilon_H. \tag{3.14}$$

The segmentation result of a human face using this formula in scale space processing is shown in Fig. 3, and more examples can be seen in [4].

4 Summary

This paper formulates the relationship between the zero thresholds of the Gaussian curvature K and the mean curvature H in multiple scale processing of 3-dimensional data in the presence of noise. It makes explicit that zero thresholds of the Gaussian and the mean curvature are not irrelevant but related via the mean curvature, which also brings the scale effects into zero thresholding. Once a proper zero threshold ϵ_H has been chosen, a consistent zero threshold ϵ_K can be given automatically by this formula. This effectively prevents the "phantom case" from occurring. Hence, the consistent zero thresholding inequality is particularly convenient for scale space processing since any zero threshold empirically imposed on K and H at a certain scale is unlikely to be available at other scales.

Acknowledgements

This work was done in the Department of Artificial Intelligence, University of Edinburgh. Thanks are given to Drs. R.B. Fisher and J. Hallam for supervision, M. Cameron-Jones for comments.

References

1. Besl, P.J. and Jain, R.C. (1986). Invariant surface characteristics for three dimensional object recognition in range images. *Computer Vision, Graphics, Image Processing, 33*, 1 (January), 33-80.

2. Cai, L.D. (1987). Diffusion smoothing on dense range data. DAI WP-**200**, Department of A.I., University of Edinburgh.

3. Cai, L.D. (1987). Some note on repeated averaging smoothing. DAI RP-**337**, Department of A.I., University of Edinburgh. Also in *Proc. of the BPRA 4th International Conference on Pattern Recognition*, Cambridge, U.K., 1988, *Lecture Notes in Computer Science*, J.Kittler (Ed.), Vol. 301, Springer-Verlag, 596-605.

4. Cai, L.D. (1988). A 'small leakage' model for diffusion smoothing of image data. DAI RP-**418** (resubmitted), Department of A.I., University of Edinburgh. Also in *Proc. of the 11th International Joint Conference on Artificial Intelligence*, Detroit, Michigan, USA, August, 1989.

5. Cai, L.D. (1988). A diffusion smoothing approach to sculptured surfaces. DAI RP-**406**, Department Of A.I., University of Edinburgh. Also in *Proc. of the 3rd IMA Conference on the Mathematics of Surfaces*, Oxford, U.K., September, 1988, 267-285.

6. De Carmo, M.P. (1976). *Differential Geometry of Curves and Surfaces*, Prentice Hall, Englewood Cliffs.

7. Ponce, J. and Brady, M. (1987). Toward a surface primal sketch. In *Three-Dimensional Machine Vision*, T.Kanade (Ed.), Kluwer Academic Publishers, 225-227.

8. Witkin, A.P. (1983). Scale-space filtering. In *Proc. of the 8th International Joint Conference on Artificial Intelligence*, 1019-1022.

9. Yang, H.S. (1988). Range image segmentation and classification via split-and-merge based on surface curvature. In *Proc. of BPRA 4th Int. Conf. on Pattern Recognition*, Cambridge, 1988, *Lecture Notes in Computer Science*, J. Kittler (Ed.), Vol. 301, Springer-Verlag, 58-67.

Optimized Parameterizations for Cubic and Bicubic Splines

T.A. Foley and R. Ekpete

Arizona State University

1 Introduction

The parametric cubic spline interpolant to points p_i is greatly influenced by the knot partition of the piecewise cubic in the parameter domain. The commonly used uniform partition and the chord length partition, see [1], often yield poorly shaped cubic spline interpolants. There have been recent advances for selecting effective partitions for curves in [7], [4], [6] and [10]. However, these methods do not easily generalize to the more difficult problem of knot selection for the bicubic spline interpolant $S(u,v)$ to surface data p_{ij}. Farin [1] and Faux and Pratt [2] note that the difficulty with this problem is that a single knot partition must be effective for all isoparametric curves in the u-direction (similarly for the v-direction).

We present a geometric optimization approach that applies to both curves and bicubic spline surfaces. The distance from the Bézier points representing the interpolating spline to the net formed by the given data is minimized over all knot sequences. For curves, the given data net is simply the polygon $p_0, \cdots, p_m$, while for surfaces, the given data net is the collection of piecewise bilinear patches that interpolate the data p_{ij}. Although this method is computationally expensive, effective C^2 curves and surfaces are obtained. Depending on the measure of distance used, we are able to obtain both visually smooth and very tight C^2 interpolating curves and surfaces.

2 Knot Selection for Cubic Spline Curves

Given $m+1$ points $p_0, \cdots, p_m$ and real parameter values $u_0 < u_1 < \cdots < u_m$, there exists a unique C^2 cubic spline $S(u)$ that satisfies $S(u_i) = p_i$, for $i = 0, \cdots, m$, and two user defined end conditions. The $m+1$ points p_i are fixed and we address that problem of how to choose the $m+1$ values u_i in an effective manner.

With $\triangle u_i = u_{i+1} - u_i$, the chord length parameterization is defined by $u_0 = 0$ and $\triangle u_i = \|\, p_{i+1} - p_i \,\|$, for $i = 0, \cdots, m - 1$. The uniform partition is simply defined by $\triangle u_i = 1$ (or some other constant). The centripetal method in [6] uses the partition spacing $\triangle u_i = \sqrt{\|\, p_{i+1} - p_i \,\|}$. A more complicated formula for $\triangle u_i$ is used in the affine invariant angle method in [4]. This method involves the affine invariant metric discussed in [9]. In addition to "distance" between points, this method involves the angles and shape of the control polygon.

It is straightforward to show that if the u_i are scaled and translated to $t_i = au_i + b$, then the same interpolating spline curve is obtained, assuming that the end conditions are scaled appropriately. That is, if $\bar{S}(t)$ is the cubic spline that satisfies $\bar{S}(t_i) = p_i$, then $S(u) = \bar{S}(t)$ when $t = au + b$. Because of this, it is often the practice to normalize the u_i partition so that $u_0 = 0$ and $u_m = 1$. With $\triangle u_i = u_{i+1} - u_i$, this yields the constraints that $\triangle u_i > 0$ for $i = 0, \cdots, m - 1$, and

$$\sum_{i=0}^{m-1} \triangle u_i = 1.$$

For the optimization techniques presented here, we find it more convenient to normalize our knot partition so that $u_0 = 0$ and $u_1 = 1$. This reduces the unknowns to $m - 1$ values $\triangle u_i > 0$, for $i = 1, \cdots, m - 1$.

The interpolating cubic spline curve $S(u)$ can be represented in many forms, such as the Hermite and B-spline forms. Farin [1] gives a thorough description of these representations, together with the corresponding tridiagonal system of linear equations that needs to be solved. (Slight adjustments are necessary when periodic end conditions are used.)

We use the Bernstein-Bézier representation for $S(u)$, which is well documented in [1] and in [2]. The cubic segment for $u_i \leq u \leq u_{i+1}$ is represented by

$$S(u) = b_{3i} B_0(t) + b_{3i+1} B_1(t) + b_{3i+2} B_2(t) + b_{3i+3} B_3(t),$$

where $t = (u - u_i)/\triangle u_i$ and $B_i(t) = \begin{pmatrix} 3 \\ i \end{pmatrix} t^i (1 - t)^{3-i}$. Since $S(u_i) = p_i$, we have $b_{3i} = p_i$, for $i = 0, \cdots, m$. The remaining "inner" Bézier points b_{3i+1} and b_{3i+2} can be easily computed from the Hermite and B-spline representations. In the Hermite representation, $S'(u_i)$ is known, and the inner Bézier points are computed by

$$b_{3i+1} = p_i + \frac{3}{\triangle u_i} S'(u_i) \quad \text{and} \quad b_{3i-1} = p_i - \frac{3}{\triangle u_{i-1}} S'(u_i).$$

Details for converting the B-spline representation to Bézier form are given in [1]. Conversion of higher degree polynomials can be affected by round-off error, but for cubic polynomials, we did not observe any problem. However, we used double precision as a safety precaution.

Two objective functions $F_1(\triangle U)$ and $F_2(\triangle U)$ are now defined that are to be minimized over the $m - 1$ variables $\triangle U = (\triangle u_1, \cdots, \triangle u_{m-1})$. Let $b_0, \cdots, b_{3m}$ be the Bézier points for the cubic spline interpolant $S(u)$ determined by the knot partition $u_0 = 0, u_1 = 1$ and $u_{i+1} = u_i + \triangle u_i$ for $i = 1, \cdots, m - 1$. For $M = 1, 2$ define

$$F_M(\triangle U) = \sum_{i=0}^{m-1} \sum_{k=1}^{2} \left[D_M(b_{3i+k}) \right]^2,$$

where b_{3i+k} are the inner Bézier points of $S(u)$ and the distance functions D_M are defined as follows. For $0 \le t \le 1$, let $L_i(t) = (1 - t)p_i + tp_{i+1}$ be the line segment from p_i to p_{i+1}. For $i = 0, \cdots, m - 1$ and $k = 1, 2$, define

$$D_1(b_{3i+k}) = \frac{min}{0 \le t \le 1} \| b_{3i+k} - L_i(t) \|$$

and

$$D_2(b_{3i+k}) = \| b_{3i+k} - L_i(k/3) \| .$$

D_1 is the distance from an inner Bézier point to its corresponding line segment $p_i p_{i+1}$, while D_2 is the distance to the points $1/3$ and $2/3$ on that line segment. If the p_i are colinear, then $F_2(\triangle U) = 0$ when $\triangle u_i = \| p_{i+1} - p_i \| / \| p_1 - p_0 \|$. The interpolant $S(u)$ will be linear in this case, hence linear precision is obtained when the objective function $F_2(\triangle U)$ is minimized.

The constraints $\triangle u_i > 0$ form an open set, thus $F_M(\triangle U)$ may not have a minimum. In practice, we constrain the $\triangle u_i$ to satisfy $0 < \triangle_{min} \le \triangle u_i \le \triangle_{max}$. The bounds could also depend on i and be multiples of the normalized chord length $d_i = \| p_{i+1} - p_i \| / \| p_1 - p_0 \|$. For example, $\triangle u_i$ could be constrained by $d_i 10^{-4} \le \triangle u_i \le d_i 10^4$. Whether or not there is a unique $\triangle U$ that minimizes F_M is an open problem. Even if there is a unique solution, it may be difficult to find numerically because there appears to be many local minima to this nonlinear optimization problem.

We used iterative optimization software in the IMSL package which generally converges to a local minimum dependent on the initial values of $\triangle u_1, \cdots, \triangle u_{m-1}$. Chord length, uniform, centripetal and the affine invariant angle knot partitions were used as the initial values for $\triangle U$. These partitions were scaled by $1/\triangle u_0$ so that $u_0 = 0$ and $u_1 = 1$. Although different initial values sometimes yielded different parameterizations, the resulting cubic spline curves were visually indistinguishable in general.

Our first example consists of eight points (x_i, y_i) where $x_i = (.2, .4, .6, .8, .8, .6, .4, .2)$ and the y_i are .45 and .55. Fig. 1a is the C^2 cubic spline interpolant to this data set using the chord length knot spacing. The interpolation points are denoted by an x and the inner Bézier points are denoted by 0. The tight curve in Fig. 1b uses the partition which

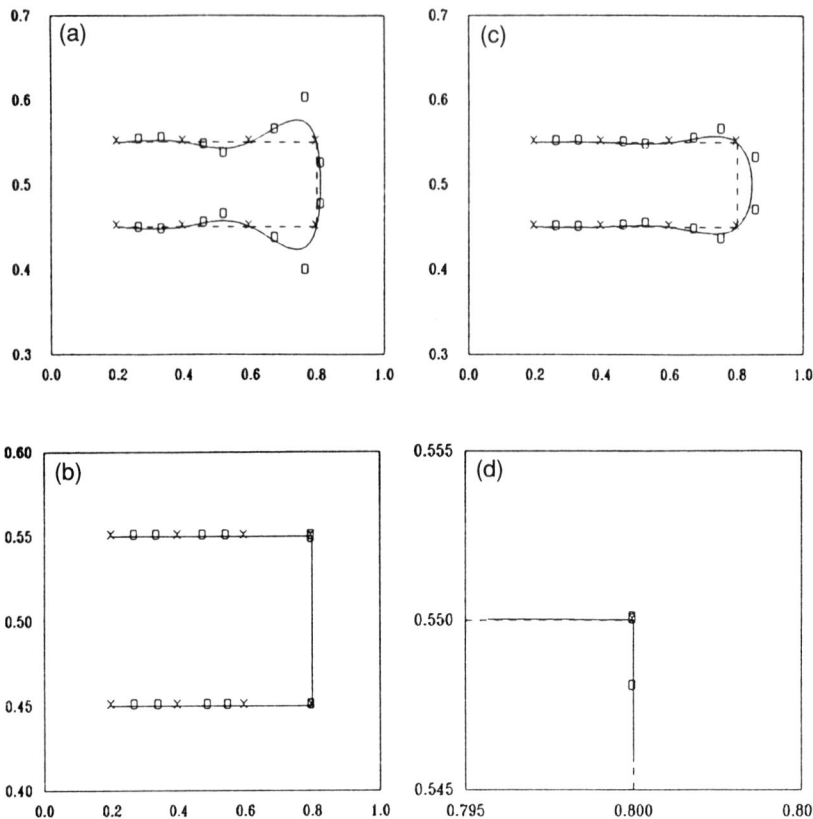

Figure 1. Cubic spline interpolants using the chord length partition in a, minimum F_1 partition in b, minimum F_2 partition in c, and an enlargement of the upper right portion of part b is in d

minimizes $F_1(\Delta U)$ and the smoother curve in Fig. 1c uses the partition which minimizes $F_2(\Delta U)$. Fig. 1d is an enlargement of the upper right portion of the C^2 cubic spline in Fig. 1b which minimizes $F_1(\Delta U)$. Our second example in Fig. 2 uses data from [5]. Fig. 2a uses the chord length partition, 2b uses the uniform partition, 2c uses the partition that minimizes F_1 and 2d uses the partition that minimizes F_2. We applied these optimization techniques to many other data sets and the results were consistent with those shown in Figs. 1 and 2. In general, the partition that minimizes F_1 yielded tight curves, while the partition that minimized F_2 generally yielded smooth well behaved curves.

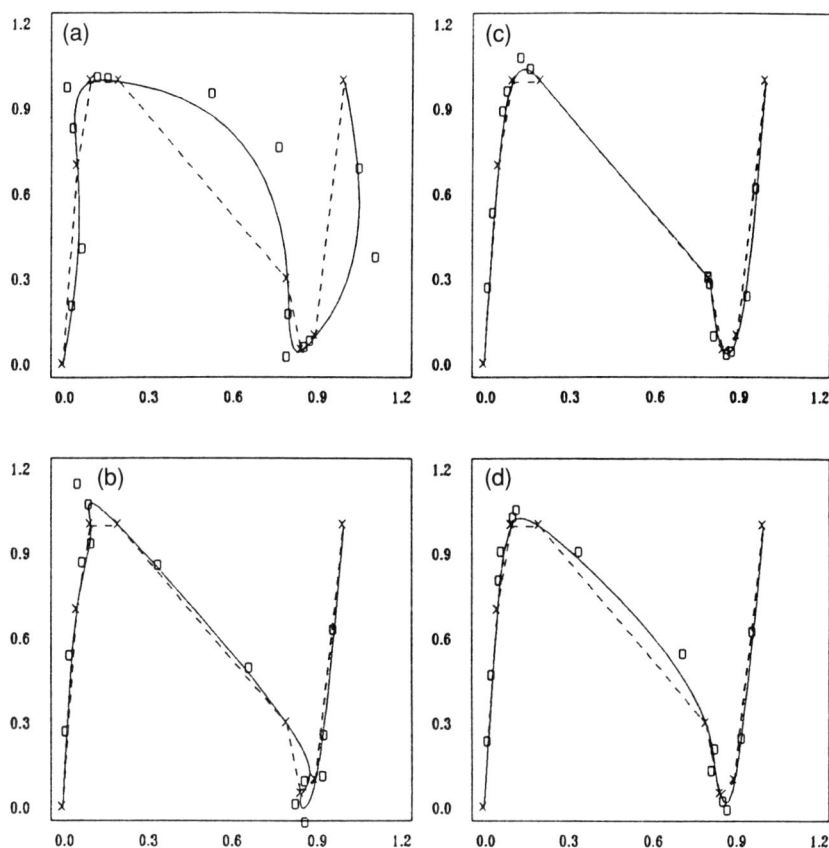

Figure 2. Cubic spline interpolants using the chord length partition in a, uniform partition in b, minimum F_1 partition in c, and the minimum F_2 partition in d

3 Knot Selection for Bicubic Spline Surfaces

One of the primary motivations for developing the knot partition for curves in the previous section was to have a method that can be extended to tensor product bicubic spline surfaces. The method presented here will minimize the distance from the Bézier points of the bicubic spline interpolant to the underlying bilinear patch determined by the data. Given points p_{ij}, for $i = 0, \cdots, m$ and $j = 0, \cdots, n$, and the two knot partitions $u_0, u_1, \cdots, u_m$ and $v_0, v_1 \cdots, v_n$, there exists a unique C^2 bicubic spline $S(u, v)$ that satisfies $S(u_i, v_j) = p_{ij}$ and appropriate end conditions.

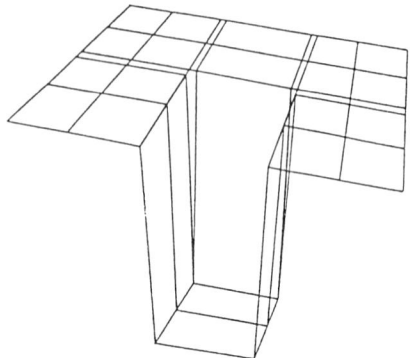

Figure 3. Data used for Fig. 4

The problem we address is how to effectively choose the knot partitions u_i and v_j. The difficulty with this problem is that the single set of parameter values $\{u_0, \cdots, u_m\}$ must be effective for all isoparametric curves in the u-direction. The same difficulty arises in the v-direction. For example, the distances between points in the first row of data in Fig. 3 differs significantly from that in the last row, yet the same knot partition u_i must be used in both cases. Since each curve $S(u, v_j)$ is a cubic spline interpolant in u, it is unlikely that the bicubic spline interpolant will be effective on different rows.

A simple solution is to simply let $\triangle u_i = 1$ and $\triangle v_j = 1$ for all i, j. This will be ineffective whenever the uniform partition has problems with any row or column of data. A common approach is to compute effective knot spacings for each row and column of data and then perform some averaging. For example, the average chord length partition is defined by

$$\triangle u_i = \frac{1}{n+1} \sum_{j=0}^{n} \| p_{i+1,j} - p_{i,j} \|$$

for $i = 0, \cdots m - 1$, and

$$\triangle v_j = \frac{1}{m+1} \sum_{i=0}^{m} \| p_{i,j+1} - p_{i,j} \|$$

for $j = 0, \cdots, n - 1$. The average centripetal partition would be similar.

In order to set up an optimization problem, we first eliminate the redundant degrees of freedom in the $m + n + 2$ values of u_i and v_j. Since the method is a tensor product, if each partition is scaled and translated independently, then the same surface is generated, assuming that derivative end conditions are scaled appropriately. That is, the partitions $s_i = au_i + b$ and $t_j = cv_j + d$ yield the same surface. This is easily seen when the bicubic spline is represented in the B-spline form

$$S(u, v) = \sum \sum d_{ij} N_i(u) \bar{N}_j(v).$$

Therefore, we can assume that $u_0 = 0, u_1 = 1, v_0 = 0, v_1 = 1$, and the remaining $m + n - 2$ unknowns are defined by $u_{i+1} = u_i + \triangle u_i$, for $i = 1, \cdots, m - 1$ and $v_{j+1} = v_j + \triangle u_i$, for $j = 1, \cdots, n - 1$. The unknown knot spacings must satisfy $\triangle u_i > 0$ and $\triangle v_j > 0$.

Given values for u_i and v_j, we represent the bicubic spline interpolant in Bézier form. For $u_i \leq u \leq u_{i+1}$ and $v_j \leq v \leq v_{j+1}$,

$$S(u, v) = \sum_{k=0}^{3} \sum_{l=0}^{3} b_{3i+k, 3j+l} B_i(s) B_j(t)$$

where $s = (u - u_i)/\triangle u_i$ and $t = (v - v_j)/\triangle v_j$. The Bézier points $b_{3i, 3j} = p_{ij}$ and the other Bézier points can be easily computed if the Hermite or B-spline representation of $S(u, v)$ is computed. Details of these conversions and software for efficiently computing the B-spline representation are described in [1].

Similar to the curve case, we define two functions $F_1(\triangle U, \triangle V)$ and $F_2(\triangle U, \triangle V)$ that are to be minimized over the $m + n - 2$ variables $\triangle U = (\triangle u_1, \cdots, \triangle u_{m-1})$ and $\triangle V = (\triangle v_1, \cdots, \triangle v_{n-1})$. Let b_{ij}, $i = 0, \cdots, 3m$ and $j = 0, \cdots, 3n$, be the Bézier points of the bicubic spline interpolant $S(u, v)$ determined by the partitions $u_0 = 0, u_1 = 1, u_{i+1} = u_i + \triangle u_i$, $v_0 = 0, v_1 = 1$ and $v_{j+1} = v_j + \triangle v_j$. For $M = 1, 2$, define

$$
\begin{aligned}
F_M(\triangle U, \triangle V) &= \sum_{i=0}^{m-1} \sum_{j=0}^{n-1} \sum_{k=1}^{2} \sum_{l=1}^{2} \left[D_M(b_{3i+k, 3j+l}) \right]^2 \\
&+ \sum_{i=0}^{m-1} \sum_{j=0}^{n} \sum_{k=1}^{2} \left[D_M(b_{3i+k, 3j}) \right]^2 \\
&+ \sum_{i=0}^{m} \sum_{j=0}^{n-1} \sum_{l=1}^{2} \left[D_M(b_{3i, 3j+l}) \right]^2.
\end{aligned}
$$

For $0 \leq s, t \leq 1$, let $BL_{ij}(s, t)$ be the bilinear interpolant to the four points p_{ij}, $p_{i+1, j}$ $p_{i, j+1}$ and $p_{i+1, j+1}$. For $0 \leq i \leq m - 1$, $0 \leq j \leq n - 1$ and $0 \leq k, l \leq 3$, define

$$D_2(b_{3i+k, 3j+l}) = \| b_{3i+k, 3j+l} - BL_{ij}(k/3, l/3) \|.$$

For $1 \leq k, l \leq 2$, define

$$D_1(b_{3i+k,3j+l}) = \begin{array}{c} min \\ 0 \leq s,t \leq 1 \end{array} \| b_{3i+k,3j+l} - BL_{ij}(s,t) \| .$$

To insure consistency over neighboring patches, we define

$$D_1(b_{3i+k,3j}) = \begin{array}{c} min \\ 0 \leq s \leq 1 \end{array} \| b_{3i+k,3j} - (p_{i,j}(1-s) + p_{i+1,j}s) \|$$

and

$$D_1(b_{3i,3j+l}) = \begin{array}{c} min \\ 0 \leq t \leq 1 \end{array} \| b_{3i,3j+l} - (p_{i,j}(1-t) + p_{i,j+1}t) \| .$$

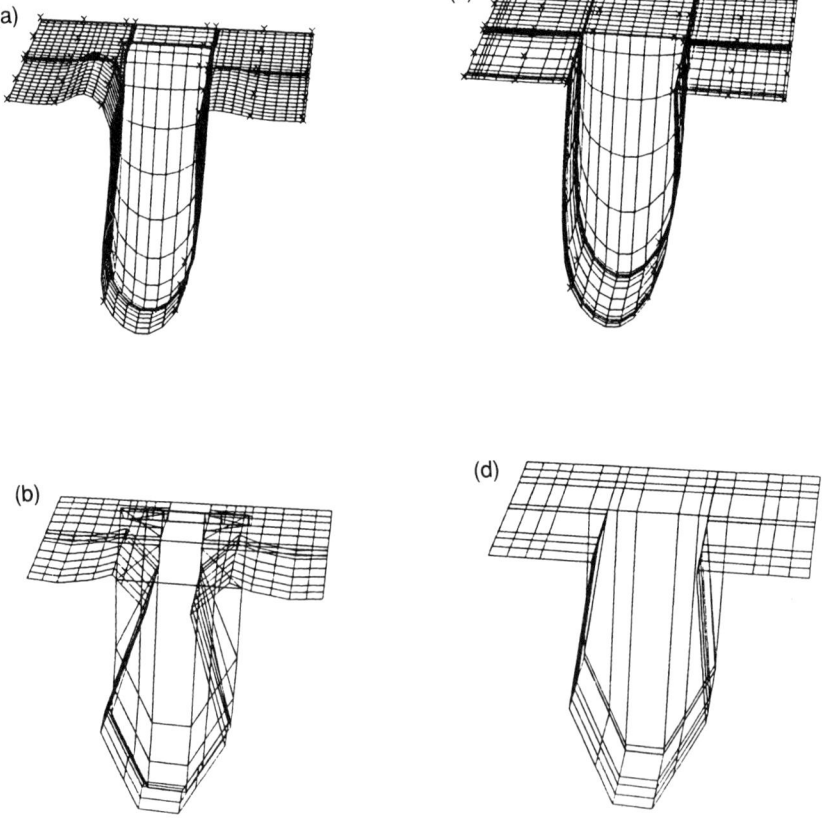

Figure 4. Bicubic spline interpolant to the data shown in Fig. 3. Part a uses the average chord partition and its Bézier net is shown in b. Part c uses the minimum F_1 partition and its Bézier net is shown in d

We constrain the knot spacing to satisfy $0 < \triangle_{min} \leq \triangle u_i, \triangle v_j \leq \triangle_{max}$, where $\triangle_{min}$ and $\triangle_{max}$ are input by the user. For the bicubic spline examples that follow, we use natural end conditions and the average chord length knot sequences are used for the initial guess of $\triangle U$ and $\triangle V$. We normalize the initial partitions by factors of $1/\triangle u_0$ and $1/\triangle v_0$ so that $u_0 = 0$, $u_1 = 1$, $v_0 = 0$ and $v_1 = 1$.

Fig. 4a is the interpolant to the data in Fig. 3 using the average chord length partition and the Bézier net of this interpolant is shown in Fig. 4b. Fig. 4c uses the partition that minimizes $F_1(\triangle U, \triangle V)$ and the Bézier net for this interpolant is given in Fig. 4d. The surface which uses the partition that minimizes $F_2(\triangle U, \triangle V)$ is not shown because it is almost identical to the average chord length method. The uniform partition also yields a surface very similar to the average chord length in Fig. 4a.

A more subtle example is given in Fig. 5, which consists of 12 points from [1]. Fig. 5a uses the average chord length partition, Fig. 5b is the Bézier net for part a, Fig. 5c uses the partition that minimizes $F_2(\triangle U, \triangle V)$ and its Bézier net is given in Fig. 5d. The partition that minimizes F_1 is not shown because it is almost identical to the one that minimizes F_2.

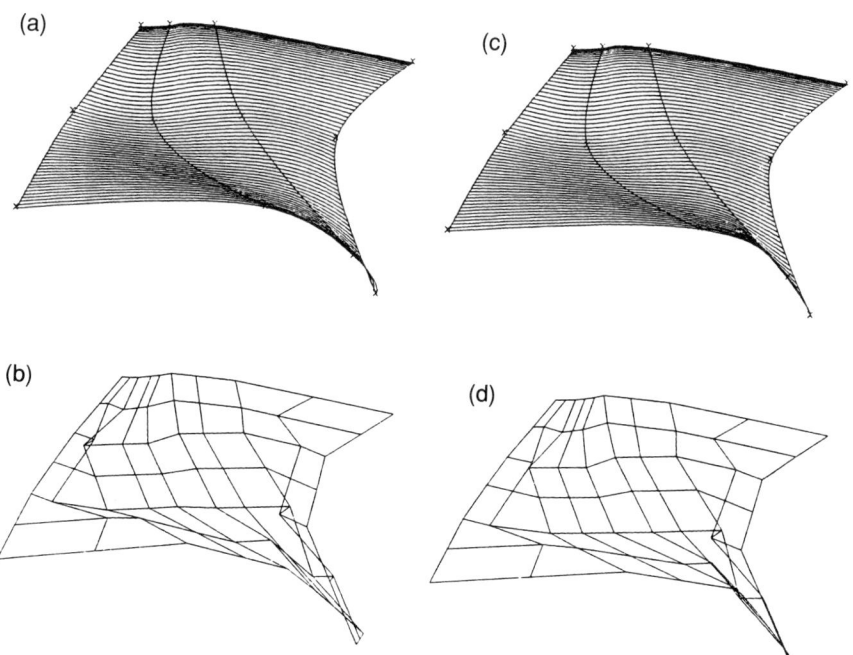

Figure 5. Bicubic spline interpolants using average chord partition in a, and the minimum F_2 partition in c. The Bézier nets for a and c are in b and d

The surfaces in Fig. 5a and c appear very similar except in the lower right corner. However, the Bézier nets in b and d show that the optimized partition has a smoother Bézier net in the lower right, the front center and the middle left portions.

4   Remarks

The optimization methods were effective on all of the curve cases and on many of the surface cases that we tested. For extreme distributions of surface data, our optimization techniques (together with uniform, average chord and average centripetal partitions) yielded poorly shaped surfaces. The major problem in these extreme cases is not with the partition, but with using the tensor product bicubic spline method. For extreme distributions, it is generally advisable to use Coons patches with geometric continuity constraints (see [1]).

Instead of minimizing the sum of squares in F_M, we also used l_1 and l_∞ vector norms on the distances from the Bézier points to the given data net. We also considered weighted versions of F_M which generally "tightened" the interpolating spline on segments with larger weights.

Since there is no guarantee that any knot sequence will yield a shape that the designer has in mind, splines with tension controls, such as [8] and [3], can be interactively used to obtain the desired shape. When using these tension splines, the default with no tension is the C^2 cubic or bicubic spline, thus a reasonably shaped initial interpolant can save the designer time. A final observation is that this technique generalizes easily to higher degree polynomial splines and to tensor product interpolants of several variables.

Acknowledgements

The first author was supported by the NSF and AFOSR grant DMS-9116930.

References

1. Farin, G. (1990). *Curves and Surfaces for Computer Aided Geometric Design*, Academic Press.

2. Faux, I.D. and Pratt, M.J. (1979). *Computational Geometry for Design and Manufacture*, Ellis Horwood.

3. Foley, T.A. and Ely, H.S. (1989). Surface interpolation with tension controls using cardinal bases. *Computer Aided Geometric Design, 6,* 97–109.

4. Foley, T.A. and Nielson, G.M. (1989). Knot selection for parametric spline interpolation. In *Mathematical Methods in Computer Aided Geometric Design*, T. Lyche and L. L. Schumaker, editors, Academic Press, 261–271.

5. Irvine, L.D., Marin, S.P. and Smith, P.W. (1985). Constrained interpolation and smoothing. *Constructive Approximation*, *2*, 129–151.

6. Lee, E. (1989). Choosing nodes in parametric curve interpolation. *Computer Aided Design*, *21*, 363–370.

7. Marin, S.P. (1984). An approach to data parameterization in parametric spline interpolation problems. *J. Approximation Theory*, *41*, 64–86.

8. Nielson, G.M. (1986). Rectangular *v*-splines. *IEEE Computer Graphics & Applications*, *6*, 35–40.

9. Nielson, G.M. and Foley, T.A. (1989). A survey of applications of an affine invariant norm. In *Mathematical Methods in Computer Aided Geometric Design*, T. Lyche and L. L. Schumaker, editors, Academic Press, 445–467.

10. Rademacher, C. and Scherer, K. (1990). Algorithms for computing best parametric cubic interpolation. In *Algorithms for Approximation II*, J.C. Mason and M. G. Cox, editors, Chapman and Hall, 193–207.

Twist Estimation for Smooth Surface Design

Hans Hagen

Universität Kaiserslautern

Abstract The methods of Computer Aided Design have arisen from the need of efficient computer representation of practical curves and surfaces used in engineering design. The generation of smooth surfaces from a set of three-dimensional data points is a key problem in this field. A new twist estimation method based upon a calculus of variation approach and a stiffness degree concept is presented. This technique generates smooth surfaces from a smooth network of curves.

Keywords: geometric modelling, design of smooth surfaces

1 Introduction

The purpose of this paper is to present a new twist estimation for smooth surface design based upon a calculus of variation approach and a stiffness degree concept.

Surfaces designed in a computer graphics environment have many applications, including the design of cars, airplanes, shipbodies and modeling robots.

The choice of the surface form depends upon the application. The two main techniques are the Bézier- and B-Spline-methods and the Gordon-Coons-type surfaces. The fundamental idea of the Bézier- and B-Spline-methods is to evaluate and manipulate the curves and surfaces by a (small) number of control points.

The boundary control points of these patch-types can be supplied by a smooth network of curves. In this paper we present a method of how to specify the inner control points of each patch in an efficient way to generate smooth surfaces.

Gordons-Coons-type surfaces can be written as a boolean sum of projectors, which are themselves interpolants to lower dimensional information. If the projectors involve derivative information, then the boolean sum involves mixed partial derivatives, the so called twist vectors. These twist vectors, which are responsible for the smoothness of the surface, can be supplied by our technique described in this paper.

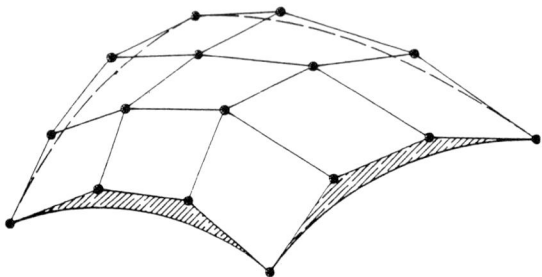

Figure 1. Bézier-patch

2   Generalized Coons Patches

This surface-patch technique was introduced by S. Coons [7] and W. Gordon pointed out [10], that such a patch can be written as a boolean sum of projectors (for more details see [2]).

Using this technique, there were two major problems: appropriate input for the twist vectors and twist incompatibilities. To remove the incompatibility of the cross partial derivatives we can use either Gregory's square [1] and [4], or Nielson's convex combination extension of the boolean sum scheme [14]:

$$X(u, w) := P_1 + P_2 - \alpha P_1 P_2 - \beta P_2 P_1 \qquad (2.1)$$

with the cubic or quintic Hermite-projector P_1 and P_2 and appropriate convex combination functions $\alpha(u, w)$ and $\beta(U, w)$. Since we are interested in curvature-input, but not in partial derivatives of curvature functions, we set $X_{uuw} = 0; X_{uww} = 0; X_{uuww} = 0$. If we use cubic or quintic Hermite interpolants as representation of the boundary curves, we get the following matrix representation of this patch:

$$\begin{bmatrix} H_0(u) \\ H_1(u) \\ \bar{H}_0(u) \\ \bar{H}_1(u) \\ \bar{\bar{H}}_0(u) \\ \bar{\bar{H}}_1(u) \end{bmatrix}^{\mathrm{T}}$$

$$\times \begin{bmatrix} X(0,0) & X(0,1) & X_w(0,0) & X_w(0,1) & X_{ww}(0,0) & X_{ww}(0,1) \\ X(1,0) & X(1,1) & X_w(1,0) & X_w(1,1) & X_{ww}(1,0) & X_{ww}(1,1) \\ X_u(0,0) & X_u(0,1) & \tilde{X}_{uw}(0,0) & \tilde{X}_{uw}(0,1) & 0 & 0 \\ X_u(1,0) & X_u(1,1) & \tilde{X}_{uw}(1,0) & \tilde{X}_{uw}(1,1) & 0 & 0 \\ X_{uu}(0,0) & X_{uu}(0,1) & 0 & 0 & 0 & 0 \\ X_{uu}(1,0) & X_{uu}(1,1) & 0 & 0 & 0 & 0 \end{bmatrix}$$

$$\times \begin{bmatrix} H_0(w) \\ H_1(w) \\ \bar{H}_0(w) \\ \bar{H}_1(w) \\ \bar{\bar{H}}_0(w) \\ \bar{\bar{H}}_1(w) \end{bmatrix} . \tag{2.2}$$

In this scheme there are no twist incompatibilities. In chapter 4 we present a solution of the twist input problem.

3 Bézier- and B-Spline Surfaces

The curves and surfaces now known as Bézier curves and surfaces were independently developed by P. de Casteljau and by P. Bézier. The underlying mathematical theory, based on the concept of Bernstein polynomials, was first introduced by R. Forrest [9]. The fundamental idea of this approach is to evaluate and manipulate the curves and surfaces by a (small) number of "control points". Since we build up Bézier-surfaces from curve-networks, we first consider Bézier-curves.

A Bézier-curve is a segmented curve. The segments $X_1(u); 1 = 0, \ldots, k$ of a Bézier-curve of degree m over the parameter interval $u_1 \leq u \leq u_{1+1}$ are:

$$X_l(u) := \sum_{i=0}^{m} b_{l \cdot m + i} \cdot B_i^m \left(\frac{u - u_l}{u_{l+1} - u_l} \right) . \tag{3.1}$$

The Bernstein polynomials $B_i^m(t) := \begin{pmatrix} m \\ i \end{pmatrix} (1-t)^{m-i} t^i; \ 0 \leq t \leq 1$ are used as blending functions.

Bernstein polynomials are special degenerated B-splines [6]. If we use B-splines as blending functions, we can generalize the whole concept to so called B-Spline-curves and surfaces [11].

B-Spline curves are similar to Bezier-curves in that a set of blending functions combines the effect of $n + 1$ control points:

$$Y(u) := \sum_{j=0}^{n} d_j N_{j,N}(u). \qquad (3.2)$$

The most important difference is the local support property of the B-Spline blending functions $N_{j,k}(u)$. Both curvetypes have the convex hull and variation diminishing property (for more details see [6]).

A Bézier-surface is a segmented surface. The segments $X_{pq}(u, w); p = 0, \ldots, k; q = 0, \ldots, r$ of a Bézier surface of degree m, n over the rectangular parameter domain $u_p \leq u \leq u_{p+1}; w_q \leq w \leq w_{q+1}$ are:

$$X_{pq}(u, w) := \sum_{i=0}^{m} \sum_{j=0}^{n} b_{p \cdot m+i/q \cdot n+j} \cdot B_i^m \left(\frac{u - u_p}{u_{p+1} - u_p} \right) B_j^n \left(\frac{w - w_q}{w_{q+1} - w_q} \right). \qquad (3.3)$$

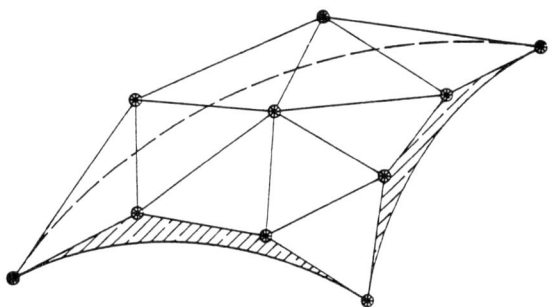

Figure 2. Triangular Bézier-patch

Instead of a control polygon a Bézier-surface(-segment) has a control polyhedron. The definition of a B-Spline-surface over a rectangular parameter domain follows directly the same pattern:

$$Y(u, w) := \sum_{i=0}^{m} \sum_{j=0}^{n} d_{ij} N_{i,M}(u) N_{j,N}(w). \qquad (3.4)$$

A triangular Bézier patch is defined by

$$X(u, w) := \sum_{I} b_{i,j,k} B_{i,j,k}^n (r(u, w), s(u, w), t(u, w)) \qquad (3.5)$$

where r, s, t are local barycentric coordinates of the triangular parameter domain and "I" denotes summation over all $i, j, k \geq 0, i + j + k = n$. The $B_{i,j,k}^n$ are generalized Bernstein polynomials of degree n given by:

$$B_{i,j,k}^n(r, s, t) := \frac{n!}{i!j!k!} r^i s^j t^k.$$

These Bernstein polynomials have properties very much like the univariate ones. Therefore they are appropriate blending functions [8].

The boundary control points of these patch-types can be supplied by a smooth network of curves. In chapter 4 we present a method of how to specify the "inner control points" of each patch in an efficient way to generate smooth surfaces.

4   Variational Twist Estimation

The functional $\int_S (k_1^2 + k_2^2) dS$ (k_1 and k_2 are the principal curvatures of surface S) is a standard fairness criterion for surfaces in engineering [15], because it is equivalent to the strain energy of flexure and torsion in a thin rectangular elastic plate of small deflection.

Hagen and Schulze used this functional for a variation formulation and solution of the surface fairing problem [12] under the assumption of orthogonal parameter lines. Just recently Farin and Hagen presented a general solution without any regularity constraints.

The strain energy of flexure and torsion in a thin rectangular elastic plate of small deflection is equivalent to the functional

$$G := \int_s (k_1^2 + k_2^2) dS. \tag{4.1}$$

Calculating the principal curvature of the surface we get

$$
\begin{aligned}
G &= \int_A \frac{(g_{11}h_{22} - 2g_{12}h_{12} + g_{22}h_{11})^2 - 2gh}{g^2} \sqrt{g} \, du \, dw \\
&= \int_A (h_{12}^2(2g + 4g_{12}^2) - h_{12}(4g_{12}(g_{11}h_{22} + g_{22}h_{11}) + (g_{11}h_{22} + g_{22}h_{11})^2 \\
&\quad - 2gh_{11}h_{22})g^{-3/2} du \, dw
\end{aligned}
$$

$X : A \longrightarrow E^3$ is a parametrisation of the surface S and g_{11}, g_{22} and g_{12} are the components of the first fundamental form; h_{11}, h_{22} and h_{12} are the components of the second fundamental form and $h := h_{11}h_{22} - h_{12}^2$ and $g : g_{11}g_{22} - g_{12}^2$.

We have a functional of the form $G := \int \int F(u, w, h_{12}(u, w)) du \, dw$, the Euler equation:

$$\frac{\partial F}{\partial h_{12}} = 2h_{12}(2g + 4g_{12}^2) - 4g_{12}(g_{11}h_{22} + g_{22}h_{11}) = 0 \qquad (4.2)$$

gives a necessary condition for the energy minimum

$$h_{12} = \frac{g_{12}(g_{11}h_{22} + g_{22}h_{11})}{g + 2g_{12}^2}. \qquad (4.3)$$

The mean curvature $H := \frac{1}{2g}(h_{11}g_{22} - 2h_{12}g_{12} + h_{22}g_{11})$ measures the deviation of a surface from a minimal surface ($H \equiv 0$). Minimal surfaces correspond to the surface formed by a soap bubble between the boundary curves.

(4.3) is equivalent to: $h_{12} = 2g_{12} \cdot H$

Theorem: A surface $X : A \longrightarrow E^3$ with

$$h_{12} = \frac{g_{12}(g_{11}h_{22} + g_{22}h_{11})}{g + 2g_{12}^2} = 2g_{12} \cdot H$$

is smooth in the sense of $\int_S (k_1^2 + k_2^2)dS \longrightarrow$ min.

This result gives "smooth normal components' of twist vectors.

Another very efficient twist estimation is the so called, Adini-twist. This twist has been introduced into CAGD through the paper by Barnhill et al. [3] based on a scheme (Adini's rectangle) from the finite element literature. The basic idea is this: the four boundary curves define a bilinearly blended Coons patch, which happens to be a bicubic patch itself. Take the corner twists of that patch as twist vectors.

In Bézier form, one may simply compute the bilinearly blended Coons patch to the four boundary Bézier polygons. This is the Bézier net of the bicubic patch with Adini-twists.

It is now a rather straightforward idea to combine the advantages of both methods by extending the stiffness-degree concept for curves [13] to surfaces.

Performing a blended optimization of the normal components of the energy, and of the Adini-twist, using the (less important) tangent components of the Adini-twist vector $Ad(u, v)$ we get:

$$\begin{aligned} X_{uw} &= (\alpha < Ad, N > + \beta h_{12}) \cdot N \qquad (4.4) \\ &+ < Ad, X_u > \cdot X_u + < Ad, X_w > \cdot X_w \end{aligned}$$

$$\alpha(u, w), \beta(u, w) \geq 0; \alpha + \beta = 1.$$

This is the Gauss-frame representation of an (in our meaning) optimal twist vector.

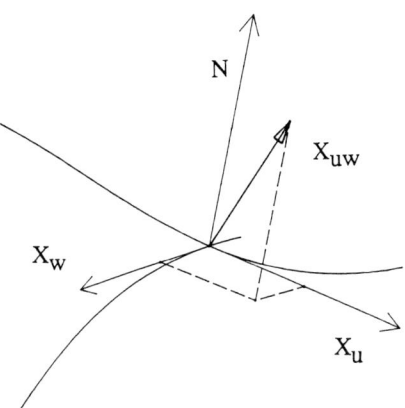

This information can also be used to specify the inner control points of Bézier-patches since the twist vectors of such a patch are given by

$$X_{uw} = (m-1)(n-1) \sum_{1=0}^{m-1} \sum_{j=0}^{n-1} \Delta^{1,1} b_{ij} B_i^{m-1}(u) B_j^{n-1}(w)$$

where $\Delta^{1,1} b_{ij} = b_{i+1,j+1} - b_{i,j+1} - b_{i,j+1} + b_{ij}$

In the case of B-spline surfaces we have the analogous situation.

Example:

We constructed a turbine blade using this method.

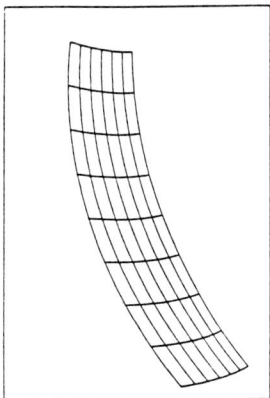

To visualize the "smoothness" and quality of this surface we use the reflection-line-method (for more details see [5]).

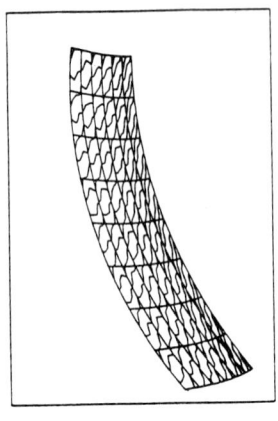

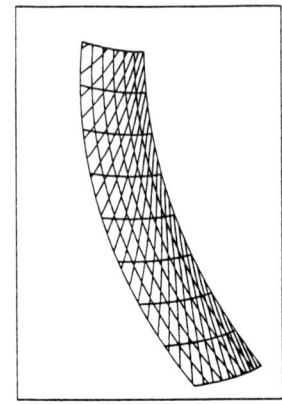

zero twist input   optimal twist input

References

1. Barnhill (1974). Smooth interpolation over triangles. In *Computer Aided Geometric Design*, Barnhill-Riesenfeld, editors, Academic Press, 45–70.

2. Barnhill (1982). Coons' patches. *Computers in Industry, 3*, 37–43.

3. Barnhill-Brown-Klucewicz (1978). A new twist in CAGD. *Computer Graphics and Image Processing, 8*, 78–91.

4. Barnhill-Gregory (1975). Compatible smooth interpolation in triangles. *J. Approx. Theory, 15*, 214–225.

5. Barnhill-Farin-Fayard-Hagen (1988). Twists, curvatures, and surfaces interrogation. To be published in *CAD*.

6. Boehm-Farin-Kahmann (1984). A survey of curve and surface methods in CAGD. *Computer Aided Geometric Design, 1*, 1–60.

7. Coons (1964). Surfaces for computer aided design. Report MAC-TR-4; Project MAC, MIT.

8. Farin (1979). *Subsplines über Dreiecken, Diss.* TU Braunschweig.

9. Forrest (1972). Interactive interpolation and approximation by Bézier polynomials. *Computer Journal, 15*, 71–79.

10. Gordon (1969). Free-form surface interpolation through curve networks. GMR–921, GM Research Laboratories.

11. Gordon-Riesenfeld (1974). B-Spline curves and surfaces. In *Computer Aided Geometric Design*, Barnhill-Riesenfeld, editors, Academic Press, 95–126.

12. Hagen-Schulze (1987). Automatic smoothing with geometric surface patches. *Computer Aided Geometric Design*, *4*, 231–235.

13. Hagen-Schulze (1990). Variational principles in curve and surface design. To be published in *Geometric Modelling-Methods and Applications*, Roller-Hagen, editors, Springer.

14. Nielson (1979). The side vertex method for interpolation in triangles. *J. Approx. Theory*, *25*, 318–336.

15. Nowacki-Reese (1983). Design and fairing of ship surfaces. In *Surfaces in CAGD*, Barnhill-Boehm, editors, North Holland, 121–134.

3D Reconstruction of Closed Objects by Piecewise Cubic Triangular Bézier Patches

Claudia Cottin[1] and Ruud van Damme[2]

[1]*University of Duisburg, Germany and* [2]*University of Twente, The Netherlands*

1 Introduction

The 3D reconstruction of closed objects from certain given data is an important issue in CAGD with many applications, *e.g.*, in medicine and biology. An extensive exposition of this subject can be found in [13]. In this context, the following problem often occurs: Given position values and normal vectors in a number of randomly distributed points on an object, one wishes to construct a 'simple' visually continuous (VC1) surface which interpolates to these data. In this context, 'simple' means above all that the interpolation method is as local as possible and that the patches which represent the surface can easily be evaluated and manipulated with a computer.

A particularly promising approach is to build up low degree polynomial patches on each element of some triangulation of the object with vertices in the given points. Methods along these lines have been developed, *e.g.*, by Farin [2], Jensen [8], Neamtu and Pfluger [9], Piper [11], and Shirman and Séquin [14]. All of them make essential use of quartic triangular patches, sometimes also in combination with quadrilateral elements. Further important references related to our work, but not relying on purely polynomial methods, are [6], [10].

Finding a good triangulation is a problem of its own (cf., *e.g.*, [12]), which we will not address in this note. We start with a given triangulation. For each of its elements, our interpolant consists of three polynomial patches, joining as indicated in Fig. 2 ('Clough–Tocher split'). Such a type of interpolants, using quartic polynomial elements, have been considered by Farin [2], and Piper [11]. We assume the reader to be familiar with the gist of these articles and the underlying theory, as expounded, *e.g.*, in [3], and in the relevant chapters of [4] or [7].

Piper motivates the use of quartics instead of cubics with the fact that, once the boundary curves are fixed, cubics do not always offer enough degrees of freedom to solve the problem; see [11] Section 3. We will show, however, that with a suitable *choice* of the boundary curves one can, to a large extent, restrict oneself to the use of cubic elements.

The organization of our paper is as follows. In Section 2 we briefly review the VC1 conditions established in [11] for the cubic case and draw some conclusions on the solvability of the involved equations. In Section 3 we present the essentials of an algorithm for the construction of a piecewise cubic VC1 interpolant. It strongly relies on an appropriate determination of the patch boundary curves. Section 4 is devoted to a more detailed discussion of this topic. The suggested procedures differ considerably from corresponding ones previously developed by other authors.

In a general setting, the basic algorithm may still render some patches which are not completely satisfactory. As a remedy, we propose several refinement strategies in Section 5, such as the readjustment of some boundary control points, the local performance of additional splits, or the fitting of some quartic elements into the surface. Some numerical examples are included in Section 6.

2 A Sufficient Condition for Equal Tangent Planes

Each polynomial patch of the interpolating surface is represented in parametric Bernstein–Bézier form

$$P(u, v, w) = \sum_{i+j+k=n} B_{i,j,k}^n(u, v, w) p_{i,j,k}, \quad u, v, w \geq 0, \ u+v+w = 1. \quad (2.1)$$

Here, n is the degree of the polynomial, $p_{i,j,k} \in \mathbb{R}^3$ are the so-called control points, (u, v, w) are the barycentric coordinates with respect to the standard domain triangle, and

$$B_{i,j,k}^n(u, v, w) := \frac{n!}{i!j!k!} u^i v^j w^k$$

are the Bernstein basis functions.

Two such patches join in a visually smooth way if the tangent planes along the common edge agree and the corresponding normal vectors point in the same direction. If the latter condition is not satisfied a so-called cusp occurs. Our first concern is the derivation of conditions for equal tangent planes only. Then additional considerations have to be performed in order to avoid cusps. However, we can give no standard method which automatically avoids cusps. This seems to be a general problem; cf., *e.g.*, also the discussion in [11].

Given two patches of the form (2.1) with control points $p_{i,j,k}$ and $q_{i,j,k}$, respectively, one has the following necessary and sufficient condition for equal tangent planes in a point $u + v = 1$ along a common edge $w = 0$: A directional derivative along this edge must lie in the plane spanned by a

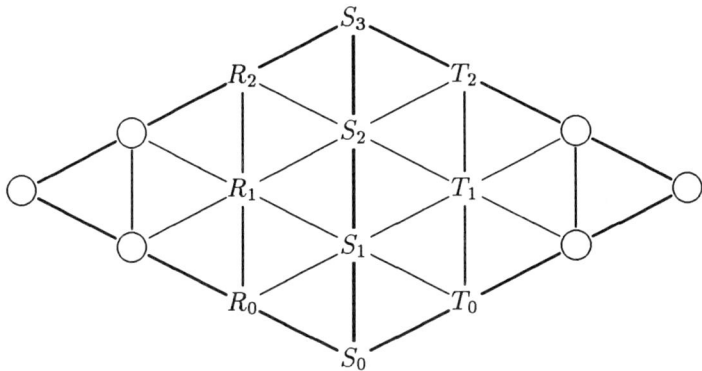

Figure 1. The control points of two adjacent cubic patches

pair of directional derivatives (one for each triangle) along the edge $v = 0$; in other words, the three vectors $J - I$, $K - I$, and $L - I$ with

$$I = \sum_{i+j=n-1} B_{i,j,0}^{n-1}(u,v,0)p_{i+1,j,0},$$

$$J = \sum_{i+j=n-1} B_{i,j,0}^{n-1}(u,v,0)p_{i,j+1,0},$$

$$K = \sum_{i+j=n-1} B_{i,j,0}^{n-1}(u,v,0)p_{i,j,1},$$

$$L = \sum_{i+j=n-1} B_{i,j,0}^{n-1}(u,v,0)q_{i,j,1}$$

are linearly dependent. In general, this condition is difficult to handle. Therefore, we will follow the approach of Piper and use the sufficient condition

$$(f_1'u + f_2 v)(J - I) + (g_1'u + g_2 v)(K - I) + (h_1'u + h_2 v)(L - I) = 0 \quad (2.2)$$

for some scalars f_1', g_1', h_1', f_2, g_2, h_2.

We consider the case of two adjacent cubic patches ($n = 3$). In order to increase readability, we set (cf. Fig. 1)

$$R_i = p_{2-i,i,1}, \quad T_i = q_{2-i,i,1}, \quad i = 0,1,2,$$
$$S_i = p_{3-i,i,0} = q_{3-i,i,0}, \quad i = 0,1,2,3.$$

Putting $e_1' := -f_1' - g_1' - h_1'$, $e_2 := -f_2 - g_2 - h_2$, (2.2) yields the following sufficient conditions for equal tangent planes along the common edge:

(i) $e_1' S_0 + f_1' S_1 + g_1' R_0 + h_1' T_0 = 0,$

(ii) $e_2 S_2 + f_2 S_3 + g_2 R_2 + h_2 T_2 = 0,$

(iii) $2(e_1' S_1 + f_1' S_2 + g_1' R_1 + h_1' T_1) + (e_2 S_0 + f_2 S_1 + g_2 R_0 + h_2 T_0) = 0,$

(iv) $(e_1' S_2 + f_1' S_3 + g_1' R_2 + h_1' T_2) + 2(e_2 S_1 + f_2 S_2 + g_2 R_1 + h_2 T_1) = 0.$

$$(2.3)$$

Note that by (2.3) *(i)* and *(ii)*, respectively, each one of the involved control points may be expressed as a barycentric combination of three other ones. Such a representation is useful in interpreting equations in terms of the coefficients $e_1', f_1', g_1', h_1',\ e_2, f_2, g_2, h_2$ and to compute these quantities.

The control points S_0 and S_3 are fixed by the interpolation condition. For a moment, we assume that the control points S_1, S_2, R_0, T_0, R_2, and T_2 were already determined. Then (2.3) *(ii)* uniquely fixes the coefficients e_2, f_2, g_2, h_2 subject to the condition $g_2 + h_2 = 1$. This condition can always be satisfied in the case of a VC1 join since then R_2, T_2 lie on strictly opposite sides of the line through S_2 and S_3 in the tangent plane at S_3.

In (2.3) *(i)*, we write $e_1' = \lambda e_1$, $f_1' = \lambda f_1$, $g_1' = \lambda g_1$, and $h_1' = \lambda h_1$ with e_1, f_1, g_1, h_1 fixed by the normation $g_1 + h_1 = 1$, and variable $\lambda \in \mathbb{R} - \{0\}$. Now (2.3) *(iii)* and *(iv)* can be written as

(i) $g_1 R_1 + h_1 T_1 = -(e_1 S_1 + f_1 S_2) - \frac{1}{2\lambda}(e_2 S_0 + f_2 S_1 + g_2 R_0 + h_2 T_0),$

(ii) $g_2 R_1 + h_2 T_1 = -(e_2 S_1 + f_2 S_2) - \frac{\lambda}{2}(e_1 S_2 + f_1 S_3 + g_1 R_2 + h_1 T_2).$

$$(2.4)$$

For each fixed λ, system (2.4) always has a solution for R_1 and T_1, provided that $g_1 \neq g_2$. In the case $g_1 = g_2$, the system is solvable iff the right-hand sides of (2.4) *(i)* and *(ii)* agree. Using (2.3) *(i)* and *(ii)*, and the relation $f_i = -e_i - 1$, $i = 1, 2$, we derive that this is true iff

$$(e_2 - e_1) \cdot [(S_2 - S_1) + \frac{1}{2\lambda}(S_0 - S_1) + \frac{\lambda}{2}(S_2 - S_3)] = 0, \qquad (2.5)$$

i.e., if $e_1 = e_2$ or if

$$C_\lambda := \lambda(S_3 - S_2) + 2(S_1 - S_2) + \frac{1}{\lambda}(S_1 - S_0) = 0. \qquad (2.6)$$

One can check that in the case of (2.5) (even if $g_1 \neq g_2$), a solution of (2.4) is given by

$$R_1^* = A - [\frac{h_1 \lambda}{2}(T_2 - R_2) + \frac{h_2}{2\lambda}(T_0 - R_0)],$$
$$T_1^* = A + [\frac{g_1 \lambda}{2}(T_2 - R_2) + \frac{g_2}{2\lambda}(T_0 - R_0)], \qquad (2.7)$$

where

$$\begin{aligned} A := \ & S_2 + e_1(S_2 - S_1) - \tfrac{1}{2\lambda}(e_2 - e_1)(S_0 - S_1) \\ = \ & S_1 + f_2(S_1 - S_2) - \tfrac{\lambda}{2}(f_1 - f_2)(S_3 - S_2). \end{aligned}$$

Obviously, this solution is invariant under the transformation

$$(S_0, S_1, S_2, S_3, R_0, T_0, R_2, T_2, g_1, g_2, h_1, h_2, e_1, e_2, \lambda)$$
$$\rightarrow (S_3, S_2, S_1, S_0, R_2, T_2, R_0, T_0, g_2, g_1, h_2, h_1, f_2, f_1, \tfrac{1}{\lambda}).$$

In many cases, (2.7) will be a very reasonable choice of R_1, T_1 since R_1 lies 'approximately' between R_0 and R_2, and T_1 between T_0 and T_2. This is especially true if λ approximately equals 1. Note that

$$T_1^* - R_1^* = \frac{\lambda}{2}(T_2 - R_2) + \frac{1}{2\lambda}(T_0 - R_0).$$

In the case $g_1 \neq g_2$, the unique solution of (2.4) (for fixed λ) is given by

$$\widetilde{R}_1 = R_1^* - \frac{1}{2}h_1 Q_\lambda, \quad \widetilde{T}_1 = T_1^* + \frac{1}{2}g_1 Q_\lambda \qquad (2.8)$$

with R_1^*, T_1^* as in (2.7), and

$$Q_\lambda = \frac{e_2 - e_1}{g_2 - g_1} C_\lambda$$

with C_λ as in (2.6). We observe that $A - \frac{h_1}{2}Q_\lambda = A' - \frac{h_2}{2}Q_\lambda$ with $A' :=$ $S_2 + e_2(S_2 - S_1) - \frac{\lambda}{2}(e_2 - e_1)(S_3 - S_2)$, such that (2.8) could be written in a more symmetric (but also more complicated) form. Solution (2.8) will only be sensible if $|Q_\lambda|$ is not too large.

3 Sketch of the Algorithm

Fig. 2 shows one triangular element with given vertices V which is split into three subtriangles. The points V, S, R, G, B, and M indicate the control points of the three cubic patches that we wish to construct.

The first step of our algorithm accomplishes the construction of the boundary curves (wire frame) defined by the vertices V and the points S. By the observations made in the previous section, the validity of (2.5) along each edge would be desirable. Since the condition $e_1 = e_2$ cannot be satisfied by means of a completely local strategy, we require that the control points should satisfy (2.6) whenever possible, or else be chosen in such a way that a 'reasonable' solution of the form (2.8) exists (see Section 4 for details). In a very general setting, this cannot be done for *all* boundaries of the given triangulation. In Section 5, we therefore suggest several strategies for the treatment of the remaining 'exceptional' boundary curves. The necessary modifications do not essentially change the procedure proposed here for the non-exceptional cases.

In the second step, the points G are chosen as one third of the sum of the neighbouring V and two S's. As a third step, the points R are computed

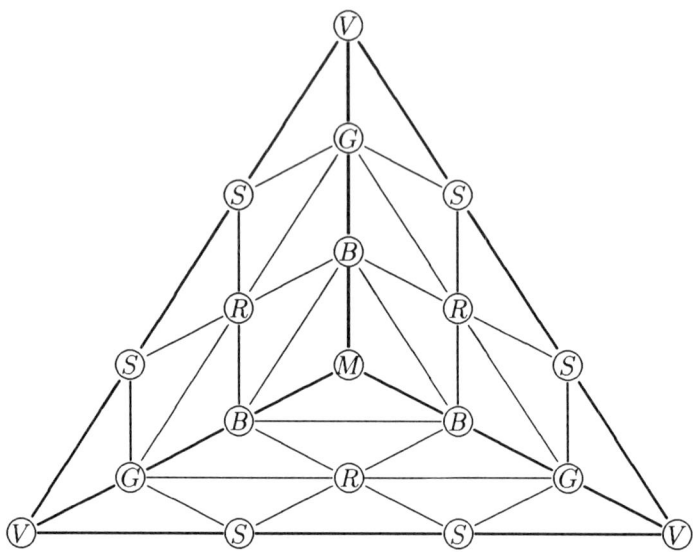

Figure 2. One macro-triangle (cubic Clough–Tocher split)

according to (2.7) or (2.8), respectively, such as to yield a VC1 transition to the neighbouring macro-triangle. In the fourth step, the points B are taken as one third of the sum of the neighbouring G and two R's. Finally M is set to one third of the sum of all B's. It is easy to see that, along the split edges, equations (2.3) *(i)-(iv)* hold with corresponding $e_i = g_i = h_i = \frac{1}{2}$, and $f_i = -\frac{3}{2}$, $i = 1, 2$, and thus provide the interior VC1 join.

4   Construction of the Wire Frame Curves

Our first objective is to determine the situations in which condition (2.6) can be satisfied for a cubic boundary curve given by control points S_0, S_1, S_2, and S_3. In case of non-parallel or equal tangent planes T_0 and T_3 at S_0 and S_3, respectively, solutions of (2.6) for S_1 and S_2 always exist. The resulting curve is obviously planar. In many situations, this is a desirable restriction on the wire frame.

The construction of feasible points S_1 and S_2 can be performed in the following way. We choose any point I not equal to S_0 or S_3 from the intersection of T_0 and T_3. Then we put $t_1 = I - S_0$ and $t_2 = I - S_3$, as well as $S_1 = S_0 + \mu t_1$, $S_2 = S_3 + \nu t_2$, $\mu, \nu \in \mathbb{R}$. Using (2.6), we arrive at the relations

$$\mu = \frac{2\lambda}{2\lambda + 1}, \quad \nu = \frac{2}{\lambda + 2}. \tag{4.1}$$

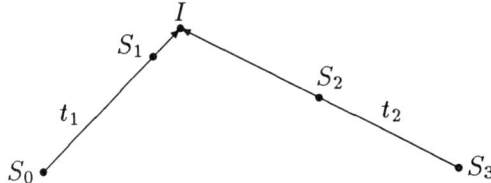

Figure 3. Construction of a boundary curve with $\lambda = 2$ ($\mu = \frac{4}{5}, \nu = \frac{1}{2}$)

If $\lambda < 0$, then either $\mu < 0$ and $\nu > 1$, or $\mu, \nu > 1$, or $\mu > 1$ and $\nu < 0$. The corresponding control polygon, and hence the resulting curve, possess an undesirable shape in these situations.

For $\lambda > 0$, we have $0 < \mu, \nu < 1$, and $\nu \to 1$, $\mu \to 0$ for $\lambda \to 0$, as well as $\nu \to 0$, $\mu \to 1$ for $\lambda \to \infty$. The corresponding control polygon is that of a convex or concave curve. In many cases, also, this second restriction (in addition to the planarity condition) seems sensible. However, a complete lack of points of inflection for the boundary curves may result in cusps along the join or in self-intersecting patches.

Therefore we proceed as follows. We determine a plane $\mathcal{T}$ in which the wire frame curve should lie such that it forms a reasonably short path between S_0 and S_3. We suggest taking the plane through S_0 and S_3 parallel to the sum of the normal vectors of $\mathcal{T}_0$ and $\mathcal{T}_3$. A more refined choice is the plane trough S_0 and S_3 which minimizes the sum of its (unsigned) angles with the normals of $\mathcal{T}_0$ and $\mathcal{T}_3$.

In this plane we compute target control points S_1^* and S_2^* without taking into account the VC1 condition for a moment. This can be done, *e.g.*, by considering the intersection of $\mathcal{T}$ with $\mathcal{T}_0$ and $\mathcal{T}_3$ which defines two tangents t_0 and t_3 through S_0 and S_3, respectively. We compute the projection P_3 of S_0 onto t_3 and the projection P_0 of S_3 onto t_0 in the plane $\mathcal{T}$. Then we set

$$S_1^* = S_0 + d(P_0 - S_0), \quad S_2^* = S_3 + d(P_3 - S_3) \tag{4.2}$$

with some scaling factor d, *e.g.*, $d = \frac{1}{3}$.

Another strategy which yields good results is sketched in Fig. 4. Let I be the intersection of t_0 and t_3. If $|I - S_0| > |I - S_3|$, we pick X on t_3 such that $|X - I| = |S_0 - I|$. Then we construct S_1^* and S_2^* as in the functional case with domain axis $X - S_0$. If $|I - S_0| < |I - S_3|$, a likewise construction is performed with X on t_0. There are always two possible choices of X, but only one is consistent with the orientation of the tangent planes. Note that the case of parallel tangent planes, when I does not exist, can be viewed as the limit case of 'almost' parallel tangent planes, yielding the same points as in (4.2) for $d = \frac{1}{3}$.

If the control polygon formed by S_0, S_1^*, S_2^*, and S_3 is convex or concave (or can be made so by small feasible changes of S_1^*, S_2^*) we use (2.6) for the

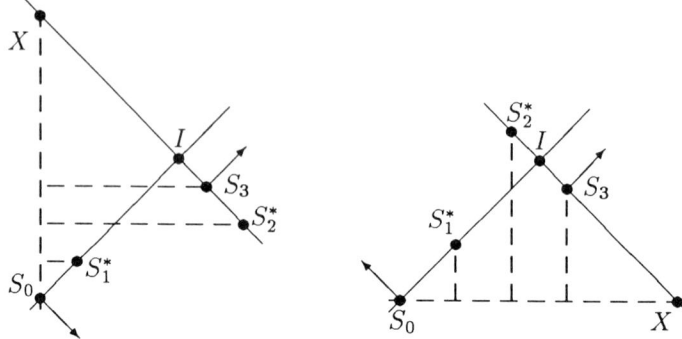

Figure 4. Boundary curve construction: the two possible choices of X

final construction of the boundary curve. For the other edges, the target points S_1^* and S_2^* could now be recomputed by relaxing the planarity condition for the curve. However, we have omitted this step in our algorithm, since we do not believe that it will yield much improvement in general.

It has to be a topic of further research to find an optimal λ in (2.6) for each wire frame curve and to determine all situations in which (2.6) can successfully be used. For our first computational results, we have taken $\lambda = 1$; *i.e.*, $\mu = \nu = \frac{2}{3}$. Then (2.6) reduces to

$$S_1 - S_2 = \frac{1}{3}(S_0 - S_3), \tag{4.3}$$

which means that the resulting curve is just quadratic.

After the boundary curves are fixed, all points G in Fig. 2 are determined according to Section 3. The points R can be computed by (2.7) for the edges which satisfy either (2.6) or the condition $e_1 = e_2$. Otherwise, they can be computed by (2.8) if the quantity Q_λ is defined and reasonably small. In the remaining cases, *i.e.*, the ones in which (2.6) does not hold and $g_1 = g_2$ and $e_1 \neq e_2$, or $g_1 \neq g_2$, but the solution (2.8) seems 'degenerate', we use some refinement strategy, as explained in the following section.

5   Refinement Strategies

One way of avoiding 'degenerate' solutions for the points R in Fig. 2 is to impose the condition $e_1 = e_2$ on all edges which cannot be made convex or concave. Suppose we have fixed a target wire frame according to one of the algorithms of the preceding section. Writing (2.3) *(i)*, *(ii)* in the form

$$\begin{aligned}
(i) \quad & S_1 - S_0 = \lambda_1(R_0 - S_0) + \mu_1(T_0 - S_0), \\
(ii) \quad & S_2 - S_3 = \lambda_2(R_2 - S_3) + \mu_2(T_2 - S_3),
\end{aligned} \tag{5.1}$$

we see that $e_1 = e_2$ is equivalent to

$$\frac{1}{\lambda_1 + \mu_1} + \frac{1}{\lambda_2 + \mu_2} = 1. \tag{5.2}$$

In general, the control points of the target wire frame will not satisfy (5.2). For each edge E not constructed by (2.6) we determine

$$C_E = \frac{1}{\lambda_1^* + \mu_1^*} + \frac{1}{\lambda_2^* + \mu_2^*} \tag{5.3}$$

with λ_1^*, μ_1^*, λ_2^*, μ_2^* chosen according to (5.1). This quantity is not well-defined in the (rare) case that $e_1 = 0$ or $e_2 = 0$ along E; but we can always suitably readjust the target points in order to avoid this condition.

We now aim at satisfying (5.2) on E by modifying the control points S_1, S_2 such that

$$\lambda_1 + \mu_1 = C_E(\lambda_1^* + \mu_1^*), \quad \lambda_2 + \mu_2 = C_E(\lambda_2^* + \mu_2^*). \tag{5.4}$$

To this end, we consider a vertex $V \equiv S_0$ in isolation; see Fig. 5. For simplicity we put this vertex in the origin. The points S^k correspond to the ones previously denoted by S_1, and the G^k to the G in Fig. 2.

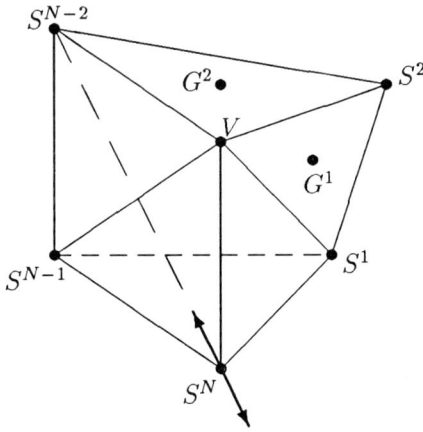

Figure 5. A vertex with $N = 5$ edges

Let us assume that there are N edges E_k at V. Putting $\Sigma^k := \lambda_1 + \mu_1$ (along E_k) and using the relation $G^k = \frac{1}{3}(S^{k+1} + S^k)$, (5.4) yields the equations

$$(3 - \Sigma^k)S^k = (\Sigma^k - \rho^k)S^{k+1} + \rho^k S^{k-1} \quad (k = 1, \ldots, N) \tag{5.5}$$

with ρ^k a free parameter. Geometrically this means that

$$\frac{Area(S^{k-1}, S^{k+1}, V)}{Area(S^{k-1}, S^k, V) + Area(S^k, S^{k+1}, V)} = \frac{3 - \Sigma^k}{\Sigma^k}. \qquad (5.6)$$

In the worst case we end up with $2N$ equations for $3N$ unknown. Remaining degrees of freedom can be fixed by demanding that the solution is as close as possible to the original target frame. Since the system (5.5) is non-linear there might still be some problems in solving it suitably. However, in very many cases at least one of the edges at V will already be fixed by (2.6), assuming this is E_1. Then one can choose the S^k, $k = 2, \ldots, N$, successively such as to satisfy (5.5) for $k = 2, \ldots, N-1$. Since it is clear from the geometrical interpretation (5.6) that a shift of S^N in the direction of S^{N-2} does not affect any of these $N - 2$ relations, (5.5) can also be fulfilled for $k = N$.

When we have found a wire frame such that (5.5) is satisfied *everywhere*, we automatically also arrive at proper solutions if the data set (normals and/or position values) is changed a little. We just have to rotate and/or shift the control points of the changed vertices. Afterwards one must, of course, recompute the control points within the involved triangles. An example of such a procedure is given in Section 6. Moreover, one can scale all control points around one vertex in any direction, since this manipulation also keeps the relations of (5.5) intact.

Another approach investigated by us is the modification and refinement of the given triangulation in such a way that all wire frame curves can be assumed to be convex or concave, and thus can be constructed consistent with (2.6). We do not give a detailed algorithm here, but only sketch a possible procedure.

If there is a non-convex/-concave edge AB we first swap the diagonal and investigate whether the edge CD (see Fig. 6) can be made convex/concave. Otherwise we determine points S_A, S_B, S_C, S_D, in the tangent planes at A, B, C, and D, respectively, and some 'midpoint' M, all lying in one plane. M is taken as a new vertex of the triangulation, and the control points (A, S_A, M), (B, S_B, M) etc. define quadratic boundary curves which hence satisfy (4.3). The points S_A, S_B, S_C, S_D, and M should be chosen in such a way that the new wire frame curves deviate as little as possible from the previously computed curves along AB and CD.

A third refinement strategy is to fit some quartic elements into the surface. After the determination of all boundary control points according to Section 4 and all corresponding G's in Fig. 2 we consider all macro-triangles with (at least) one 'exceptional' edge (*i.e.*, (2.6) does not hold, and solution (2.8) does not exist or is degenerate). The points R in Fig. 2 have been computed if they do not belong to such an edge. We proceed by constructing three quartic polynomial patches with control points indicated in Fig. 7 (a) which fit into the overall surface in a VC1 fashion.

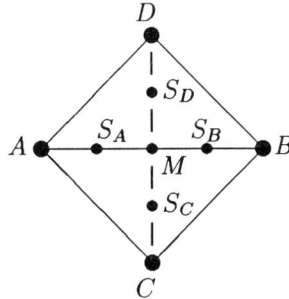

Figure 6. Modification of the triangulation

All points S and G in Fig. 7 (a) are computed based on the corresponding ones in Fig. 2 by the degree elevation algorithm. The same is done with the points R for the non-exceptional edges. We observe that, as before, G is still one third of the sum of the three surrounding control points, and that the degree elevation process obviously does not influence the VC1 transition along the non-exceptional edges.

Along each exceptional edge the control points x in Fig. 7 (b) are determined such as to yield a VC1 transition to the adjacent macro-triangle. This step can be performed, for example, as was done in a similar construction in [11]. The control points F in Fig. 7 (a) are free parameters and can be used to model the shape of the interpolant. In the final step, the points C are taken to be one third of the sum of the three surrounding ones in order to achieve the interior VC1 joins.

For the implementation of a universal algorithm one can also appropriately combine all three strategies presented in this section. The fitting in of quartics will unproblematically work for all exceptional edges. On the other hand, the necessary computations will be quite costly, and in some instances we may end up with a large number of quartic patches if we only use this strategy. The other two refinement procedures, in the form presented above, have to be performed for all non-convex/-concave edges (even for the non-exceptional ones). Of course, we can then omit a prior computation of Q_λ and of the solutions (2.8). Both strategies could also be carried out in such a way that they keep the non-exceptional edges unchanged if they do not turn into exceptional ones by the manipulations on the other edges; but a corresponding algorithm becomes more involved.

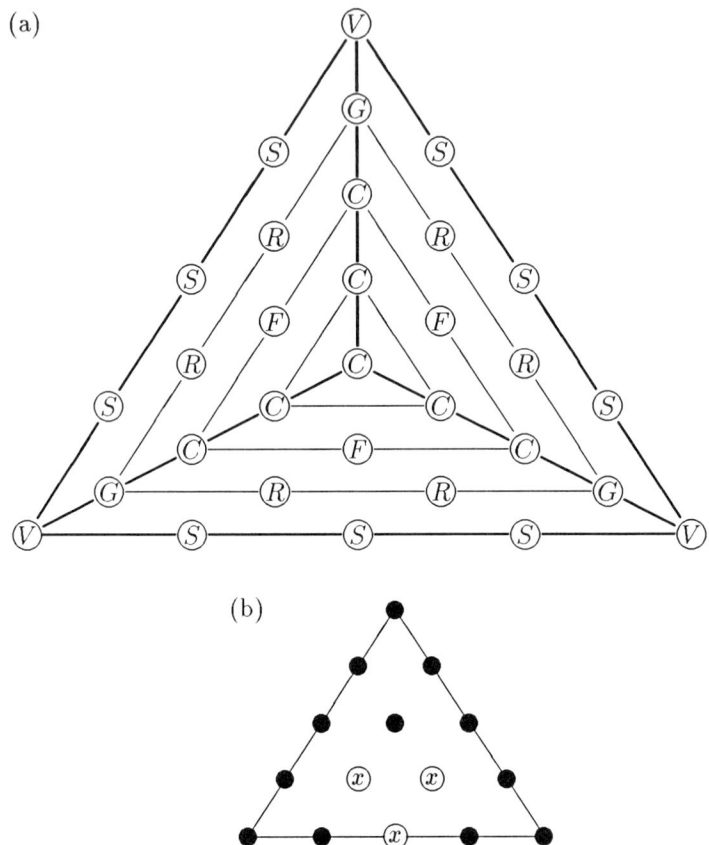

Figure 7. (a) One macro-triangle (quartic Clough–Tocher split), (b) One quartic micro-triangle

6　Numerical Examples

In this section we present some pictures produced by our algorithm. As refinement strategy we used the readjustment of the control points according to (5.5). Whenever there existed a vertex without any convex or concave boundary curve attached to it, we solved (5.5) with a simple relaxed Newton method, minimizing the sum over the distances of the new control points from the target points.

Figures 8 and 9 show the Franke function (see [5]) on an 8 ∗ 8 regular grid and on the standard 36 scattered data set (see, *e.g.*, [1] or [15]). Of course, we never used the fact that the data set was drawn from a *function*.

In Figure 10 the data set consists of $5*5$ planar data points with all normals perpendicular to the plane, except the middle one, which lies within this plane. We use the idea mentioned in Section 5 and start with the plane itself as a Lagrange interpolant, hence ignoring the normals. Choosing the control points as in the functional case, we immediately satisfy (2.4), since this equation is always valid in the functional case. A simple rotation of $\pi/2$ of the control points at the middle vertex yields the Hermite interpolant shown in Figure 10.

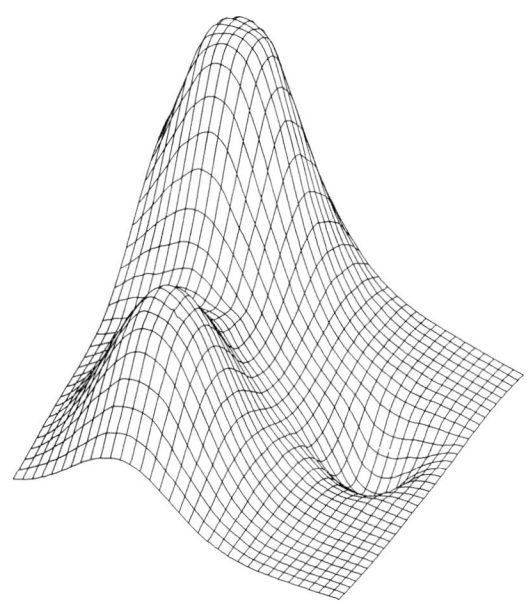

Figure 8. The Franke function on an $8*8$ regular grid

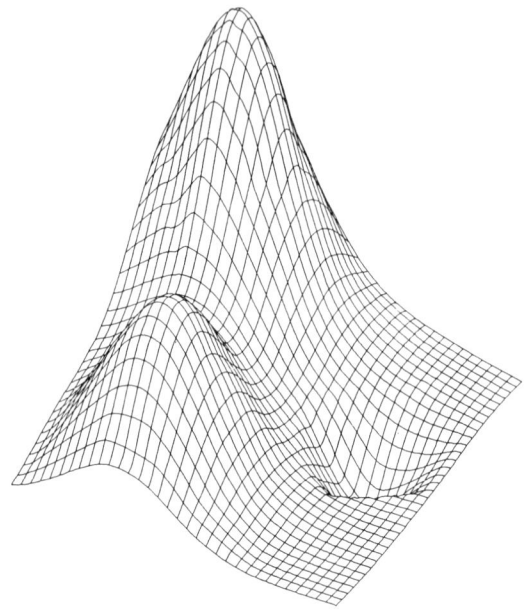

Figure 9. The Franke function on the standard 36 scattered data set

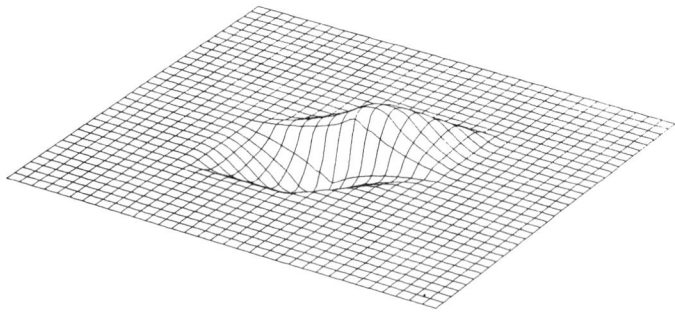

Figure 10. $5 * 5$ planar data points with perpendicular normals, except the middle one (see text)

Acknowledgements

The first author was assisted by DFG Postdoctoral Grant Co 153/3-1. Moreover, she acknowledges partial support by the University of Twente (Faculty of Applied Mathematics).

References

1. Alfeld, P. (1985). Derivative generation from multivariate scattered data by functional minimization. *Computer Aided Geometric Design*, *2*, 281-296.

2. Farin, G. (1983). Smooth interpolation to scattered 3D data. In *Surfaces in Computer Aided Geometric Design*, R.E. Barnhill and W. Boehm eds., North Holland, 43-63.

3. Farin, G. (1986). Triangular Bernstein–Bézier patches. *Computer Aided Geometric Design*, *3*, 83-127.

4. Farin, G. (1988). *Curves and Surfaces for Computer Aided Geometric Design*, Academic Press.

5. Franke, R. (1982). Scattered data interpolation: test of some methods. *Math. Comp.* , *38*, 181-200.

6. Herron, G. (1985). Smooth closed surfaces with discrete triangular interpolants. *Computer Aided Geometric Design*, *2*, 297-306.

7. Hoschek, J. and Lasser, D. (1989). *Grundlagen der geometrischen Datenverarbeitung*, Teubner.

8. Jensen, T. (1987). Assembling triangular and rectangular patches and multivariate splines. In *Geometric Modeling*, G. Farin ed., SIAM, 203-220.

9. Neamtu, M. and Pfluger, P.R. On visually smooth interpolation schemes in IR^3. Submitted for publication.

10. Nielson, G.M. (1987). A transfinite, visually continuous, triangular interpolant. In *Geometric Modeling*, G. Farin ed., SIAM, 235-246.

11. Piper, B. (1987). Visually smooth interpolation with triangular patches. In *Geometric Modeling*, G. Farin ed., SIAM, 221-233.

12. Schumaker, L.L. (1987). Triangulation methods. In *Topics in Multivariate Approximation*, C. K. Chui, L.L. Schumaker, and F.I. Utreras eds., Academic Press, 219-232.

13. Schumaker, L.L. (1990). Reconstructing 3D objects from cross-sections. In *Computation of Curves and Surfaces*, W. Dahmen, M. Gasca, and C. A. Micchelli eds., Kluwer, 275-309.

14. Shirman, L.A. and Séquin, C.H. (1987). Local surface interpolation with Bézier patches. *Computer Aided Geometric Design*, *4*, 279-395.

15. Whelan, T. (1986). A representation of a C^2 interpolant over triangles. *Computer Aided Geometric Design*, *3*, 53-66.

Wireframe Conversion for Surface Modelling

T. Hermann

Computer and Automation Institute, Hungarian Academy of Sciences, Budapest

1   Introduction

In surface modelling a common way of defining a surface is by using a wireframe. In a lot of modellers it is required that the wireframe has to be in the form of a rectangular mesh. The rectangular mesh has several advantages: it is easy to define; there are only four sided patches; and the theory of the four-sided patches is well-developed. Nevertheless, there is a whole range of surfaces which cannot be adequately represented by a rectangular mesh. Typical examples are the so-called suitcase corner (Fig. 1) and the branching of a tree-like volume (Fig. 2) where the surface does not have a rectangular topology. This leads to the introduction of such a curve network (or wireframe) where it is no longer required that a patch should be four sided or that a point of the surface may belong to only two curves. Such a curve network can contain n-sided patches (where it is required that $n > 2$) and it is allowable that a segment's end point may be an inner point of another segment (T-node). See e.g. [1], [3], [5], [6], [11], [12], [14], [15] and [17], concerning interpolation of n-sided patches.

In our institute we built an experimental surface modeller called FFS-GT (Free Form Surfaces with General Topology). This modeller's philosophy is that the user may define a wireframe with free topology (allowing n-sided patches and T-nodes) and then the modeller automatically generates the surface. (See [13], [7], [8].) A typical curve network can be seen in Fig. 3.

Here we encountered the following problem: how to identify the patches of the curve network? This problem is far from trivial, and in addition without some restrictions to the wireframe, the problem has no unique solution (see e.g. [10]). In the following we present two different algorithms which have different restrictions on the wireframe.

The structure of the paper is as follows: in the next section we give our basic assumptions about the curve network and define our terminology. In section 3 we present our first algorithm which we called *local*, in section 4 the other algorithm called *global* and in section 5 we sum up our results.

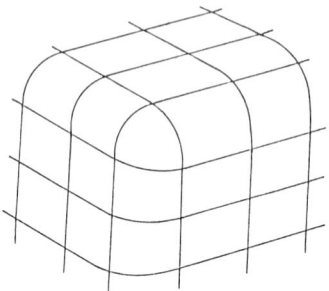

Figure 1.

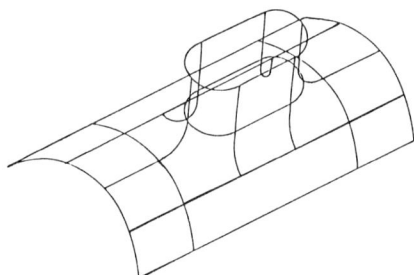

Figure 2.

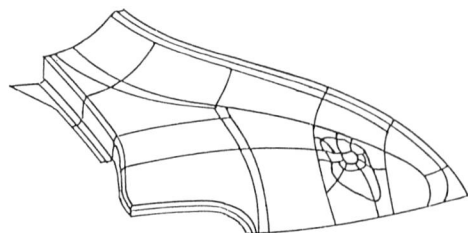

Figure 3.

2   Basic Assumptions and Terminology

Our aim is to model a two-sided G^1 continuous surface with the help of a wireframe.

There are points where two or more curves meet. We shall call these *vertices*. A curve consists of a series of *segments*. A segment is a parametric arc (polynomial or rational) and both end points of a segment are vertices. The wireframe can be considered to be a graph so that the vertices of the graph are the vertices of the curve network and the edges are the segments.

Now we need some definitions from the graph theory. The *degree* of a vertex is the number of edges incident to it. If v and w are two vertices from a graph then a *path* from v to w is a sequence of vertices and edges leading from v to w. A path is a *cycle* if all its edges are distinct and no vertex occurs twice in it with the exception of the first and the last one, which are identical. A path is *simple* if all its vertices are distinct. In the following when we speak about paths then it should be understood that we mean a simple path. A graph is *connected* if every pair of vertices of the graph is connected by a path. A graph is *k-connected* if every pair of vertices of the graph is connected by at least k paths so that any two such paths do not contain common vertices (with the exception of the first and the last vertices of the paths). We require the wireframe to be a biconnected (i.e. 2-connected) graph. (It is easy to see that this requirement is natural for every surface.) We suppose the curve network is given with the following data structure. We have a list of the vertices and every vertex contains a sublist of the corresponding segments. We suppose that the boundaries of the curve network (of the surface) are given in the form of lists. If the surface is closed then these lists are empty.

In our algorithms we need a systematic way of exploring a graph. We will use a method called *depth-first search*. To carry out a depth-first search on a graph, start from some vertex v and choose an edge leading v to follow. Traversing the edge leads to a new vertex. Continue in this way, at each step selecting an unexplored edge leading from the most recently reached vertex which still has unexplored edges otherwise backtrack. If the graph is connected, each edge is traversed exactly twice (once exploring a new vertex and once when backtracking).

As the surface is G^1 continuous we require from the curve network that in each vertex there has to be a tangent plane which is tangential to all the segments which contain the vertex.

3   The Local Algorithm

What is the difficulty in identifying the patches? When all the curves lie in a given plane then the answer is trivial. If we start from one corner of

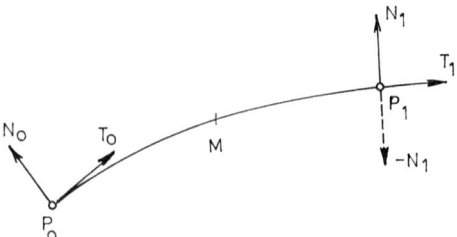

Figure 4.

a patch and at every corner we turn to the left then we go around a patch and so we determine its border. The problem is that in space we cannot tell which is the "left" direction. Let us consider the surface to be oriented. If we knew the sign of the normal vector corresponding to the tangent plane then we would be able to tell what is the "left" direction and so we would be able to solve our problem. Hence one way to present our problem is that we need an algorithm which determines the signs of the normals. This reasoning led to our first algorithm which we call "local" because it uses only local information from the wireframe. The local algorithm is the following. We explore the graph by depth-first search. At the first vertex we choose the sign of the normal arbitrarily. At the depth-first search we always proceed from a vertex where we have already determined the sign of the normal to a vertex where we still have not determined it. Then we determine (by an intelligent guess) the sign of the normal of the new vertex. This happens in the following way. Let us name the vertex from where we arrived as P_0 and the other vertex as P_1. The corresponding normals are N_0 and N_1, resp. The unit tangent vectors at the endpoints of the segment leading from P_0 to P_1 are T_0 and T_1, the midpoint of the segment is M. (See Fig. 4.)

We distinguish two cases. In the first case let us suppose that P_0, P_1 and M determine a unique plane. Let us denote the normal of this plane as N and the unit vectors in the direction of $N \times T_0$ and $N \times T_1$ are denoted as B_0 and B_1. Then let us consider the following equation

$$N_0 = xN + yT_0 + zB_0.$$

Obviously x, y and z are uniquely determined. With these values let us make the following vector

$$V = xN + yT_1 + zB_1.$$

Now we choose from N_1 and $-N_1$ that normal which is nearer to V.

In the second case P_0, P_1 and M do not determine a unique plane (i.e. they lie on a straight line). Then we choose between N_1 and $-N_1$ that one which is closer to N_0.

The algorithm above is based on the assumption that the patches corresponding to the curve network do not have too large torsion. The basic idea of the algorithm is due to T. Várady [18]. Originally he suggested the use of the Frenet frames as local coordinate systems. The present form of the algorithm is due to the author. The algorithm is implemented and tested in the FFS-GT experimental surface modeller. It has proved to be fast and reliable.

Still, we feel that the problem cannot be considered as solved. It is easy to construct an example where the algorithm fails (even if our counterexample is rather artificial). e.g. when a segment is a straight line and N_0 and N_1 are orthogonal to each other. Our main concern is that although this failure happens only very rarely (and in general due to some error by part of the user) we have no information about where it happened.

4   The Global Algorithm

To solve this problem we want an alternative algorithm which is fast and has the advantage that in case of failure we can detect automatically the reason of that failure. We attribute the shortcomings of the previous algorithm to the fact that it used only local information from the curve network. So our second algorithm is based on global information, namely on the graph of the wireframe.

Our idea is based on the fact that there are efficient (linear) algorithms for embedding planar graphs into the plane [16] and to decompose graphs into their triconnected (i.e. 3-connected) components [9]. In the following we shall refer to them as *Embed* and *Tricon*. These algorithms are based on some tricky multiple applications of depth-first search. To use them we have to make an additional supposition about the wireframe, namely that the surface is homeomorphic to a plane, which may contain some holes. We do not feel this restriction to be too severe because one always can cut the surface to be modelled into parts which satisfy this requirement.

So we can suppose that the graph of our wireframe is planar. Every embedding of the graph of the wireframe into the plane determines a patch structure. Of course a planar graph can be embedded into the plane in several ways. Two embeddings are considered to be equivalent if they

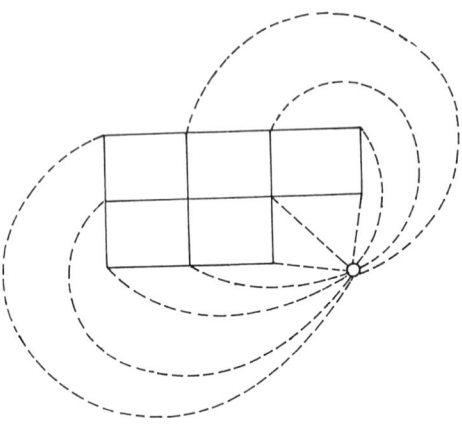

Figure 5.

determine the same patch structure. In the following we shall consider the equivalent embeddings to be identical. It is well-known that in the above sense a planar graph has only one embedding if and only if it is triconnected [19]. So if our graph is triconnected then we apply Embed and in this way identify the patches. [1] But in general a curve network is not triconnected. e.g. even the rectangular mesh is only biconnected. Now we have not only the graph of the wireframe but other information too. Our idea is to put this additional information into the graph by transforming it. (Of course in the transformations we preserve our original graph and mark the new vertices and segments.)

The first information that we use is the boundary of the surface. The boundary (which is a cycle of the graph) can be considered as the boundary of a new patch which closes the surface. We call this patch an *outer* patch. Additionally the user may declare any patch to be empty i.e. a hole (if it is not bordering another patch already declared to be empty). For the outer patch and for each hole let us add a new vertex to the graph which we consider to be an inner point of that patch and let us connect it to every vertex of the patch (Fig. 5).

Another piece of information available is the following. At every vertex we know the cyclic order of the segments. (This is the point where we use the condition that the wireframe is G^1 continuous.) Now we substitute such vertices of the graph whose degree n is greater than three with a subgraph: a cycle containing n vertices corresponding to the cyclic ordering (Fig. 6).

---

[1]In [4] the authors solve the problem of wireframe conversion for certain solids. In the case of solids the triconnectivity is a more natural requirement than for surfaces.

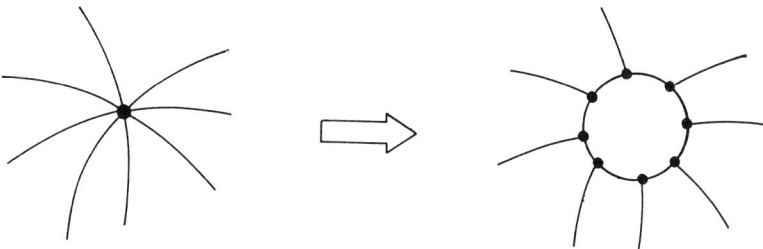

Figure 6.

After the transformations we check our graph, with the help of Tricon, to see if it is triconnected or not. If it is triconnected then we embed it with the help of Embed. (Our experience is that for the great majority of wireframes the transformed graph is already triconnected. Where it is not, this happened due to some error on the part of the user.) If the graph is not triconnected then Tricon divides it into triconnected components and presents them to the user with the request that additional edges be inserted to make it triconnected.

5 Conclusion

We have presented two algorithms to identify patches. The local algorithm proved to be very fast and reliable. In our experience the only problem with it was that in the case of failure the user needed a rather long process to detect and correct the error. This necessitated our second algorithm. The global algorithm is much more complex and slower. Due to the intricacies of Embed and Tricon we have not fully implemented our second algorithm but we intend to do so for the sake of completeness.

Acknowledgements

I would particularly like to thank my research group colleagues: Gábor Lukács, Gábor Renner, Zoltán Rozgonyi and Tamás Várady for many helpful and relevant discussions during the course of this work.

References

1. Catmull, E.E. and Clark, J.H. (1978). Recursively generated B-spline surfaces on arbitrary topological meshes. *Computer Aided Design, 10*, 350–355 .

2. Chiyokura, H. and Kimura, F. (1984). A new surface interpolation method for irregular curve models. *Computer Graphics Forum, 3,* 209–218.

3. Doo, D.W.H. and Sabin, M.A. (1978). Behaviour of recursive division surfaces near extraordinary points. *Computer Aided Design, 10,* 356–360.

4. Dutton, R.D. and Brigham, R.C. (1983). Efficiently identifying the faces of a solid. *Comput. & Graphics, 7,* 143–147.

5. Gregory, J.A. (1986). N-sided surface patches. In *The Mathematics of Surfaces*, J. A. Gregory, editor, Oxford University Press,

6. Gregory, J.A. and Charrot, P. (1980). A C^1 triangular interpolation patch for computer-aided geometric design. *Computer Graphics and Image Processing, 13,* 80–87.

7. Hermann, T. and Renner, G. (1989). Subdivision of n-sided regions into four-sided patches. In *The Mathematics of Surfaces III*, D. C. Handscomb, editor, Oxford University Press, 347–357.

8. Hermann, T., Renner, G. and Várady, T. (1990). Surface Interpolation Based on a Curve Network of General Topology. (manuscript).

9. Hopcroft, J.E. and Tarjan, R. (1973). Dividing a graph into triconnected components. *SIAM J. Comput., 2,* 135–157.

10. Hojnicki, J.S. and White, P.R. (1988). Converting CAD wireframe data to surfaced representations. *Computers in Mechanical Engineering, 7,* 19–25.

11. Jones, A.K. (1988). Nonrectangular surface patches with curvature continuity. *Computer Aided Design, 20,* 325–335.

12. Renner, G. (1991). Polynomial n-sided patches. In *Curves and Surfaces*, P. J. Laurent, A. Le Méhauté, L. L. Schumaker (eds.), Academic Press, Boston, 407-410.

13. Renner, G. (1990). Designing complex surfaces based on curve-networks with irregular topology. *Geometric Modelling Studies, 3,* Comp. Aut. Inst. Hung. Acad. of Sci.

14. Sarraga, R.F. (1987). G^1 interpolation of generally unrestricted cubic Bézier curves. *Computer Aided Geometric Design, 4,* 23–39.

15. Shirman, L.A. and Séquin, C.H. (1987). Local surface interpolation with Bézier patches. *Computer Aided Geometric Design, 4,* 279–295.

16. Tarjan, R. (1971). An Efficient Planarity Algorithm. STAN-CS-244-71, Comput. Sci. Dep., Stanford U.

17. Várady, T. (1991). Overlap patches: a new scheme for interpolating curve networks with n-sided regions. *Computer Aided Geometric Design*, 8, 7-27.

18. Várady, T. (1988). Private communication.

19. Whitney, H. (1932). Congruent graphs and the connectivity of graphs. *Am. J. Math.*, *54*, 150–168.

How to Construct the Skeleton of CSG Objects

Christoph M. Hoffmann

Computer Science Department, Purdue University

1 Introduction

Consider a bounded solid T in $\mathbf{R}^3$ with compact boundary, such as a CSG object. For every point p in $\mathbf{R}^3$, we define its *distance* from the boundary of T as the minimum Euclidean distance $d(p, q)$ where q is on the boundary of T. For every p, there is always at least one point q in the boundary of T such that $d(p, q)$ is minimum, and such a point q will be called a *footpoint* of p on T.

The *interior skeleton* of T consists of all points p that are interior points of T and have more than one footpoint on T, and the limits of point sequences in this set; i.e., the relative closure of the set of all interior points with multiple footpoints. Similarly, the *exterior skeleton* of T consists of all points p that are exterior to T and have more than one footpoint, along with their limit points. An example in two dimensions is sketched in Fig. 1. In the literature, skeletons are also referred to as medial-axis transforms and Voronoi diagrams; e.g. [1, 23].

The interior skeleton can be used in pattern recognition, e.g., [1, 12]; and in finite-element mesh generation [16, 17]. For example, as argued in [17], knowing the skeleton allows one to answer a number of basic shape interrogations including detecting constrictions and their length scales, extracting holes, and decomposing complex shapes into topologically simple subdomains. Such queries facilitate generating finite-element meshes of high quality. The exterior skeleton can be used in motion planning, e.g., [25]; and for mesh generation for computational fluid dynamics.

It is well-known that the interior skeleton can represent a solid as follows. With each skeleton point, associate its distance from the boundary of T. Consider the set of spheres centered on the points of the skeleton and of radius equal to the center's distance from the boundary. Then the envelope of these spheres is the boundary of T.

Previously published work on the skeleton has considered the problem in two dimensions, [3, 11, 12, 16, 17, 18, 23, 24]. Although an extension

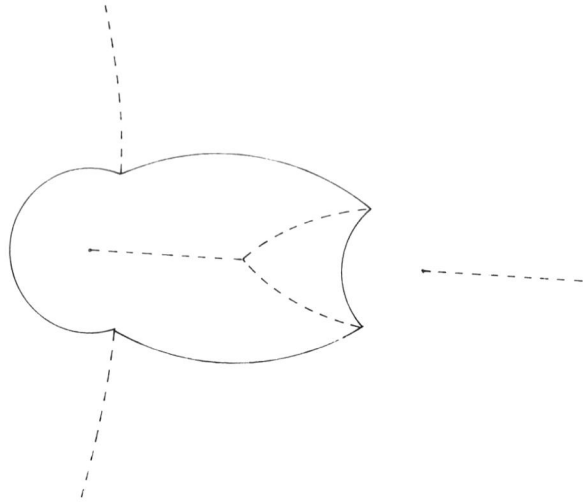

Figure 1. Interior and exterior skeletons of a simple domain

to three dimensions has been suggested early-on [1], very little has been published to date on this problem, unless we admit for T sets of discrete points, and include the extensive work on Voronoi diagrams; see, e.g., [19]. In [15], it is proposed to decompose three-dimensional shapes into spheres. The centers of these spheres can be thought of as a discrete approximation of the interior skeleton. [14] proves that the bisector of a linearly separable set in any dimension is a manifold.

In this paper we explain how to construct the interior skeleton of three-dimensional solids. We restrict the solids to CSG objects [20]. Doing so simplifies some of the analytical computations necessary to determine critical points and curves of the skeleton. However, except for the determination of critical points our algorithm applies unchanged to solids with curved boundaries as long as the surfaces involved are twice differentiable.

In [7] we have investigated the geometry of the surfaces that arise in the construction of the skeleton of CSG objects. Here, we study how to find initial points on the curves and surfaces comprising the skeleton and give a general algorithm for constructing the vertices, edges, and faces of the skeleton. The algorithm constructs all parts of the skeleton by increasing distance from the boundary, in a four-dimensional (x, y, z, r)-space. The resulting skeleton, in 4D, is then a complete representation of the original solid, as explained before.

2 Skeleton Construction

In [16, 17], Patrikalakis and Gursoy construct the interior skeleton of planar domains bounded by circle and line segments as follows:

1. Considering pairs and adjacent triples of boundary elements, determine a set of initial vertices in the skeleton that lie closest to the domain boundary.

2. Offset the boundary into the domain interior so that the closest initial vertices in the skeleton are on the offset boundary.

3. Repeat Steps 1 and 2 until the boundary has zero area.

Note that a boundary element is either an edge or a vertex.

The algorithm of Patrikalakis and Gursoy is a grass-fire algorithm. The principal technical reason for restricting boundary elements to circle and line segments is that the reduction in step 2 then yields a smaller domain that is again bounded by circle and line segments. Thus, no new types of boundary elements are introduced. A second reason for choosing these shape primitives is that all edges of the skeleton must be conic and line segments, so that the computation determining skeleton vertices is equivalent to intersecting two conics, a simple algorithmic problem. More complex curve segments such as boundary elements generate intersection problems that would be much harder to solve.

Neither property generalizes to CSG objects. The offset of a CSG object is not necessarily a CSG object; for, while the offset of a natural quadric is again a natural quadric, the tubular offset surfaces of certain edges are not natural quadrics. Note, however, that each tube could be approximated by a CSG object using the techniques of [21, 22]. Thus, an approximate skeleton might be constructed with an algorithm that retains many of the characteristics of the planar version literally. On the other hand, it is not really necessary to construct the offset solid explicitly, for all proximity computations can be expressed in terms of the original boundary elements. Even so, determining closest bisecting points between boundary elements requires considerably more machinery than solving two bivariate quadratic equations. Note that in the three-dimensional problem, the boundary elements are the faces, edges, and vertices of the CSG object.

We use the following algorithm for computing the skeleton of a CSG object:

1. By considering all pairs of boundary elements, determine for each pair the points that are equidistant from both elements and have minimum distance.

2. Sort the points by their distance from the boundary.

3. Processing the points by increasing distance, construct the skeleton by tracing the arising edges and faces.

Steps 1 and 3 are explained in some detail in the following sections. Note that the algorithm is structurally very similar to the two-dimensional version. In contrast to the two-dimensional method, however, not all critical points of the skeleton are found. In consequence, step 3 of our algorithm includes a local adjacency analysis that can determine all adjacent edges and faces for a given skeleton point.

Note that certain equidistant (bisecting) points, found in step 1, form curves or surfaces. For example closest bisecting points between two parallel cylindrical faces are on a line, and nearest equidistant points between concentric spherical faces lie on a sphere.

In principle, our algorithm can be used for solids with more complex boundaries. However, complex curved surfaces introduce algebraic complexities into step 1. In consequence, the generalization would necessitate introducing techniques for solving general systems of algebraic equations.

A different approach can be used to construct an *approximate* skeleton as follows. First, sample the solid's surface obtaining a set of surface points. Using a robust point-set Voronoi algorithm, construct the Voronoi diagram of this point set. Then, eliminate certain Voronoi edges and surfaces thereby obtaining an approximate skeleton. For details and a two-dimensional example see [4].

3 Determining Nearest Bisecting Points

Given two boundary elements, we show how to determine nearest bisecting points between them. The boundary elements are the vertices, edges, and faces of the solid. Certain boundary elements have nearest bisecting points with respect to themselves. For instance, the center of the sphere, the central circle of a torus and its center of symmetry are such points. Likewise, the center of an edge that is a circular arc must be considered.

Edges and faces are subsets of space curves and of surfaces, respectively. We call the corresponding curve or surface the *carrier* of the boundary element. The problem of determining nearest bisecting points between two boundary elements is reduced to the problem of determining nearest bisecting points between the corresponding carriers. Briefly, if F is a face, we first determine nearest bisecting points with respect to its carrier f. Those nearest bisecting points with footpoints outside F are discarded. Next, nearest bisecting points are determined with respect to the bounding edges of F. Again nearest bisecting points with footpoints not on the edge are discarded. Finally, nearest bisecting points with respect to the vertices of F are determined. Edges are handled analogously. In consequence, determining nearest bisecting points between two (nonadjacent) faces may require determining nearest bisecting points between all edge pairs bounding the faces. Since we should determine nearest bisecting points between all pairs of boundary elements, essentially no additional work is required apart from testing whether their footpoints lie on edges or faces.

In order to find nearest bisecting points between the carriers f and g, we find pairs of footpoints p and q, that is, points p on f and q on g such that $d(p,q)$ is minimum. Such a pair will be called a *closest approach pair*. Clearly then the bisector of the pair is a nearest bisecting point.

The geometric basis for identifying a closest approach pair (p,q) is the well-known observation that the connecting line $\overline{pq}$ must be perpendicular to f at p and perpendicular to g at q. In fact, this condition also describes point pairs at which the Euclidean distance has local extrema, both minima or maxima. So, we determine *all* extremal point pairs and compute their distances discarding all that do not have minimum distance.

In the following, we describe how to compute extremal approach pairs with the surfaces and curves that are carriers of faces and edges in CSG solids. We will omit a number of special cases. For example, the closest approach pairs between two intersecting surfaces are the points on the surface intersection; the closest approach pairs between two parallel cylinders lie on corresponding generators; and so on.

3.1 A generic procedure for finding closest approach pairs

In order to determine closest approach pairs (p,q) between two carriers f and g, we can formulate and solve a system of algebraic equations. The system has the following structure:

$$\begin{aligned}
f(p) &= 0 \\
g(q) &= 0 \\
\mathsf{perp}&(p,q,f) \\
\mathsf{perp}&(q,p,g).
\end{aligned} \tag{3.1}$$

The first two lines assert that p is on f and that q is on g. If f is a surface given implicitly, then the first line is a single equation. If f is a curve that is the intersection of f_1 and f_2, then the line represents the two equations $f_1(p) = 0$ and $f_2(p) = 0$. Finally, if f is a point, then the line is not an equation, but the coordinates of p are the given coordinates of the point f. Similar considerations apply to the second equation.

The third line asserts that the connecting line $\overline{pq}$ is normal to f at p. With f an implicit surface, this can be expressed by three equations, as explained in [8, 9, 10]. If f is a curve, the third line represents the condition that $\overline{pq}$ is perpendicular to the tangent to f at p, [9], and if f is a point the third line expresses no condition.

For CSG objects, f and g are algebraic and therefore (3.1) is a system of algebraic equations whose variety should be zero-dimensional. While this approach will not generate many cases, it has the disadvantage that the system (3.1) is algebraically unnecessarily complex. This makes the system harder to solve and generates unnecessary candidate pairs. Therefore, we consider the arising cases in greater detail in an effort to simplify matters. In fact, many cases will require solving only a single quadratic or quartic equation, and can therefore be handled very easily.

3.2   Closest approach pairs between CSG surfaces

We consider closest approach pairs for two algebraic surfaces f and g, each of which is either a plane, a natural quadric, or a torus. We organize the various cases by the type of g, considering only the interesting surfaces f.

It is helpful to consider a point to be a sphere of radius zero, to consider a line to be a cylinder of radius zero, and to consider a circle to be a torus of minor radius zero. For example, determining the closest approach pair of two cylinders is equivalent to determining the closest approach pair between their axes. This reduces the number of cases to be considered.

Closest approach pairs with a plane

Let g be a plane, given in implicit form by

$$a_1 x + b_1 y + c_1 z + d = 0$$

and assume that $a_1^2 + b_1^2 + c_1^2 = 1$. Most CSG surfaces f pose uninteresting problems, except the torus. So, let f be a torus, choosing the coordinate system as shown in Fig. 2, such that the axis of rotation coincides with the z-axis and the major radius is 1. Moreover, without loss of generality we assume that $b_1 = 0$.

Let $q = (a_0, b_0, c_0)$ be a point on the plane g, and let C be the interior skeleton of the torus, a circle of radius 1 in the xy-plane. The point q has a corresponding point of extremal approach on the torus if we can draw a line L from q, perpendicular to g, through the circle C to meet the z-axis, for then L will intersect the torus normally. Since $b_1 = 0$, the line L cannot be perpendicular to the plane g unless it lies in the plane $y = 0$, so we know that $b_0 = 0$. In order to pass through the circle C, furthermore, L must pass through the point $(1, 0, 0)$ or $(-1, 0, 0)$. Let $z = \lambda_0$ be the intercept of L with the z-axis. Then the point $(\lambda_0 a_0/(\lambda_0 - c_0), 0, 0)$ is the intersection with the xy-plane. So, L must satisfy the following equations

$$
\begin{aligned}
a_0 a_1 + c_0 c_1 + d &= 0 \\
a_1 c_0 - a_0 c_1 - a_1 \lambda_0 &= 0 \\
\lambda_0 a_0 &= \pm(\lambda_0 - c_0).
\end{aligned}
\tag{3.2}
$$

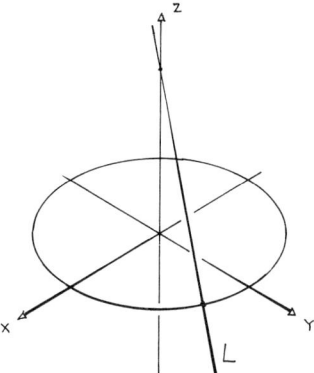

Figure 2. Torus skeleton and one torus normal

The first equation states that q is on the plane g, the second that the line intercepting the z-axis at $(0, 0, \lambda_0)$ is normal to the plane, and the last one that this line passes through $(1, 0, 0)$ or $(-1, 0, 0)$. The unknowns are a_0, c_0, and λ_0.

For each sign choice, the system (3.2) consists of two linear equations and one quadratic equation, with two solutions. Each solution will generate two extremal points on the torus. A distance computation settles which ones have minimum distance from the plane. Summarizing, we have

Theorem 1. *Determining closest approach pairs between a plane and a torus requires solving two systems of three equations in three unknowns, each system consisting of two linear equations and one quadratic equation.*

Closest approach pairs with a sphere

From a geometric point of view, closest approach determination with a sphere is essentially a two-dimensional problem. Since all CSG surfaces are surfaces of rotations, the closest approach points will lie in a plane through the axis of rotation that contains the center of the sphere. In consequence, we expect very simple equations.

Without loss of generality, we assume that g is the point $q = (a, b, c)$, and is either a vertex or the center of a sphere. The closest approach to a cylinder f is determined by the orthogonal projection of q onto the line that is the cylinder's axis. This requires solving a linear equation and is trivial.

Assume next that f is the cone

$$x^2 + y^2 - u^2 z^2 = 0. \tag{3.3}$$

The normals to the cone lie on cones with the equation

$$x^2 + y^2 - (z - \lambda_0)^2/u^2 = 0. \qquad (3.4)$$

Here, λ_0 is the z-coordinate of the intersection of the normals with the cone axis. To find a closest approach pair we determine λ_0 so that the point (a, b, c) lies on the corresponding cone of normals, i.e., we solve a quadratic equation in λ_0.

If f is a torus, we align the coordinate system as described in Section 3.2. We seek a line through the unit circle in the xy-plane, intercepting the z-axis at λ_0, and passing through the point $q = (a, b, c)$. Those lines form cones with the equation

$$x^2 + y^2 - (z - \lambda_0)^2/\lambda_0^2 = 0. \qquad (3.5)$$

So, we determine for which value of λ_0 this cone contains the point q, i.e., by solving a quadratic equation in λ_0. In summary, we have

Theorem 2. *Determining closest approach pairs between a point or a sphere and a line or a cylinder requires solving a linear equation, and determining closest approach pairs between a point or a sphere and a cone or a torus requires solving a quadratic equation.*

Closest approach pairs with a cylinder

Without loss of generality we replace the cylinder g with its axis, and assume that the axis is given as the parametric line

$$(a_0 + a_1\mu,\ b_0 + b_1\mu,\ c_0 + c_1\mu) \quad \text{where} \quad a_1^2 + b_1^2 + c_1^2 = 1. \qquad (3.6)$$

If f is another cylinder, we assume a coordinate system in which the cylinder axis of f is $(0, 0, \lambda)$. We assume that the two lines are skew, for otherwise the problem is uninteresting. A closest approach pair is found along the line of constriction; that is, along the common normal of the two lines. Perpendicularity to the two given lines is expressed by two linear equations. If λ_0 and μ_0 are the two (unknown) parameter values that specify the closest approach points on the two lines, the direction of the line of constriction is the vector

$$(a_0 + \mu_0 a_1,\ b_0 + \mu_0 b_1,\ c_0 + \mu_0 c_1 - \lambda_0). \qquad (3.7)$$

So, we must solve the linear system

$$\begin{aligned}
c_0 + \mu_0 c_1 - \lambda_0 &= 0 \\
\mu_0 - c_1\lambda_0 + (a_0 a_1 + b_0 b_1 + c_0 c_1) &= 0
\end{aligned} \qquad (3.8)$$

each equation stating perpendicularity to one of the lines.

If f is a cone, we assume a coordinate system in which the cone's axis coincides with the z-axis. So the cone equation is (3.3). The surface normals form the cones (3.4), and we seek λ_0 such that the corresponding cone of normals has a generator that intersects the line g at a right angle. Let μ_0 be the parameter value specifying the intersection of the cone of normals with the line g. The generator through this point has the direction (3.7). Hence, the unknowns μ_0 and λ_0 are found by solving the system

$$
\begin{aligned}
(a_0 + \mu_0 a_1)^2 + (b_0 + \mu_0 b_1)^2 - (c_0 + \mu_0 c_1 - \lambda_0)^2/u^2 &= 0 \\
\mu_0 - c_1 \lambda_0 + (a_0 a_1 + b_0 b_1 + c_0 c_1) &= 0
\end{aligned}
\tag{3.9}
$$

consisting of a linear and a quadratic equation. Here the first equation states that the line g intersects the cone of normals at the point specified by μ_0, and the second equation states that the corresponding generator is perpendicular to g.

If f is a torus, we assume a coordinate system as described before, in Section 3.2 and illustrated in Fig. 2, so that the exterior skeleton is the z-axis and the interior skeleton the unit circle in the xy-plane. We seek a line L through the unit circle, intercepting the z-axis at $z = \lambda_0$, such that L intersects the line g in a right angle, at a point with parameter value μ_0. The two unknowns λ_0 and μ_0 are found by solving

$$
\begin{aligned}
(a_0 + \mu_0 a_1)^2 + (b_0 + \mu_0 b_1)^2 - (c_0 + \mu_0 c_1 - \lambda_0)^2/\lambda_0^2 &= 0 \\
\mu_0 - c_1 \lambda_0 + (a_0 a_1 + b_0 b_1 + c_0 c_1) &= 0.
\end{aligned}
\tag{3.10}
$$

The first equation fixes the intersection with g and is quartic. The second equation states perpendicularity and is linear. We summarize the findings with

Theorem 3. *Let g be a line or a cylinder. Determining the closest approach pairs with a cylinder or line requires solving a linear system, determining closest approach pairs with a cone requires solving a quadratic equation, and with a torus it requires solving a quartic equation.*

Closest approach pairs with a cone

Assume that f and g are both cones. We assume that one of the cones, say f, has the equation (3.3), and that the axis of g is given by the line (3.6). Moreover, the axes of the two cones are assumed to be skew. We identify a normal cone of f and a normal cone of g such that each cone contains the apex of the other. The normal cones to g can be expressed as a quadratic form $\hat{g}(x, y, z, \mu_0)$, where μ_0 specifies the apex position on the axis of g. Thus, we must solve a system with the variables λ_0 and μ_0 of the form

$$
\begin{aligned}
(a_0 + \mu_0 a_1)^2 + (b_0 + \mu_0 b_1)^2 - (c_0 + \mu_0 c_1 - \lambda_0)^2/u^2 &= 0 \\
\hat{g}(0, 0, \lambda_0, \mu_0) &= 0.
\end{aligned}
\tag{3.11}
$$

The system consists of two quadratic equations.

If f is a torus, we assume a coordinate choice as before with the axis of rotation the z-axis and the interior skeleton of the torus the unit circle in the xy-plane. The axis of g is again assumed to be given by (3.6), and the normal cones by $\hat{g}(x, y, z, \mu_0)$. Here, we solve the system

$$\begin{aligned}
(a_0 + \mu_0 a_1)^2 + (b_0 + \mu_0 b_1)^2 - (c_0 + \mu_0 c_1 - \lambda_0)^2/\lambda_0^2 &= 0 \\
\hat{g}(0, 0, \lambda_0, \mu_0) &= 0.
\end{aligned} \qquad (3.12)$$

Thus we must solve a quadratic and a quartic equation. Summarizing the situation, we have

Theorem 4. *Let g be a cone. Then the closest approach pairs with a skew cone are found by solving a system with two quadratic equations, and the closest approach pairs with a torus are found by solving a system with a quadratic and a quartic equation.*

Closest approach pairs with a torus

Let f be a torus, where the coordinate system is aligned as before, and assume that g is another torus whose axis is specified by (3.6). We assume that the two axes are skew. We must find a line intercepting the axis of f at $z = \lambda_0$ and the axis of g at a point specified by μ_0 that also intersects the interior skeletons of g and of f. Such lines are determined by two cones of normals, one consisting of normals to f, the other of normals to g, where each cone contains the apex of the other cone. We recall that the normal cone to a torus has an implicit equation of degree 4 when the apex position is a variable, so that we must solve a system of two quartic equations.

Theorem 5. *Let f and g be two tori. Then the closest approach pairs are determined by solving a system of two quartic equations.*

3.3   Closest approach pairs involving curves

Edges of CSG objects have carriers that can be represented as the intersection of two surfaces g and h. The surfaces have degree one, two, or four. If p is a point on the carrier, it satisfies the implicit equations of g and of h. Moreover, the normal plane to the curve, at p, is spanned by the gradients to g and h, evaluated at p and denoted N_g and N_h, respectively. The tangent vector to the curve at p is given by the cross product of the normals, i.e., by $N_g \times N_h$.

In the following, we assume that g and h intersect transversally at p. That is, p must not be a singular curve point of $g \cap h$. We formulate the system with the implicit equations of g and h. If the intersection curve is parametrizable, for example when g is a plane and h is a quadric, the parametric form of the curve can be used instead, and doing so simplifies the systems by lowering the number of variables to be determined. The details are routine and we will not discuss this variant further.

The curve/surface case

We explain how to find closest approach points between the curve $g \cap h$ and the surface f. If p is a point on f and q is on $g \cap h$, then they are a closest approach pair only if the connection line $\overline{pq}$ lies in the normal plane to $g \cap h$ at p.

If f is a plane, let $z = 0$ be its implicit equation. We seek a curve point p at which there is a linear combination of the surface normals that is perpendicular to f. Let $p = (x_0, y_0, z_0)$ be a point on the curve at which such a linear combination exists. Then p is found by solving

$$
\begin{aligned}
g(x_0, y_0, z_0) &= 0 \\
h(x_0, y_0, z_0) &= 0 \\
u_1 g_x + u_2 h_x &= 0 \\
u_1 g_y + u_2 h_y &= 0.
\end{aligned}
\tag{3.13}
$$

The system has one degree of freedom accounted for by the fact that only the ratio of u_1 and u_2 is important. We could therefore proceed by solving the system (3.13) first with $u_1 = 1$, and then solving it again with $u_1 = 0$ and $u_2 = 1$. Note that the equations do not exceed the maximum degree of g and h.

If f is a sphere, let $q = (a, b, c)$ be its center. We need to find a curve point $p = (x_0, y_0, z_0)$ in whose normal plane we can find a line that contains q:

$$
\begin{aligned}
g(x_0, y_0, z_0) &= 0 \\
h(x_0, y_0, z_0) &= 0 \\
u_1 N_g + u_2 N_h &= (x_0 - a, y_0 - b, z_0 - c).
\end{aligned}
\tag{3.14}
$$

The third equation states that a certain linear combination of the two normals is equal to the vector from the curve point to center of the sphere. It is equivalent to three scalar equations.

If f is a cylinder, assume its axis is given by the line (3.6). We seek a curve point $p = (x_0, y_0, z_0)$ whose normal plane contains a line intersecting the cylinder axis at a right angle. We assume that the z-axis is the cylinder axis. Thus we have to solve the system

$$
\begin{aligned}
g(x_0, y_0, z_0) &= 0 \\
h(x_0, y_0, z_0) &= 0 \\
u_1 N_g + u_2 N_h &= (x_0, y_0, z_0 - \lambda_0) \\
(x_0, y_0, z_0 - \lambda_0) \cdot (a_1, b_1, c_1) &= 0.
\end{aligned}
\tag{3.15}
$$

Here, $(0, 0, \lambda_0)$ is the intersection of the normal line with the cylinder axis. The third equation again corresponds to three scalar equations.

If f is a cone, we seek a normal line that lies on a cone of normals to g. Assuming that the cone g has the equation (3.3). We solve the following system

$$
\begin{aligned}
g(x_0, y_0, z_0) &= 0 \\
h(x_0, y_0, z_0) &= 0 \\
u_1 N_g + u_2 N_h &= (x_0, y_0, z_0 - \lambda_0) \\
x_0^2 + y_0^2 - (z_0 - \lambda_0)^2 / u^2 &= 0.
\end{aligned}
\tag{3.16}
$$

For the torus, finally, we assume a coordinate system as in Section 3.2. We seek a line normal to the curve that intersects the z-axis at $(0, 0, \lambda_0)$ and also intersects the unit circle in the xy-plane. This line has the direction $(x_0, y_0, z_0 - \lambda_0) = u_1 N_g + u_2 N_h$. We solve the following system:

$$
\begin{aligned}
g(x_0, y_0, z_0) &= 0 \\
h(x_0, y_0, z_0) &= 0 \\
u_1 N_g + u_2 N_h &= (x_0, y_0, z_0 - \lambda_0) \\
x_0^2 + y_0^2 - (z_0 - \lambda_0)^2 / \lambda_0^2 &= 0.
\end{aligned}
\tag{3.17}
$$

Note that the last equation expresses perpendicularity of the line to the torus.

In case the space curve is the intersection of two quadrics, we obtain systems whose nonlinear equations have degree 2, and, in the case of the torus, an additional quartic equation.

The curve/curve case

Let $g_1 \cap h_1$ and $g_2 \cap h_2$ be the two curves on which we seek closest approach pairs (p, q). With $p = (x_1, y_1, z_1)$ and $q = (x_2, y_2, z_2)$, we find the points subject to the condition that each lies on the normal plane of the other, or, equivalently, that the connection line is in both normal planes.

$$
\begin{aligned}
g_1(x_1, y_1, z_1) &= 0 \\
h_1(x_1, y_1, z_1) &= 0 \\
g_2(x_2, y_2, z_2) &= 0 \\
h_2(x_2, y_2, z_2) &= 0 \\
u_1 N_{g_1} + u_2 N_{h_1} &= (x_1 - x_2, y_1 - y_2, z_1 - z_2) \\
v_1 N_{g_2} + v_2 N_{h_2} &= (x_1 - x_2, y_1 - y_2, z_1 - z_2).
\end{aligned}
\tag{3.18}
$$

Note that the degrees of the equations do not exceed the degree of the surfaces involved.

4   Local Skeleton Analysis

Given a bisecting point, we develop a criterion to test whether the point is on the skeleton of the solid T, and if so, whether it is a vertex, on an edge, or on a face of the skeleton. Let p be a point of closest approach to the two elements E_1 and E_2 of the boundary of T. We determine the minimum distance of p to every other boundary element E_3. If p is closer to some E_3 than to E_1 and E_2, then p is not on the skeleton of T, otherwise it is.

As a result of the admissibility test, for each point of closest approach we find s distinct boundary elements of equal minimum distance. This is recorded as the structure

$$(p, r, (E_1, p_1), ..., (E_s, p_s))$$

where p is the point, r the minimum distance of p from the boundary of T, and, for $1 \leq k \leq s$, p_k is a footpoint of p and E_k is the boundary element on which p_k lies.

Now let $(p, r, (E_1, p_1), ..., (E_s, p_s))$ be a skeleton point so determined. Then p is on a face, edge, or vertex of the skeleton according to whether the Jacobian of the r-offset surfaces of $E_1, ..., E_s$, considered as intersecting hypersurfaces in $\mathbf{R}^n$, has rank $n - 2$, $n - 1$, or n, assuming transversal intersection. Here, n is determined by the number of variables needed to formulate the equation, and is in the simplest case 4 [7]. So, the dimension of the tangent space at p of the r-offset intersection determines the topology of the skeleton in the neighborhood of p.

Suppose we are at a point on an edge or vertex of the skeleton, and wish to find the adjacent faces and edges of the skeleton not yet known. We proceed as follows to find the carriers of adjacent faces and edges:

1. Let $E_1, ..., E_s$ be the boundary elements whose r-offsets intersect at p and contain the footpoints of p. Let $n - k$ be the rank of the Jacobian, where $k = 0$ or $k = 1$.

2. Select all subsets of the E_k such that the Jacobian of the r-offsets, at p, has rank at most $n - k - 1$ but is not smaller than $n - 2$. The intersection of these r-offsets, for each subset, forms an adjacent Voronoi surface or curve and is one of the carriers we seek.

For each carrier so found, we then determine the edge or face that lies on it. This is done as follows.

If p is on an edge, we need to determine the direction in which the adjacent face lies. This is done by finding a point on either side of the edge and determining its distance from the boundary of T. If the distance is less than r, then the point is either not on the skeleton or else is on one of the faces we already know. Otherwise, the point is on a new face. If p is a vertex, that is, if the rank of the Jacobian is n, then the procedure for

finding adjacent edges is analogous. In this case, we do not find adjacent
faces directly. Rather, we first locate the adjacent edges, and then locate
adjacent faces from points on the edges.

5 Building a Skeleton Section

We assume we are given all points of closest approach between boundary
element pairs, sorted by their minimum distance from the boundary. Let
the corresponding distance sequence be $r_0, r_1, ..., r_m$. By a *section* of the
skeleton we mean those skeleton points whose distance from the boundary
is between r_{k-1} and r_k, where $1 \leq k \leq m$. There is one additional skeleton
segment with points at distance greater than r_m. In step 3 of the algorithm,
the skeleton is constructed section by section.

The first section is constructed beginning with all points at distance r_0.
If the boundary of T contains locally convex edges, then these edges are
the initial set and $r_0 = 0$. The skeleton faces and edges are now evaluated
with the techniques of [6], in a four-dimensional (x, y, z, r)-space, where r
is the distance of a skeleton point to the boundary of T. Using a first-
order approximant, in 4-space, allows us to cut an approximant at the
next critical distance r_1 if needed. Similarly, when constructing subsequent
sections, we limit approximants to extend up to but not beyond the next
critical distance.

By using the marching-cubes approach of [6], we can detect whether
faces or edges are about to intersect. The intersection of skeleton faces
can be done either by intersecting the defining equation systems, or by
intersecting a face with its trimming surface, also specified as a system of
equations [7].

When building a skeleton section, we thus construct vertices, edges and
faces ordered by their distance from the boundary. At any time, several
edges and faces have been partially evaluated, and these partial evaluations are the current *frontiers*. When reaching the next critical distance
r_k, new frontiers are added to the list of current frontiers, namely those
corresponding to the closest approach points at distance r_k. Eventually,
each frontier closes. This happens when several frontiers come together
without generating a new frontier. That is, the approximants at two or
more frontiers join, and at their intersection the local approximants are
not increasing in distance from the boundary, nor are there subsets of the
nearest element set that have such approximants.

6 Summary

We have sketched a general algorithm for constructing the (interior) skeleton of three-dimensional solids. The algorithm first determines points of

closest approach, sorts them by distance, and then constructs the skeleton by increasing distance. The restriction to CSG impacts fundamentally only the determination of closest approach points. In section 3 we show how this restriction can be exploited by deriving very simple algebraic systems that can be solved without expensive machinery. Given a good equation solver, however, the algorithm is easily extended to solids with more general boundaries.

The face and edge construction of the skeleton is done numerically, based on the generic techniques developed in [5, 6, 9]. The advantage is a uniform code that applies unchanged to all types of curved boundary elements, not only those arising in CSG. By constructing approximants in a four-dimensional space, moreover, we can determine robustly when the skeleton construction is locally completed, by monitoring the current distance from the boundary.

Acknowledgements

This research has been supported in part by the National Science Foundation under grants CCR 86-19817 and DMC 88-07550, and by the Office of Naval Research under contract N00014-90-J-1599. Helmut Pottmann and Kokichi Sugihara suggested a number of valuable improvements to the manuscript. Thanks also to Ching-Shoei Chiang, Pamela Vermeer, and Jianhua Zhou for their suggestions. Malcolm Sabin since pointed out to me the work by Cecil Armstrong at the Queen's University in Belfast. Armstrong computes the skeleton from a Delaunay triangulation of points on the boundary, and uses the skeleton to generate finite element meshes in 2D and 3D.

References

1. Blum, H. (1967). A transformation for extracting new descriptors of shape. In *Models for the Perception of Speech and Visual Form*, Weiant Whaten-Dunn, eds., MIT Press, Cambridge, MA, 362–380.

2. Blum, H. and Nagel, R. (1978). Shape description using weighted symmetric axis features. *Pattern Recognition 10*, 167–180.

3. Bookstein, F.L. (1979). The line skeleton. *Comp. Graphics Image Proc. 11*, 123–137.

4. Boots, B., Okabe, A. and Sugihara, K. (1991). *Spatial tesselations — concepts and applications of Voronoi diagrams*. John Wiley and Sons, Chichester, England, 1991.

5. Chandru, V., Dutta, D. and Hoffmann, C. (1990). Variable-radius blending with cyclides. In *Geometric Modeling for Product Engineering*, 39–58, K. Preiss, J. Turner, M. Wozny, eds., North Holland.

6. Chuang, J.-H. (1990). Surface Approximations in Geometric Modeling. PhD Dissertation, Comp. Sci., Purdue University.

7. Dutta, D. and Hoffmann, C.M. (1990). A geometric investigation of the skeleton of CSG objects. *Proc. ASME Conf. Design Automation*, Chicago, IL.

8. Hoffmann, C. (1989). *Geometric and Solid Modeling, an Introduction*, Morgan Kaufmann, San Mateo, CA.

9. Hoffmann, C. (1990). A dimensionality paradigm for surface interrogation. *CAGD*, to appear.

10. Hoffmann, C. (1990). Algebraic and numerical techniques for offsets and blends. In *Computations of Curves and Surfaces*, M. Gasca, W. Dahmen, S. Micchelli, eds., Kluwer Academic Publ., 499–528.

11. Lee, D.T. (1982). Medial axis transformation of a planar shape. *IEEE Trans. Pattern Anal. Mach. Intell.*, PAMI-4, 363–369.

12. Montanari, U. (1969). Continuous skeletons from digitized images. *JACM 16*, 534–549.

13. Nackman, L.R. (1982). Curvature relations in three-dimensional symmetric axes. *Comp. Graphics Image Proc. 20*, 43–57.

14. Nackman, L.R. and Srinivasan, V. (1989). Bisectors of linearly separable sets. *Res. rept. RC14140*, IBM Yorktown Heights, NY.

15. O'Rourke, J. and Badler, N. (1979). Decomposition of three-dimensional objects into spheres. *IEEE Trans. Pattern Anal. Mach. Intell.*, PAMI-1, 295–305.

16. Patrikalakis, N.M. (1989). Shape interrogation. *Proc. 16th Annl. MIT Sea Grant College Program Lecture and Seminar*, Hemisphere Pbl., Cambridge, MA, 83–104.

17. Patrikalakis, N.M. and Gursoy, H. (1989). Shape interrogation by medial axis transform. *Design Lab. Memo. 90-2*, Sea Grant College Program, MIT.

18. Preparata, F.P. (1977). The medial axis of a simple polygon. *Proc. 6th Symp. Math. Foundations of Comp. Sci.*, 443–450.

19. Preparata, F. and Shamos, M. (1985). *Computational Geometry*, Springer Verlag, New York.

20. Requicha, A. and Voelcker, H. (1977). Constructive solid geometry. *Tech. Memo 25*, Prod. Automation Proj., University of Rochester.

21. Rossignac, J. and Requicha, A. (1984). Constant-radius blending in solid modeling. *Comp. Mech. Engr. 3*, 65–73.

22. Rossignac, J. and Requicha, A. (1984). Offsetting operations in solid modeling. *CAGD 3*, 129–148.

23. Srinivasan, V. and Nackman, L.R. (1987). Voronoi diagram of multiply connected polygonal domains. *IBM J. Res. Dev. 31*, 373–381.

24. Srinivasan, V., Nackman, L.R. and Meshkat, S. (1989). Automatic mesh generation by symmetric partitioning of polygonal domains: I. algorithm. *Res. Rept.*, IBM Yorktown Heights, NY.

25. Stifter, S. (1990). The Roider method: A method for static and dynamic collision detection. In *Issues in Robotics and Nonlinear Geometry*, C. Hoffmann, ed., JAI Press.

Interference Checking in the 5-axis Machining of Parametric Surfaces

E. McLellan[1], G.M. Young[2], R.J. Goult[1] and M.J. Pratt[3]

[1]*LMR Systems, Newport Pagnell, Buckinghamshire,* [2]*School of Electronic Systems Design, Cranfield Institute of Technology, Bedford and* [3]*Design and Manufacturing Institute, Rensselaer Polytechnic Institute, New York, USA*

1 Introduction

In the numerically controlled machining of 'sculptured' parametric surfaces a cutting tool is driven so that (subject to specified numerical tolerances) it remains in contact with the desired surface, removing unwanted material as it passes. There are several possible strategies for choosing the cutter contact paths. The choices most commonly used are (i) a series of isoparametric lines on the surface, and (ii) a set of intersection curves of the surface with an equidistantly spaced set of parallel planes. In the first case a parameter increment in the transverse direction is specified to determine what is known as the *stepover* between successive cutter passes; in the second case the spacing between planes fulfils the same function. The actual cutter contact paths are approximated in terms of a sequence of small linear or circular arc motions of the cutter, depending upon the capabilities of the machine tool control unit [1]. The lengths of the individual steps (known as *stepout*) are usually determined from an assessment of local curvature of the cutter contact paths.

Cutting tools have a cylindrical shaft and a cutting head whose geometry may take a variety of forms. In machining sculptured surfaces a ball-end cutter is commonly used. This has a hemispherical cutting surface, and its use leads to comparatively simple cutter control computations, since the centre of the cutting head moves on a surface which is a constant parallel offset from the one being machined. More complex is the corner-radius cutter, which has a flat end and a toroidal region giving a smooth transition onto the cylindrical side of the cutting head. In the limiting case where the corner radius is zero a flat-ended cutter results. In either of the last two cases the computation of the motion of the tool reference point (the centre of the flat end) with respect to the surface being machined is much more complex than for the ball-ended cutter. However, as explained below, the use of these more general cutter forms can result in distinct practical advantages. Other yet more complex cutter geometries are increasingly being used in practice.

As mentioned earlier, the motion of the cutter is governed by the machine tool controller. The input to this controller is generated by computer means. For simple parts this may be done using the computational facilities of the controller itself, by the machine operator on the shop floor. For more complex parts, however, particularly those having complex surfaces of the type considered here, some form of 'part program' is usually written off-line. This consists of a section describing the geometry of the part and a section specifying, at a high level, motion of the cutter with respect to that geometry. Nowadays the geometry is often derived directly from a part model generated by a CAD system. The part program is processed by the computer, the result being the specification of a large number of low-level tool motions together with other machine tool control information, expressed in a standard format known as CLDATA (cutter location data) whose logical structure is specified in ISO 3592-1978 [2]. This must then be further postprocessed to tailor it for input to a specific machine tool controller.

At the completion of this process, the machine control data must be tested. The most obvious procedure is actually to use it to drive the machine tool in the machining of a test part. This is expensive, for several reasons. Firstly, it takes the machine tool out of productive operation for a while. Secondly, it may show up deficiencies in the machining program; in particular, it may be found that the cutter gouges the part, i.e. machines away material in regions where this is not intended. And thirdly, it may even cause damage to the machine tool itself. For these reasons it is highly desirable, before going to the shop floor, to simulate the effect of the machining program on the computer. The simulation will highlight problems in the machining strategy, which can then be rectified before the program is actually put into productive operation. Most previous work in this area [3-7] has been concerned with the simulation of a complete machining program, any subsequent correction being manually performed. The present paper is however concerned with simulation at a more detailed level; unwanted interferences of the cutting tool are detected at each tool position in the computed path, and corrective action is taken automatically whenever possible. In many cases this will avoid the necessity for repeated runs of the simulation program and subsequent manual corrections to the input data. Little has previously been published on solutions to this problem, though Jerard et al. [5] indicate that they have been investigating an alternative approach to that described here, and Choi and Jun [8] have developed an algorithmic procedure for detecting and avoiding interference when machining with a ball-end cutter.

The main geometric factors affecting the machining accuracy and surface finish that can be achieved when milling sculptured surfaces are the curvatures of the surface and the shape of the tool. This is true whether the part is being manufactured on a 3 or 5 axis machine tool. Under 3 axis control the motion of the tool reference point is controlled, enabling the tool to follow any general trajectory in cartesian space whilst maintaining the axis of rotation of the tool in a fixed direction, typically parallel to the Z axis of the machine tool. On a 5 axis machine tool, operating under multi-axis control, the orientation of the tool axis may be additionally adjusted by rotary motions about two of the cartesian axes.

There are considerable advantages to be gained from using a flat-ended or corner-radius cutter under multi-axis control to machine sculptured surfaces. This is particularly true where the area of the surface is large, reducing the total machining time by enabling a wider region to be machined with a single pass across the surface and eliminating hand finishing. The drawback is that if the surface has regions concave with respect to the tool side of the surface, there may be problems of accessibility. This leads to the possibility of unwanted penetration of the surface or adjacent surfaces by parts of the cutter remote from the intended contact area. This paper discusses the advantages of 5-axis machining and describes an approach to overcoming the problem of unwanted penetrations of the surface.

Whether a ball-ended, corner-radius or flat-ended cutter is used to machine a sculptured surface under 3 axis control, cusps will be left between adjacent passes across a surface. The heights of these cusps are dependent on the surface curvature, tool geometry and the stepover between cuts. However if 5 axis control is available the full benefits of using a flat-ended cutter (with or without a corner radius) can be realized by maintaining the tool axis parallel to the surface normal at the contact point [9,10]. In practice, the tool axis is normally inclined at a slight angle to the surface normal to achieve better cutting conditions.

If the tool side of the surface is always convex, then the complete surface can be machined to the desired tolerance without any unwanted penetrations by maintaining the tool axis parallel to the surface normal. However should there be any concave or saddle regions on the surface gouging will almost certainly occur unless some avoiding action is taken.

The procedures that have been developed for detecting these undesired penetrations and avoiding them require as input a previously determined tool location and axis orientation and the surface parameters at the point of contact. The surface geometry is accessed via a parametric evaluator which returns position and values of first and second derivatives. The method is therefore easily implemented with any type of parametric surface.

2 Method

Since for large multi-patch surfaces most of the surface patches will be
remote from the tool at any instance a two stage process is used. The first
stage determines which patches are sufficiently remote from the tool for
there to be no possibility of any interference, and the second stage computes
the maximum penetration, if any, of the remaining suspect patches.

   The procedures that have been developed for detecting and avoiding undesired penetrations of parametric surfaces during machining assume that
an ideal tool location t_e and axis orientation t_a have been determined,
where the tool is in tangential contact with the surface at a known parametric location. This approach has been taken so that there is no restriction
on the basic machining strategy that may be employed.

   For example, the most common strategy used for 5-axis machining of a
surface is to maintain the point of contact of the tool with the surface on
an isoparametric line of the surface and align the tool axis at a specified
angle to the surface normal at the point of contact, in the plane defined
by the surface normal and the direction of forward motion. Alternatively,
the point of contact could be determined by projecting a point, not lying
in the surface, in a specified vector direction onto the surface. Yet another
method could be to constrain the tool axis to lie in a given plane whilst
maintaining the tool in tangential contact with the surface with its axis
at a specified angle to the projection of the surface normal at the point of
contact onto the plane.

   In the initial implementation the tool geometry has been restricted to
the set of tools that can be described in terms of diameter and corner
radius, namely ball-ended, corner-radius and flat-ended cutters (Fig. 1).
However, the basic algorithms are equally applicable to any tool geometry.

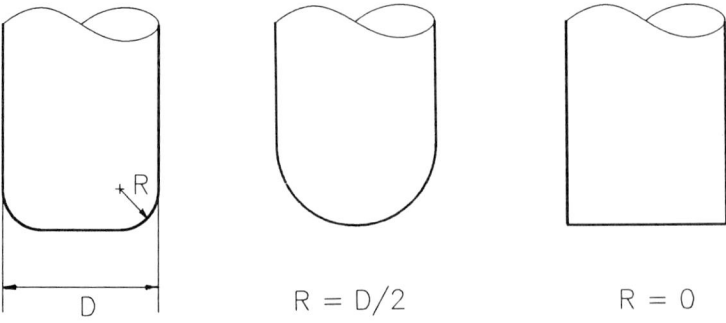

Figure 1. Tool geometry

The only restriction on surface geometry is that it must be addressable via a parametric evaluator, a software package which for any given (u, v) pair of parameter values returns the cartesian coordinates of the corresponding surface point, first and second derivatives and the value of the normal at this point. Parametric evaluators have been written at Cranfield for VDAFS standard surface data [11], for Bézier surfaces of arbitrary degree and for many of the parametric surfaces included in the STEP specification [12].

The procedures followed for each tool location and axis orientation are:

1. Scan all the surface patches except the patch that the tool is intended to be machining to determine any suspect patches where there could possibly be penetration.

2. Check all suspect remote patches for actual interference.

3. Check the patch being machined for actual interference, other than at the cutting point.

Before any of these procedures are invoked a preliminary analysis of the surface is performed.

2.1   Preliminary analysis of surface

To facilitate the scanning of the surface to determine suspect patches, it was decided that some method of creating a set of simple analytic surfaces 'boxing' each patch was required. The technique eventually used is to compute a set of five spheres for each patch. In order to determine these spheres a mesh of points (typically 5*5 for each patch) and their associated derivatives is evaluated by invoking the appropriate parametric evaluator for the surface.

The five spheres are:

(a) The sphere of minimum radius that fully encloses the patch.

(b) Up to 4 'edge' spheres. An 'edge' sphere is one that just encloses a set of evaluated points lying on a patch boundary curve, whose centre lies on the perpendicular bisector of the chord between the two end points which also intersects the line through the end points of the opposite boundary. No sphere is computed for a degenerate edge.

The last four spheres are intended to be as close a fit as possible to the actual surface boundaries. In practice, the sphere radii are increased by a small percentage to compensate for the fact that only sample data points are used.

Although the computational effort required to calculate these enclosing spheres is significant, it need only be made once during the machining of a particular surface. The subsequent checking against these spheres to determine suspect patches is a simple series of tests.

The principal curvatures at the sample points in each patch are also calculated during this preliminary analysis of the surface and the maximum and minimum determined. In addition limits for stepout in u and v for each patch are calculated. These are used to test for convergence in the subsequent iterative procedures involved in determining the presence or absence of surface penetration by the tool. These limits depend upon the tolerance requirements and the relative magnitude of the geometric and parametric ranges of the surface.

2.2   Elimination of remote patches

The scanning of remote surface patches, performed at each tool location, is speeded up by determining the infinite tool cylinder (the cylinder whose axis is coincident with the tool axis and whose diameter is that of the cylindrical part of the tool) and three spheres enclosing the tool:

1. The tool contact sphere which is determined by the tool geometry and the angle between the tool axis, t_a, and the surface normal r_n at the point of contact (Fig. 2). This sphere has its centre at the intersection of r_n and the tool axis.

2. The large shallow sphere tangential to the tool tip and containing the cutting portion of the tool, effectively a plane for flat ended cutters.

3. The minimum radius sphere enveloping the cutting portion of the tool (Fig. 3).

The non-current patches are then scanned for possible interference using the following schedule for each patch.

1. Test for interference of the infinite tool cylinder with the minimum sphere enclosing the patch.

2. Test for interference of the minimum radius sphere with all spheres associated with the patch.

3. Test for interference of the shallow tool sphere with all spheres associated with the patch.

4. Test for interference of the tool contact sphere with all spheres associated with the patch.

5. If the cutter is not ball-ended, test for interference of the actual tool with all spheres associated with the patch.

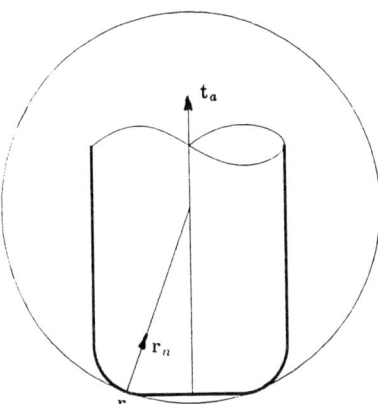

Figure 2. Tool contact sphere

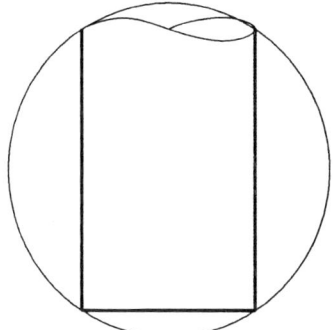

Figure 3. Tool enveloping sphere

To minimize the computation a patch is cleared from being suspect as soon as a negative result occurs in one of the above tests. Any patch that gives positive results in all of the tests is added to the list of suspect patches which are examined in detail in the next stage.

2.3 Final stage

There are two separate tasks in the final stage of the interference checking procedure. The first is to test the suspect remote patches for actual penetration; the second is to examine the current patch, with which the tool is in tangential contact.

Interference with remote patches

For each suspect remote patch, iterative procedures to determine the penetration of the tool into (or the minimum distance of the tool from) the patch are performed. If interference was detected with the patch at the previous tool position, the (u, v) parameters of that penetration point are taken as the iteration start values (u_0, v_0), otherwise the parametric midpoint of the patch is used.

The iterative search for interference of a tool with a patch is performed using the following algorithms. Given (u_0, v_0), evaluate

$$\mathbf{r}(u_0, v_0)$$

$$\mathbf{r}_u(u_0, v_0); \quad \mathbf{r}_v(u_0, v_0)$$

$$\mathbf{r}_n(u_0, v_0) = \frac{\mathbf{r}_u \times \mathbf{r}_v}{|\mathbf{r}_u \times \mathbf{r}_v|}.$$

Compute

d penetration distance (+ve indicates a hit)
p associated closest point on tool
p_n tool surface normal at p

and

t point on tool surface having tool normal t_n parallel to $\mathbf{r}_n$.

Values of δu and δv to give a better estimate of the penetration point are then determined. In general, two algorithms are used, illustrated in Figs. 4 and 5, the ultimate objective being to align p_n and $\mathbf{r}_n$. Note that if t lies on the flat base of the cutter the use of Algorithm 2 is forced.

Algorithm 1:
Given a surface point $\mathbf{r}$ *and surface normal* $\mathbf{r}_n$ *and a point on the tool* t *with the parallel tool normal* t_n, *compute the point* $\mathbf{b}$ *where this* t_n *intersects the tangent plane at* $\mathbf{r}$, *then by induced parameterization on the tangent plane determine the stepout in* δu *and* δv *on the surface,* $(\delta u_1, \delta v_1)$.

Algorithm 2:
Given an initial surface point $\mathbf{r}$, *and the tool normal* p_n *at the 'nearest' point on the tool* p, *using a local quadratic approximation to the surface determine* $\delta u, \delta v$ *for the point* s *with the same normal* p_n, $(\delta u_2, \delta v_2)$.

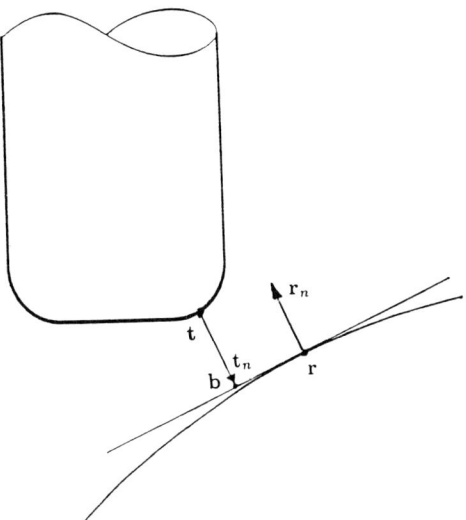

Figure 4. Geometry for algorithm 1

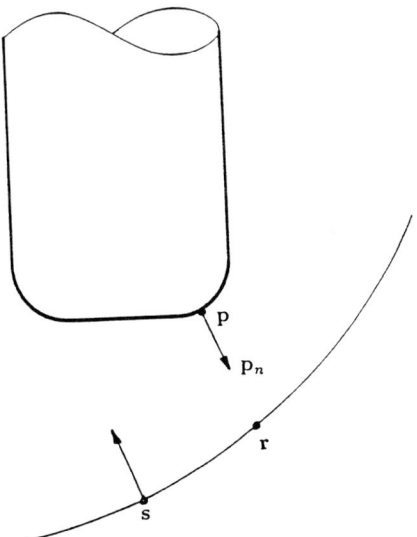

Figure 5. Geometry for algorithm 2

Whenever possible the results of both algorithms are combined using the formula below, otherwise the results of the appropriate algorithm are used.

$$\delta u = \frac{\delta u_1 \times \delta u_2}{\delta u_1 + \delta u_2}, \qquad \delta v = \frac{\delta v_1 \times \delta v_2}{\delta v_1 + \delta v_2}.$$

The values of δu and δv are limited to keep $\mathbf{r}(u, v)$ within the current patch.

The resultant values of δu and δv are compared with the limiting values computed in the preliminary analysis in order to test for convergence. If no convergence has been achieved after a given number of iterations (typically 20), three more iterations are performed and the results of the last four iterations are interrogated to determine the worst penetration detected.

On convergence the surface point, tool point and normal in addition to the penetration and the (u, v) parameters of the surface point where the penetration is deepest are recorded. If there is no convergence then the worst case results are saved.

When a positive penetration is detected, indicating interference, the details are stored for subsequent interrogation to determine the avoidance strategy.

Interference with the current patch

Once the checking for penetration of remote patches has been completed, the current patch is examined. This patch is handled differently because the tool is known to be in contact with it and the search is to determine if the tool penetrates the surface elsewhere within the patch. The following test schedule is used.

1. Test for the patch being convex or 'pseudo' convex when viewed from the tool by comparing the maximum sampled curvature for the patch with a curvature tolerance dependent on the current machining tolerance value and the tool diameter. This amounts to treatment of shallow concave patches as if they were convex in the case where the cutter geometry indicates that any interference would be within the permitted machining tolerance. If the patch is convex or 'pseudo' convex then no local interference should occur.

2. Given

    κ_t = Curvature of the minimum spherical enclosure for the tool with the cutting point as a diametral point of contact.

    κ_p = Maximum sampled curvature of the patch.

    If $\kappa_p \leq \kappa_t$ no local interference can occur.

3. Otherwise the same algorithms are used as for the remote patches to search for interference on the patch, using a number of starting points arranged in a pattern about the contact point. (The contact point cannot be used as a starting point for these searches, since it is a solution). Five start points are actually used. One is in the forward direction, one to each side, and two in the rearward direction with respect to the point of contact, at distances approximately D/4 and 3D/4, where D is the tool diameter.

All non-coincident penetration points are stored for subsequent consideration. If the algorithms converge to the same point as a result of different start points only one solution will be saved.

Finally, if any interference is detected with either a remote patch or the current patch, the largest penetration is determined and its location recorded for use in determining the avoidance strategy to be employed.

3 Avoidance Strategy

There are three possible avoidance strategies:

1. tilt the tool axis,

2. lift off normal to the surface,

3. lift off in the direction of the tool axis.

In the present implementation the user is allowed to select the preferred strategy.

If the first strategy has been selected then an attempt is made to avoid interference by tilting the tool in the most appropriate direction, either rolling about or pitching towards or away from the forward direction perpendicular to the tool axis, $\mathbf{t}_a$. This is done by first determining the most appropriate direction, then calculating the tilt angle α required for minimal avoidance of the surface at the worst point of interference. The direction is indicated by the local axis $\mathbf{u}^*$ perpendicular to $\mathbf{t}_a$ that defines the plane in which the tilt is to occur, the sign of the angle being selected accordingly. An additional constant angle γ, specified by the user, is added to ensure a minimum clearance of the surface to allow free rotation of the tool. The resultant angle β can be constrained to be within user specified limits.

The new tool axis vector $\mathbf{t}_a$ that would result from tilting the tool through β in the direction $\mathbf{u}^*$ is computed from the equation

$$\mathbf{t}_a = \mathbf{t}_a \cos \beta + \mathbf{u}^* \sin \beta.$$

Also calculated are $\mathbf{t}_e$, the new tool-end coordinates, and the centre and radius of the new tool contact sphere.

A series of checks is then made in order to confirm that interference has been sucessfully avoided.

First a check is made for possible local interference of the tool with the surface at the point of contact as a result of an increase in the effective radius of the tool caused by tilting. If interference is detected then the maximum angle of tilt β_{max} is computed such that the tool will fit the surface at the point of contact. This may not be possible. If a suitable angle is computed it is again constrained to the permitted limits and yet again a new $\mathbf{t}_a$ and $\mathbf{t}_e$ are computed, otherwise the original $\mathbf{t}_a$ and $\mathbf{t}_e$ are restored and the tool will be lifted off as the only means of avoiding interference.

Finally, the new tool position and orientation ($\mathbf{t}_e$ and $\mathbf{t}_a$) are checked for possible interference by calling the interference checking procedure again, using the previously found interference points as start points. If avoidance has been successful these start points are stored for use as start data for the next tool location.

If tilting has not sucessfully avoided interference or lift off alone was selected, then an attempt is made to avoid the unwanted penetration by lifting off in the specified direction.

The lift off distance is taken as

$$\Delta = \delta + \epsilon + \tau$$

where

$$\delta = \frac{d}{\mathbf{r}_n \cdot \mathbf{v}}$$

and the various symbols have the following meaning:

d maximum penetration.
$\mathbf{r}_n$ unit surface normal at the point of maximum penetration.
$\mathbf{v}$ unit vector in direction of desired liftoff.
$\epsilon = 0$ if surface is convex at contact point.
$\epsilon = \sqrt{r_c^2 - d^2 + \delta^2} - r_c$ if surface is concave at contact point.
r_c minimum principal radius or curvature.
τ user specified tolerance.

A new tool location $\mathbf{t}'_e = \mathbf{t}_e + \Delta\mathbf{v}$ is determined and $\mathbf{t}_a$ is unchanged. If the liftoff selected was normal to the part surface the interference checking procedure is again called to confirm that no new interference has been generated as a result of taking avoiding action.

Note: Only the tilting strategy will permit the original specified surface to be machined. Whenever liftoff is invoked some unwanted material will remain uncut; this is indicated by warning messages in the output.

4   Results and Conclusions

A number of real and simulated cases have been processed through the software and known interference points detected and successfully avoided as illustrated in Figs. 6 and 7.

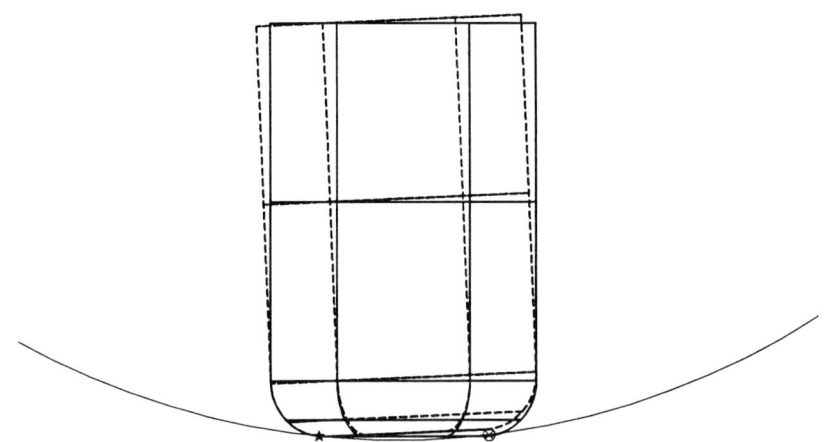

Figure 6. Avoidance by adjusting tool axis

Although every effort has been made to speed up interference checking, it is still a lengthy process and should only be invoked when there is known to be the probability of unwanted penetration. In general the algorithms converge well, though they can become unstable where the curvature of the surface is close to the corner radius of the tool. In this case there is a tendency for the 'nearest' point **p** on the tool to oscillate between two locations.

The work to date has assumed large multi-patch surfaces such as those frequently used in the motor industry. We propose to extend the concept by reparameterizing the surface to give parametric areas of the same order of magnitude as the cutter footprint, in order to improve the performance, particularly on large single patch surfaces. We also intend to implement more general tool geometry so that the method can handle the conical tools often used in practice.

Acknowledgement

The work described was originally carried out under contract to Computer Aided Manufacturing International, Inc. (CAM-I) for the Sculptured Sur-

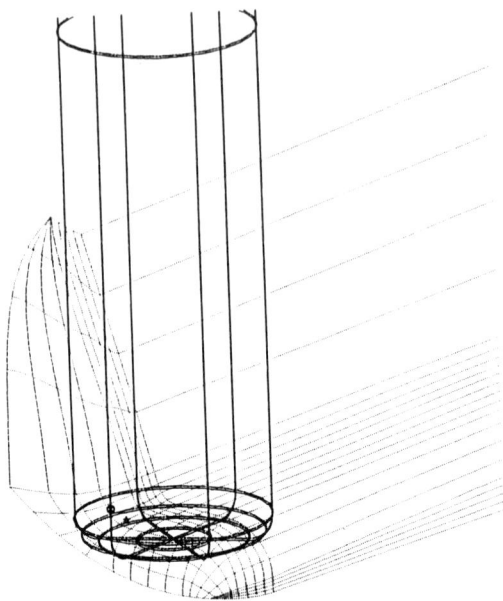

Figure 7. Avoidance by lifting off along tool axis

faces Program, and the software is implemented in the SSV3 version of the CAM-I APT IV Processor [13].

References

1. Koren, Y. (1983). *Computer Control of Manufacturing Systems*. (International Students Edition). McGraw-Hill International Book Company Japan.

2. ISO 3592 (1978). Numerical control of machines - NC processor output - Logical structure and major words (International Standard).

3. Drysdale, R.L., Jerard, R.B., Schaudt, B. and Hauck, K. (1989). Discrete simulation of NC machining. *Algorithmica*, *4*, 33-60.

4. Fridshal, R., Cheng, K.P., Duncan, D. and Zucker, W. (1982). Numerical control part program verification system. *Proc. Conf. on CADCAM Technology in Mechanical Engineering*; MIT Press, Cambridge, MA.

5. Jerard, R.B., Hussaini, S.Z., Drysdale, R.L. and Schaudt, B. (1989). Approximate methods for simulation and verification of NC machining programs. *The Visual Computer*, *5*, 329-348.

6. Wallis, A.F. and Woodwark, J.R. (1984). Creating large solid models for NC toolpath verification. *Proc. CAD84 Conf., Brighton, UK,* Butterworths.

7. Wang, W.P. and Wang, K.K. (1986). Geometric modelling for swept volume of moving solids. *IEEE Computer Graphics and Applications,* 8-17.

8. Choi, B.K. and Jun, C.S. (1989). Ball-end cutter interference avoidance in NC machining of sculptured surfaces. *Computer-aided Design, 21,* 371-378.

9. Klass, R. and Schramm, P. (1991). Numerically-controlled Milling of CAD Surface Data. *Geometric Modelling: Methods and Applications* (eds. H. Hagen and D. Roller). Springer-Verlag.

10. Vickers, G.W. and Quan, K.W. (1989). Ball-mills versus end-mills for curved surface machining. *Trans ASME: Journal of Engineering for Industry,* 111, 22-26.

11. DIN (1984). Verbund der Automobilindustrie Flächenschnittstelle (VDA-FS), DIN 66301 (German National Standard).

12. ISO DS 10303 (1990). Industrial Automation Systems - Exchange of Product Model Data (Draft International Standard).

13. CAM-I (1990). SSV3 APT IV Processor, *Report No. PS-90-SS-01,* CAM-I Inc., Arlington, Texas.

Conical Sails

F.D. Hales

*Department of Transport Technology, Loughborough
University of Technology*

1 Introduction

This paper on yacht sails deals with the characteristics and development of
surfaces that are designed to meet a simple (but optimised) aerodynamics
criteria, [1]. Four sail outlines are considered:

1. Simple triangular sails.

2. Quadrilateral sails.

3. Mainsails.

4. Jib sails.

Sails have been in use for some thousands of years and have largely
evolved to their present state with essentially the combination of craft,
skill and shrewd observation that lies behind many a sucess story.

Whether commercial or pleasure interests have produced the most, or
best, contributions to sail technology, does not concern us today, but it is
useful to list the important factors in present day sails as:

1. materials

2. outline shape

3. aerodynamic requirements of sail surface

4. aerodynamic and tension loading

5. construction.

The need for strong, durable, and rotproof materials is fairly self-
evident. It led naturally to the replacement of natural fibres by terylene
and its treatment by calandering and resinating to achieve reduced porosity,
smoothness, strength, and an approach to isotropy that was inconceivable

with natural fibre fabrics. Even before the use of "kevlar" and "mylar" the notion that sails would settle down, stretch into shape, or "run in" was clearly untenable. Even more so with these later materials. Sails now stay the shape they are made. As the stiffness of the materials increases, the difficulties that arise with significant departures from single curvature surfaces also increase.

The outline sail shapes of modern yachts are dictated by a number of design and rule factors. Many of these parameters are essentially scaling factors and as such are not of concern here, but typically jibsails have a concave leech (back edge) and mainsails a convex leech, or roach, with actual dimensions closely controlled by the designer or class rules.

Sail aerodynamics are complex in detail, but a first approximation can be easily specified by noting that a constant lift section coefficient and constant downwash would lead to a situation of maximum normal force and minimum drag which can be shown to be optimal [1]. Given an appropriate sail planform, the simplest sail section profiles which achieve these conditions are those in which:

1. constant incidence

2. constant section shape

are maintained throughout the height of the sail.

The constant incidence and section sail shape is the required final shape including deflections arising from both aerodynamic and sail control loadings. The sail deflections, due to the aerodynamic loadings (arising from range of wind speeds) over which any individual sail operates, will vary as speed squared. While sail-trimming techniques can attenuate the effect of these deflections it may be expected that significant overall shape changes will occur within the operational envelope of a sail. Whilst the detailed section shape of sails may be complex it may be described primarily in terms of camber (depth/chord) with values in the range (say) 0.05 to 0.15.

Sails of basically triangular shape are restrained normal to 1 or 2 edges (jibs and mainsails) and have high loads applied to the corners. Remembering that the material is often non-isotropic, constructional techniques may be used to diffuse these high corner loads and to provide greater stiffness in the directions of greater stress.

The purpose of this paper is to take as a starting point the concept of a sail defined as a surface having constant aerodynamic incidence and section shape. Then to determine whether and how such a shape can be achieved for the sail planforms that are current, using single curvature surfaces.

The single curvature solution has several important uses in that: if deflections are unimportant then it represents an approximation to the aerodynamically required shape, or if deflections are important then it represents a known zero stress surface to which loadings may be applied and the resulting deflections determined.

In practice it appears likely that the first of these situations is common.

2 Simple Triangular Sails

The simplest sail planform is a triangle with a horizontal foot, Fig. 1. Clearly a segment of the surface of a cone, with its straight line generators, originating at the head of the sail, meets the criteria of both constant incidence and constant section shape. In addition, the single curvature surface of the cone means that the shape can be produced without generating any stress in the surface.

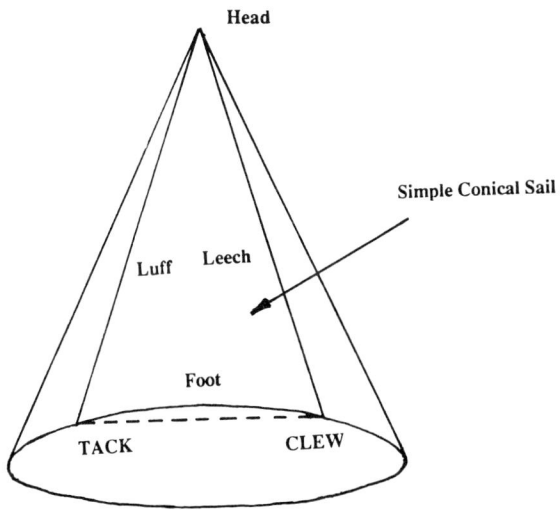

Figure 1. Simple conical sail

The outline of the process, see [1], that carries out these calculations is, see Fig. 2:

1. establish axes

2. specify sail corners and camber

3. solve sail triangle

4. determine cone camber and geometry

5. cone location and orientation

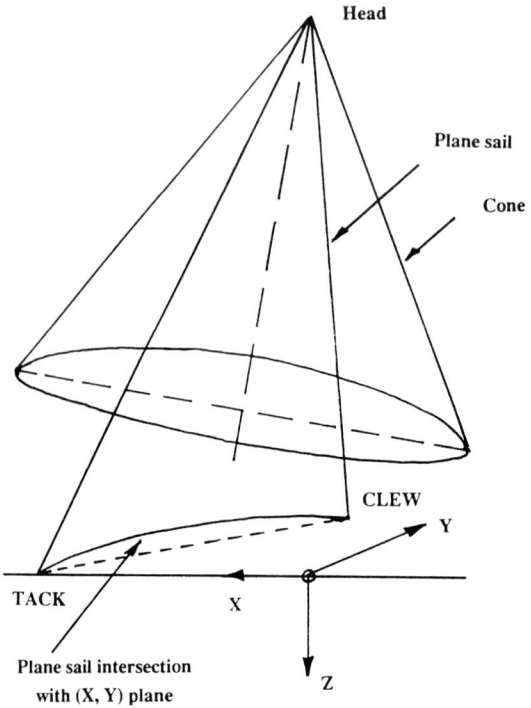

Figure 2. Genoa geometry

6. cone equation

7. cone-horizontal place intersection

8. development of cone segment into a flat surface.

Not all sails have a horizontal foot Fig. 3. A raised foot sail lying on the single cone considered (vertex 0_1 at B) would clearly not meet incidence and shape criteria along the foot of the sail. However, the introduction of a second cone (vertex 0_2 at A) at the tack provides a means of controlling incidence and shape in the foot region. The computational process outlined above needs to be applied to both the upper and lower cones, and then extended by an additional stage (see [2]):

7a) upper-lower cone intersection

yielding 2 segments from the developed cones. An illustration of the projection of this intersection onto the sail plane is shown in Fig. 4.

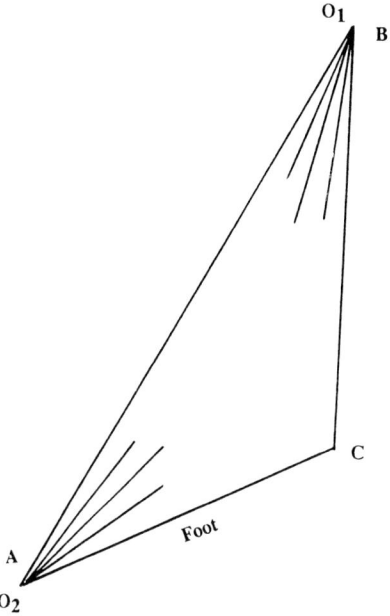

Figure 3. Raised foot triangular sail

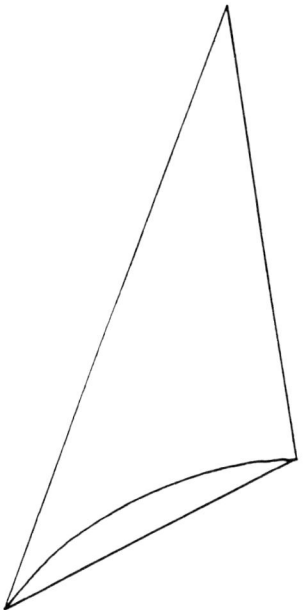

Figure 4. Upper-lower cone intersection

3 Quadrilateral Sails

Examples of quadrilateral sails are not as commonly seen as sails of superficially triangular planform. They do, however, provide not only an interesting study in relation to gaff-rigged boats and spritsail barges, but also provide the bridge to creation of surfaces for actual mainsail and jib planforms.

Consider the quadrilateral sail $ABCD$ of Fig. 5, as before constant incidence and constant shape criteria will be adopted.

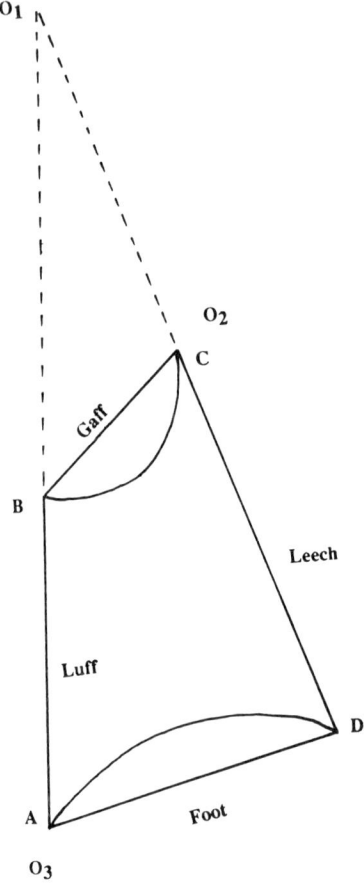

Figure 5. Quadrilateral sail

Again, the required surface, bounded by the quadrilateral $ABCD$, is generated by segments of appropriate cones in this case three cones. An illustration of the resulting cone intersections between cones 1 and 2 and

1 and 3 is shown in Fig. 5. The primary cone being located with vertex at 0_1 and secondary and tertiary cones with vertices 0_2 at pt. C and 0_3 at pt. A. In this case the cone generators for cone 1 lie along the luff and leech, for cone 2 along the gaff peak and leech, and for cone 3 along the luff and foot of the quadrilaterial sail. The calculation process is virtually unchanged (in principal) needing only

3a) solve sail quadrilateral $ABCD$ and sail triangle 0_1AD yielding 3 segments from the developed cones.

4   Mainsails

Although modern mainsails and jibsails are notionally triangular their actual shapes have significant differences from the simple triangular shapes already considered. Considering first mainsails (see [3]), Fig. 6 indicates the differences which are; the presence of a headboard, essentially horizontal foot attached to a boom, and a significant convex curvature of the leech to form a "roach". Typically also mainsails have a set of battens (to aid the roach to "set") equally spaced along the leech.

These batten locations can be used as the nodes for a piecewise linearisation of the leech. The resulting linear segments of the leech together with the sail luff can then be used to generate a family of 5 cones ($C1, ...C5$), Fig. 7. The surface generated by these five cones and the sail boundaries constitute a single curvature surface with specified incidence and shape. The near horizontal sail foot attached to a boom cannot however meet the aerodynamic shape constraint.

Closure of the surface may then be achieved by a sixth foot cone located at the tack devised to permit load transfer from the sail surface to the boom, the total surface defined by the intersections of the sail boundary and these six cones constitutes the mainsail.

Again the calculation process is virtually unchanged (in principle) needing only:

3b) solve sail octagon and the associated set of sail triangles

The intersection between cones 1 and 6 is straightforward, being that of the two cone jib. The intersections between cones 1 to 5 are more complex. A set of primary intersections may be defined as:

$$C5/C4$$

$$C4/C3$$

$$C3/C2$$

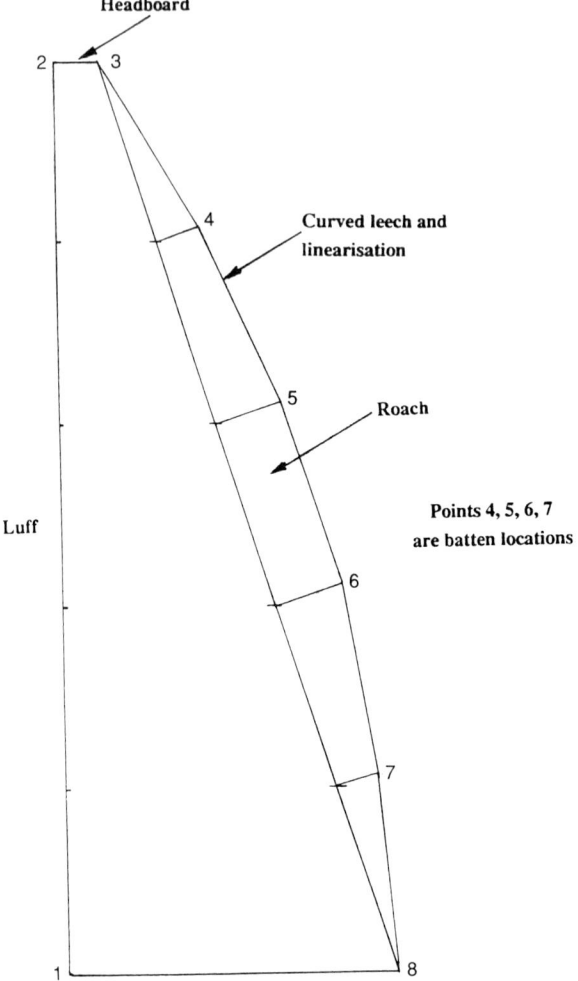

Figure 6. Mainsail

$$C2/C1$$

$$C1/C6$$

The projection of these intersections onto the sail plane is shown in Fig. 8.

To actually generate such a surface the surface segments need to be developed and the set of segments corresponding to the projection in Fig. 8 are given in Fig. 9.

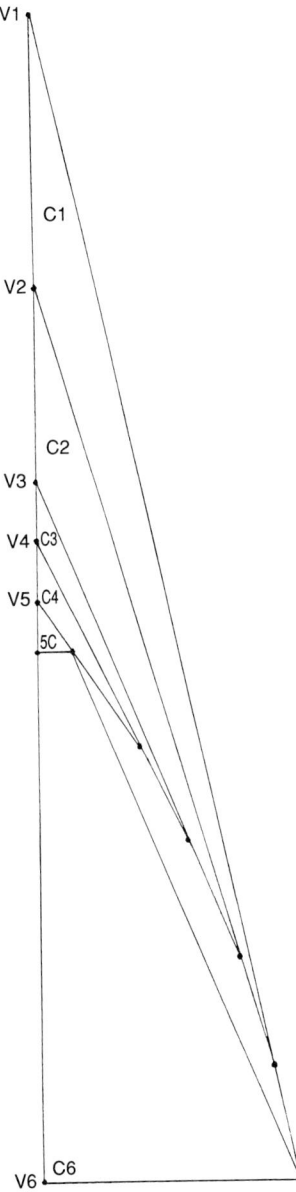

Figure 7. Mainsail generation by a family of cones

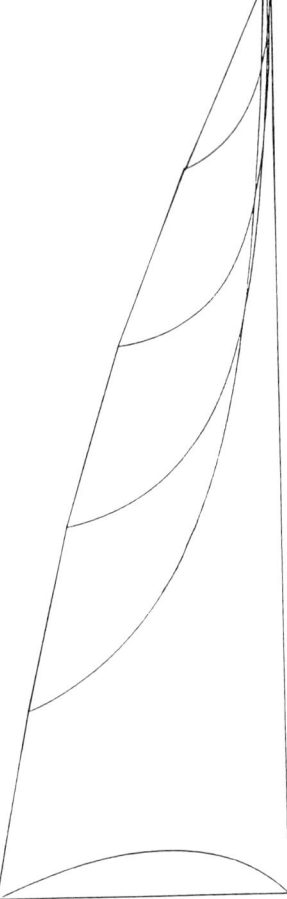

Figure 8. Projection of cone primary intersections onto the sail plane

Overlapping of these primary intersections onto other segments at the sail head may be seen in Fig. 8. Because the segment arising from cone $C1$ dominates the surface, the overlapping segments fall onto $C1$. The effects may be considered by studying the secondary intersections with cone $C1$:

$$C5/C1$$

$$C4/C1$$

$$C3/C1.$$

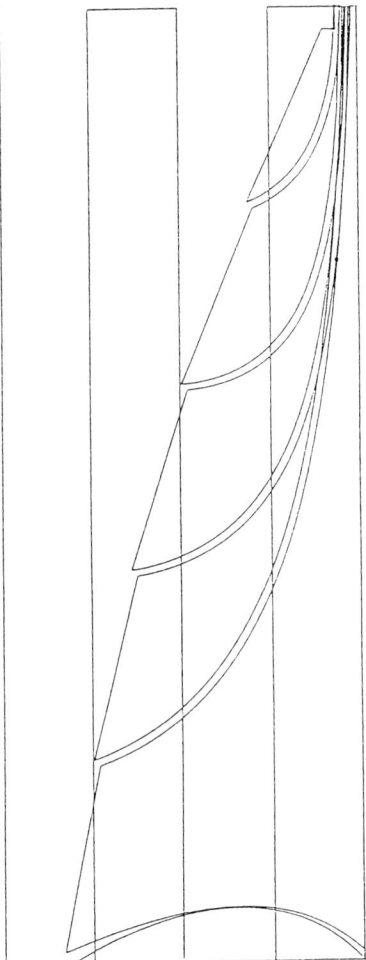

Figure 9. Developed cone segments (primary intersections)

The developed cone segments from these secondary intersections are shown in Fig. 10. Further processing could develop the compound segment shapes in this overlapping region. From the pragmatic viewpoint the overlaps occur in only a small area and where the generating cones are approaching each other to meet at their common line along the luff of the sail so that in constructing such a surface only simple corrections need be made.

Finally, because sail cloth is anisotropic, benefits can be found by aligning the axis of the cloth with radial lines on the sail (if they exist). With

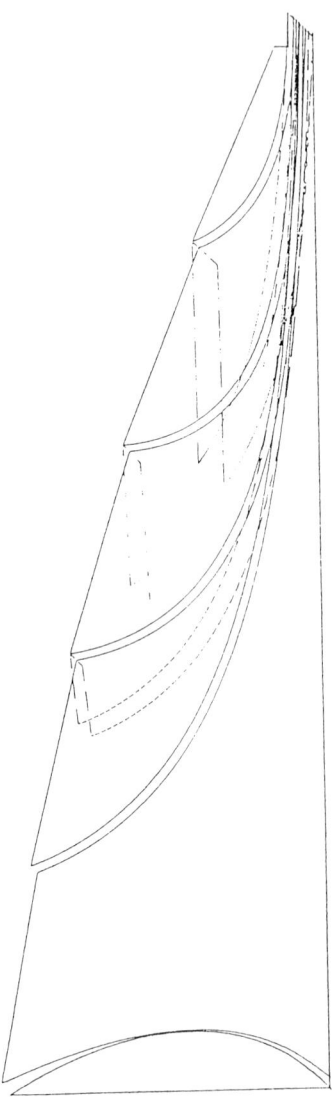

Figure 10. Developed cone segments (secondary intersections)

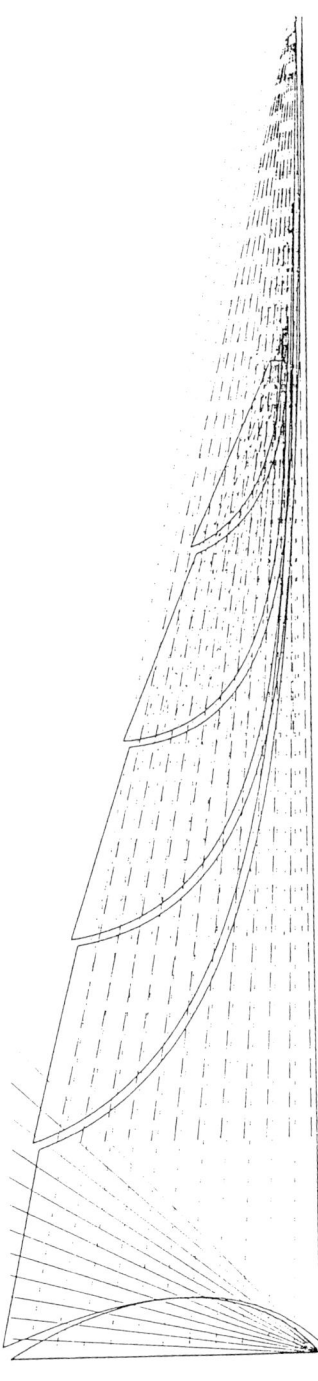

Figure 11. Radial line generation of the sail segments

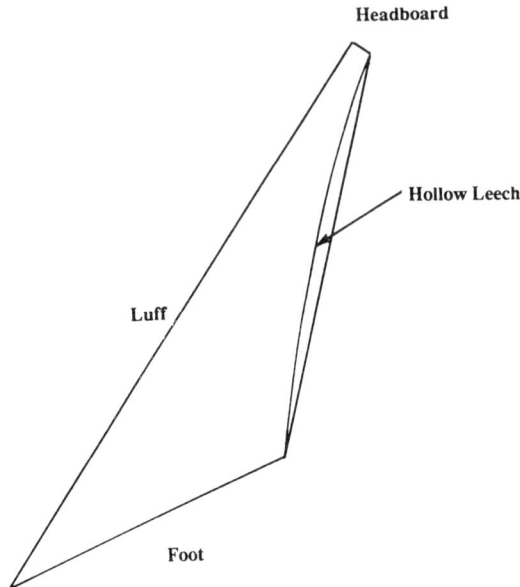

Figure 12. Jibsail

a sail generated in this way, each segment has its set of radial generators. These may be developed and used to improve structural strength. Such a development is shown in Fig. 11.

5 Jibsails

The jibsail is the last of the sail planforms to be considered. Fig. 12 shows a typical shape with a hollowed leech and head board. The same methodology as that used for the mainsail is again appropriate. Fig. 13 illustrated the linearisation of the leech and the identification of a family of 6 cones which constitute the generators for the sail surface.

In principle the method of solution is that already discussed and the projection of the primary intersections onto the sail plane is shown in Fig. 14. Finally the development of the surface segments is shown in Fig. 15.

6 Discussion

In any process of this type the immediate question that arises is - "but does it actually work"? - and if "yes" the next is - "how good are the sails

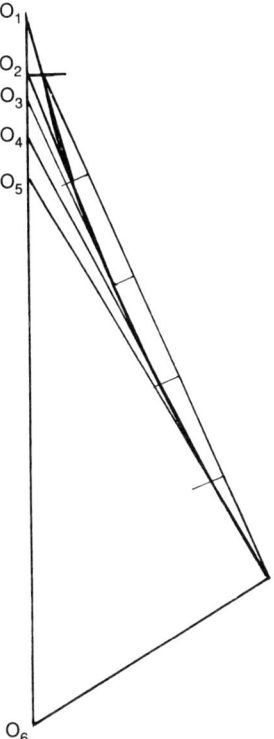

Figure 13. Jibsail generating conics

designed in this way"? Before assessing the questions, a short review of what this approach achieves is useful.

From aerodynamic study a specification for sail incidence and shape throughout the height of the sail can be derived. Because sails operate over a wide range of relative wind speed and direction that specification (itself) may be a composite (weighted) design case. Many sail controls are available to modify both sail incidence and shape to optimise the performance of any actual sail for the current prevailing conditions. Hence for skillful sailors, provided the local optimal shape is attainable as a result of use of sail controls, the sail itself need not *a priori* possess an optimal shape. Indeed, because sail controls have considerable authority over some aspects of shape and incidence, some deviations of a sail from optimum shape have little significance. On the other hand a great many sailors find complex sail adjustments difficult to do and may easily fail to find the optimum.

Against this modern high performance sails are manufactured from materials with considerable stiffness, thus a close approximation with single curvature surfaces to the required final shape has considerable benefit. The

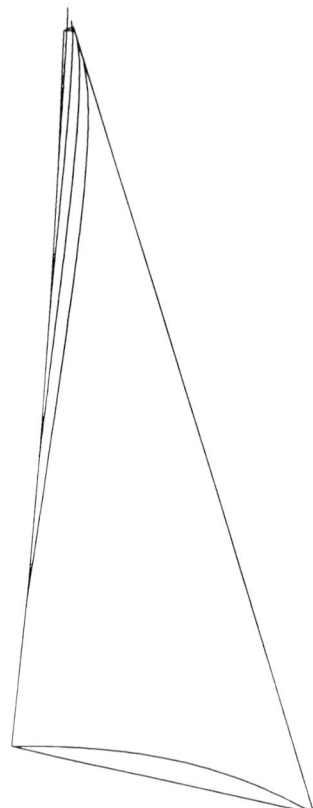

Figure 14. Projection of primary intersections and jibsail plane

datum shape being achieved with zero loading requires lower sail trimming loads to modify and leaves greater reserves of material strength and durabilty for induced aerodynamic loadings.

Another benefit is the perfect geometry of the solution and the availability of full scale patterns that, fitted together, create a sail of the correct size and shape which will set without creases. The ease of generating a sophisticated sail design by merely specifying the co-ordinates of a simple polygon and associated cambers and then running a couple of computer programmes has a great deal to commend it.

A development feature of course is that the output data files could as easily be the input for a laser cutter as a plotter, providing a very significant computer aided design and manufacturing segment to the yacht sail industry.

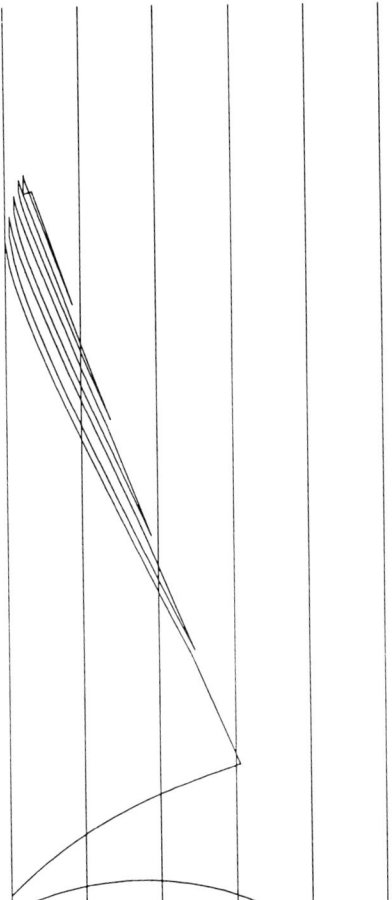

Figure 15. Jib development cone segments

To return to the simple questions:

"yes - the process does work"

"yes - the sails are good - they set without creases, and with low loads"

Whether they perform better than good sails made by the traditional (or computerised traditional) techniques has not been assessed - yet.

References

1. Hales, F.D. (1989). Yacht sail aerodynamics. Parts I, II and III, Report Numbers: TT8903, TT8904 and TT8905, Department of Transport Technology, Loughborough University of Technology, ISBN: 0-904947-15-7, 0-904947-16-5, 0-904947-17-3.

2. Hales, F.D. (1978). The geometry of plain sails. Parts I and II, Report Number: TT7803, Department of Transport Technology, Loughborough Univesity of Technology, ISBN: 0140-9751.

3. Hales, F.D. (1990). The conceptual design of yacht mainsails. Report Number: TT9008, Department of Transport Technology, Loughborough University of Technology, ISBN: 0-904947-22-X.